Approximate Conversion Tables

English to Metric

English	×	= Metric
inches	2.54	centimeters
feet	0.305	meters
yards	0.91	meters
miles	1.61	kilometers
square inches	6.45	square centimeters
square feet	0.09	square meters
square yards	0.84	square meters
acres	0.40	hectares
square miles	2.6	square kilometers
cubic inches	16.4	cubic centimeters
cubic feet	0.028	cubic meters
cubic yards	0.76	cubic meters
cubic miles	4.19	cubic kilometers
ounces	28.3	grams
pounds	0.45	kilograms
tons	0.9	tons
fluid ounces	30	milliliters
quarts	0.95	liters
gallons	3.8	liters

Metric to English

Metric	×	= English
centimeters	0.39	inches
meters	3.28	feet
meters	1.09	yards
kilometers	0.62	miles
square centimeters	0.15	square inches
square meters	11	square feet
square meters	1.20	square yards
hectares	2.47	acres
square kilometers	0.38	square miles
cubic centimeters	0.06	cubic inches
cubic meters	35.31	cubic feet
cubic meters	1.31	cubic yards
cubic kilometers	0.24	cubic miles
grams	0.04	ounces
kilograms	2.20	pounds
tons	1.1	tons
milliliters	0.033	ounces
liters	1.06	quarts
liters	0.26	gallons

Energy

1 barrel of crude oil = 42 gallons

7 barrels of crude oil = 1 metric ton = 40 billion BTUs

1 metric ton of coal = 28 million BTUs

1 gram U_{235} = 2.7 metric tons of coal = 13.7 barrels of crude oil

1 BTU (British Thermal Unit) = 252 calories = 0.0002931 kilowatt-hour

1 kilowatt-hour = 860,421 calories 3412 BTU

Metric conversions in the text are approximate.

Geology of California

Geology of California

Robert M. Norris, Professor of Geology
Robert W. Webb, Professor of Geology, Emeritus

University of California, Santa Barbara

JOHN WILEY & SONS

New York • Chichester • Brisbane • Toronto

Produced by Ken Burke & Associates
  Designer: Christy Butterfield
  Copy editor: Shirley Bruce Henderson
  Illustrations: Carl Brown
  Cover and title page photos: Spence Air Photos, courtesy of Department
                                of Geography; University of California,
                                Los Angeles
  Typesetting: Holmes Composition Service

Library of Congress Cataloging in Publication Data

Norris, Robert Matheson, 1921–
  Geology of California.

  Includes bibliographies.
  1. Geology—California. I. Webb, Robert Wallace,
1909–   joint author. II. Title.
QE89.N67   557.94   76-27281
ISBN 0-471-61566-8

Printed in the United States of America

85 86 87 10 9 8

To Virginia Norris and Elaine Webb,
whose patience made this book possible.

Preface

This book is written primarily for those interested in the geologic aspects of the environment of California. In the past two decades particularly, the volume of information on the geology of California has increased tremendously. Accordingly, it is important to select carefully from the welter of facts with which students are often deluged. This can present a problem, because students may have varied, sometimes minimal, preparation. A single general education course in physical geology is the expected norm for users of this book. For review, however, the text includes brief appendices that cover some of the basic concepts of physical geology.

Students are frequently concerned about the relevance of college courses. How does one make the study of California geology relevant? Perhaps the solution is to encourage students to pursue an understanding of the processes and designs characteristic of the landscape, emphasizing insights from observations rather than the simple reporting of observations themselves. Field trips can contribute substantially to this, but unfortunately field experience is not always possible. We believe that the best substitute for unlimited field trips is the problem approach. For example, in discussing the Whitney (Muir) Crest of the Sierra Nevada, the question should be not so much what the crest is like, but why and how did it get that way. The problem approach *involves* the class, and in our experience modern students interpret relevance primarily as involvement.

The material in this text has been winnowed from the studies of many geologists, to whom we gratefully acknowledge our obligation. We are particularly indebted to the California Division of Mines and Geology and the contributions of Olaf Jenkins, Gordon Oakeshott, Ian Campbell, and Mary Hill. The United States Geological Survey, Menlo Park, granted access to original materials, and works by many Survey members have been especially helpful. Our colleagues John Crowell, Arthur Sylvester, George Tilton, and Donald Weaver offered valuable criticism and advice.

Illustrations are acknowledged beyond, but special credit is due to Malcolm Clark, who suggested important sources of illustrative material, and David Doerner, who prepared many of the photographs for reproduction.

Shirley Bruce Henderson provided expert evaluation and editorial criticism and assistance. Her discernment and constructive suggestions improved the manuscript immeasurably. Her accommodation to our personal foibles is appreciated.

The geologic science departments of the University of California, Santa Barbara, and Dalhousie University, Halifax, supplied facilities and special courtesies.

We take full responsibility for the shortcomings of this volume, and we hope colleagues and students alike will be generous in offering their criticisms. We welcome comments.

Robert M. Norris
Robert W. Webb

Santa Barbara, California
31 December 1975

Contents

Chapter One

Geologic Setting

All observation must be for or against some
view if it is to be of any service.

Charles Darwin

California is a state of geologic contrasts. Of the 48 contiguous states, it contains the highest and lowest elevations only 80 miles (130 km) apart, plus a variety of rocks, structures, mineral resources, and scenery equaled by few areas of the world. Furthermore, because much of the state is arid or nearly arid, its rock sequences are exposed with unusual clarity.

California's rocks vary from ancient Precambrian to presently forming sediments, and several of the state's formations are type examples for North America and the world. The geologic map of California (Figure 1-1) shows general rock ages and Figure 1-2 locates geomorphic provinces. The extent and relationships of the provinces should be compared with the data of the geologic map. Precambrian gneiss and schist, associated granitic and basic intrusives, and metasedimentary sections are confined to southern California, specifically the San Gabriel and San Bernardino mountains, the Basin Ranges, and the Mojave Desert. Paleozoic sedimentary and metavolcanic rocks are found in the northern Sierra, the Basin Ranges, and the Mojave; Paleozoic metamorphic and metasedimentary rocks occur in the southern Sierra Nevada and the Klamath Mountains. Mesozoic sedimentary rocks abound in the Coast Ranges and the western Sacramento Valley; they also are present in the northwestern Sierra Nevada, the Basin Ranges, and much of southern California. Most granitic rocks of California are Mesozoic. They are especially abundant in the Sierra Nevada and the mountains of southern California, but also occur sporadically in the Coast Ranges and the Klamath Mountains. Another chiefly if not entirely Mesozoic rock unit is the important Franciscan formation, which is widespread in the Coast Ranges and also occurs in a few areas of coastal southern California and on Santa Catalina Island.

California's Cenozoic marine sedimentary rocks are particularly significant for man because of their rich petroleum resources. Coastal California, especially south of San Francisco Bay, contains great

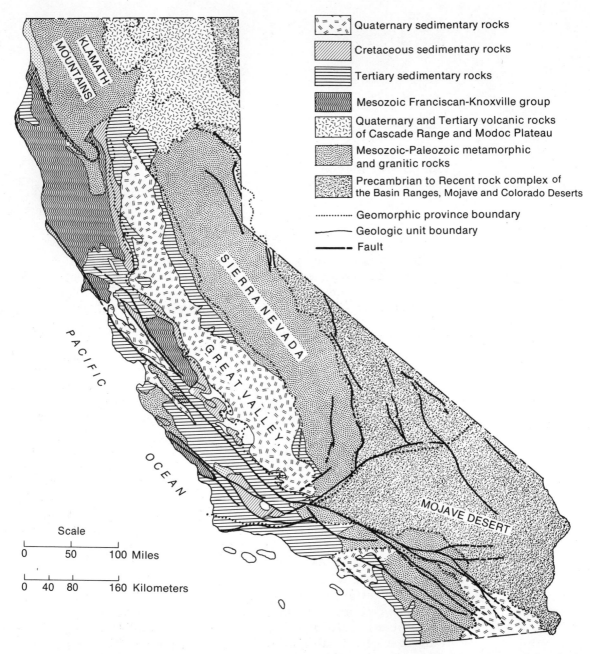

Quaternary sedimentary rocks

Cretaceous sedimentary rocks

Tertiary sedimentary rocks

Mesozoic Franciscan-Knoxville group

Quaternary and Tertiary volcanic rocks of Cascade Range and Modoc Plateau

Mesozoic-Paleozoic metamorphic and granitic rocks

Precambrian to Recent rock complex of the Basin Ranges, Mojave and Colorado Deserts

.... Geomorphic province boundary

———— Geologic unit boundary

——— Fault

Figure 1-1. Geologic map of California, showing principal faults and generalized geologic units. (Source: California Division of Mines and Geology)

thicknesses and extensive exposures that provide most of the sources and traps for petroleum. Such rocks also are distributed widely in the Great Valley (especially the San Joaquin) and are present in limited quantity in the Imperial Valley. Presumably there is no petroleum in eastern California, where the Cenozoic rocks are

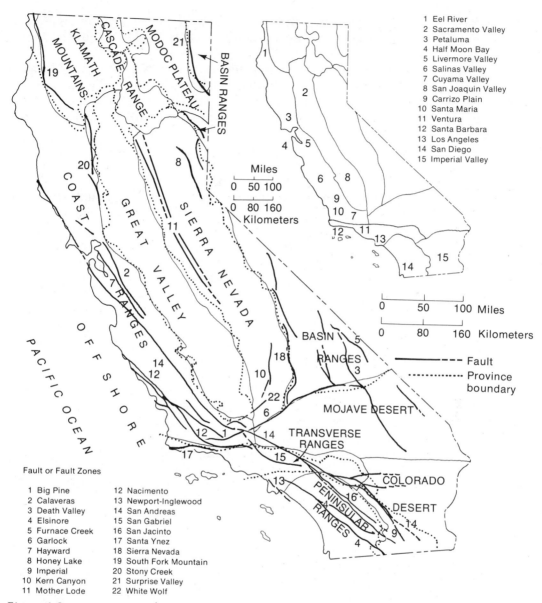

1 Eel River
2 Sacramento Valley
3 Petaluma
4 Half Moon Bay
5 Livermore Valley
6 Salinas Valley
7 Cuyama Valley
8 San Joaquin Valley
9 Carrizo Plain
10 Santa Maria
11 Ventura
12 Santa Barbara
13 Los Angeles
14 San Diego
15 Imperial Valley

Fault or Fault Zones

1 Big Pine
2 Calaveras
3 Death Valley
4 Elsinore
5 Furnace Creek
6 Garlock
7 Hayward
8 Honey Lake
9 Imperial
10 Kern Canyon
11 Mother Lode

12 Nacimento
13 Newport-Inglewood
14 San Andreas
15 San Gabriel
16 San Jacinto
17 Santa Ynez
18 Sierra Nevada
19 South Fork Mountain
20 Stony Creek
21 Surprise Valley
22 White Wolf

Figure 1-2.

Geomorphic provinces and principal faults. Insert (upper right) shows principal marine sedimentary basins. (Source: California Division of Mines and Geology)

almost entirely nonmarine. In the Los Angeles, Ventura, Santa Maria, and San Joaquin basins, the middle and late Cenozoic marine rocks are tens of thousands of feet thick. The Ventura Basin contains sedimentary rocks that aggregate over 50,000 feet (15,000 m) and incorporate one of the most complete Cenozoic sedimentary records in the world. Thick marine sections also are found in the Eel River embayment in the northern Coast Ranges. California's principal

Cenozoic marine sedimentary basins are represented in Figure 1-2. These geographic features should be considered in conjunction with the geologic map.

The building of the ancestral Sierra Nevada—the Nevadan Mountains—in middle and late Mesozoic time expelled the marine waters occupying the Paleozoic Cordilleran seaway, allowing deposition of thick, nonmarine Cenozoic deposits like those in the Mojave and Colorado deserts. Restricted continental basins, including those occupied by the extinct Pleistocene Lakes Searles, Owens, and Manly (Death Valley), reveal the sedimentary, climatic, faunal, and floral history of the late Cenozoic.

Active faulting is an important feature of California's structural pattern. Examination of the fault map shows the dominance of the San Andreas fault and its branches. This fault, of right-slip movement, has been crucial in California's geologic history since at least the Miocene and in some opinions since the Jurassic. Cumulative displacement has been estimated from 1 to 350 miles (1.6–560 km). Other important faults are the Calaveras and Hayward in the San Francisco Bay area, the Nacimiento in the southern Coast Ranges, the San Jacinto of the Peninsular Ranges, the Sierra Nevada in eastern California, and the Garlock, which separates the Mojave Desert from the Sierra Nevada and the Basin Ranges.

California's landscapes are extremely varied and range from the broad, nearly flat floor of the Great Valley to the jagged, glaciated Sierra Nevada. Twelve geomorphic provinces are recognized (Figure 1-2): the Sierra Nevada, the Klamath Mountains, the Cascade Range, the Modoc Plateau, the Basin Ranges, the Mojave Desert, the Colorado Desert, the Peninsular Ranges, the Transverse Ranges, the Coast Ranges, the Great Valley, and the Offshore. The geomorphic provinces are topographic-geologic groupings of convenience based primarily on landforms and late Cenozoic diastrophic and erosional history.

GEOMORPHIC PROVINCES

Sierra Nevada

When Forty-Niners moved west across the interior desert, the formidable barrier of the 400-mile (640 km) long Sierra Nevada blocked their path. Those who arrived in the colder months and who made frontal assaults on the Sierran wall endured hardships of snow and storm; those choosing to circumvent it were parched by the arid wastes of the deserts of the south and southwest. The highest and most massive of California's topographic features, the Sierra Nevada plays a key role in making California habitable. It squeezes moisture from clouds moving inland from the Pacific, stores winter snows and rainfall in its soils, lakes, and forests, and provides large rivers that water the agriculturally rich Central Valley and the San Francisco Bay area.

Parenthetically, the name Sierra Nevada is derived from Spanish: *sierra* meaning "jagged range" and *nevada* meaning "snowed upon." The plural form "Sierras" is a common but redundant usage, as is "Sierra Nevada Mountains."

Great Valley

Following descent from the crest of the Sierra Nevada through deep canyons and the foothills of the Mother Lode belt, a traveler enters the flat-floored trough known as the Great Valley. Drained by the Sacramento and San Joaquin rivers, the Great Valley extends nearly 500 miles (800 km) north and south, separating the Sierra Nevada from the Coast Ranges by an average of 40 miles (64 km). From Red Bluff to Bakersfield, the monotonous floor is interrupted only by the Sutter (Marysville) Buttes. Much of the valley's elevation is close to sea level (Sacramento +30 ft or +10 m); even Bakersfield (comparatively high) has an elevation of only about 400 feet (120 m). Beneath the valley's silt and gravel cover, a thick sedimentary sequence carries important petroleum and natural gas deposits.

Northern Provinces

Three major geologic units occur in northern California. The broad, forest-covered Modoc Plateau and the volcanic peaks of the Cascade Range are the southern and western extensions of topographies belonging more typically to provinces of the northwestern United States. The third northern province, relatively inaccessible and the least known geologically, is the Klamath Mountains. All three provinces close the Great Valley on the north.

Southern Provinces

Terminating the Great Valley on the south are several units of the Transverse Ranges, which are composed of many overlapping mountain blocks of nearly east-west trend in contrast to the northwest-southeast lineation of the Coast Ranges and the Sierra Nevada. The Transverse Ranges consist of several major parallel and subparallel ranges and intervening valleys. Chief among the ranges are the Santa Ynez, Santa Susana, Santa Monica, San Gabriel, and San Bernardino, with intervening, sediment-filled valleys or basins such as the Santa Ynez, Ventura, Ojai, Santa Clara, Simi, San Fernando, San Gabriel, and Santa Barbara Channel. Between the Transverse Range and Peninsular Range provinces lies the Los Angeles Basin, with thousands of feet of post-Jurassic sediment overlying crystalline basement rocks. Southward, the Peninsular Ranges incorporate the San Jacinto, Santa Ana, and other ranges in the hinterland of San Diego and are dominated by rock types prevalent in the Sierra. Much of the Peninsular Range province falls outside California, continuing south nearly 800 miles (1290 km) as the peninsula of Baja California.

Southeastern Provinces

The Mojave Desert, the Colorado Desert, and the Basin Ranges constitute California's desert regions. The Basin Ranges extend from Utah's Wasatch Mountains to the Sierra Nevada. In these desert regions, borates and saline minerals are derived from modern playa and ancient lake basins, and many of the mountains carry mineral deposits. Elevations range from 14,246 feet (4345 m) at White Mountain in the Inyo-White Mountains to −283 feet (−86 m) near Badwater in Death Valley. Great differences in relief provide some spectacular landscapes, for instance, where San Jacinto Peak (10,805 ft or 3296 m) overlooks Palm Springs (475 ft or 145 m) and Salton Sea (−235 ft or −72 m). Such varied features as Death Valley National Monument, Salton Sea, and the Colorado River delta are included in the southeastern provinces.

Coast Ranges

West of the Great Valley the Coast Ranges extend 550 miles (880 km) north and south and are divided by the San Francisco Bay system, a network of waterways and straits produced by drowning of river-cut and block-faulted valleys. The Coast Ranges show strong northwest-southeast trends, induced by folds and faults of the same strike. The chain contains dominantly sedimentary rocks underlain by two unlike kinds of basement rocks that are mostly of middle Mesozoic age. One of these, the Franciscan formation, is widespread and figures prominently in today's interpretations of the roles of sea-floor spreading and plate tectonics in developing the continental margin. The other type is a granitic sequence with associated metasediments. These rocks are exposed in several areas, and the granitic sequence probably correlates with granitic units of the Sierra Nevada. Geologic history of the Coast Ranges is intricately interwoven with tectonics of the San Andreas and other major faults, particularly those of the western part of the state. Sedimentary units include the Great Valley sequence exposed along the boundary between the Coast Ranges and the Great Valley, plus Cenozoic basin deposits of thousands of feet of clastic sediments. In some places a narrow, discontinuous coastal plain faces the Pacific. More often, however, especially in the south, the Coast Ranges rise abruptly from the sea to almost 6000 feet (1800 m), forming a western barrier almost as continuous as the Sierra Nevada to the east.

Offshore

The Offshore province is composed of two main regions. North of Point Arguello, the province appears related to the Coast Ranges and is characterized by continental shelf and slope topography. East and south of Point Arguello, the province may correlate with the Transverse and Peninsular ranges. This southern section contains elevated blocks and ridges that are occasionally expressed as islands, with deep intervening closed basins.

CALIFORNIA IN THE CONTINENTAL FRAMEWORK OF NORTH AMERICA

California's geologic setting is only one aspect of the complex patterns of rocks, structure, and history composing the western Cordillera of the United States and the geologic features circumscribing the Pacific Basin. The mountains of southern Alaska, western Canada, western continental United States, western Mexico, Central America, the Andes, and parts of Antarctica all seem to share roughly similar geologic histories. Since the renascence of the concept of continental drift and development of the ideas of plate tectonics and sea-floor spreading, it is important to understand the eastern Pacific. For example, we now appreciate that the building of the Sierra Nevada during the Mesozoic Nevadan orogeny was only a minor incident in the major event of the Cordilleran orogeny, which involved most of western North America.

Sophisticated isotopic age dating now permits correlating such California events as the Nevadan and Coast Range orogenies with episodes like the Basin Range Sevier and Rocky Mountain Laramide orogenies. Consequently, geologists have established that such local deformations were merely parts of a major event of the eastern Pacific–western North American margin, with tectonic pulsations and intervals of quiescence.

CURRENT GEOLOGIC STUDIES

Geologic maps are the principal products of geologic field work, and thorough understanding of a region's geology is possible only when a geologic map can be examined. Accordingly, most countries have organized geologic surveys. The U.S. Geological Survey, which employs several hundred geologists, is responsible for all facets of the federal government's geologic mapping program. In addition, most states have their own geologic surveys and cooperate with the federal survey in undertaking many geologic studies, including map publication. In California, the responsible agency is the Division of Mines and Geology, a branch of the California Department of Natural Resources.

Although geologic mapping of California is proceeding systematically, much remains to be done. Large areas are incompletely studied, and sometimes pertinent geologic maps are unavailable or outdated. Surface geology was summarized originally as the Geological Map of California, published by the Division in 1916 on a scale of approximately 1 to 750,000. It was an extremely generalized representation, drawn from the vague reconnaissance studies that appeared before World War I. Up to that time the Division's primary concern had been mining and mineral-resource problems.

The Division's emphasis changed gradually, however, with the 1929 appointment of Olaf P. Jenkins. Jenkins initiated a second geologic map, which appeared in 1938 on a scale of approximately 1 to 500,000. The geologically unknown areas of California were glaringly apparent since more than one-third of the state appeared as "unmapped." Jenkins was joined in 1948 by Gordon B. Oakeshott. Together they sponsored the detailed mapping required as the base for solving California's geologic problems. The necessity of educating the public and the government to the importance of the geologic environment was also apparent. As Deputy Chief Geologist, Oakeshott focused on geologic education and for almost a quarter of a century was one of the state's leading earth science educators. The first two sheets of the present geologic atlas appeared as Jenkins closed his administration in 1958. Ian Campbell then was appointed State Geologist and Chief of the Division, and under his direction the third geologic map was completed by 1969. It is an atlas of 27 sheets, on a scale of approximately 1 to 250,000 (Figure 1-3). During this period, the orientation of the state program shifted toward detailed geologic studies of environmental problems while still encompassing research on mineral commodities and mining potential.

SYSTEMATIC STUDY OF CALIFORNIA GEOLOGY

The geology of California is exceedingly complex and becomes more so as the consequences of plate tectonics are evaluated. Where, then, to start, and why start at a particular place in developing a logical, interesting, and yet necessarily limited coverage of this fascinating geology?

We shall begin with the Sierra Nevada because it provides a basis for understanding the bedrock geology of the entire state. Other regions differ from the Sierra primarily in events and rocks subsequent to the middle Mesozoic. The plutonic rocks throughout the state are overwhelmingly products of the granitic magmatic invasions that accompanied the Nevadan orogeny. In addition, the Sierra is the topographic backbone of the state and has shed debris since the middle Mesozoic, forming many of the state's younger rock units. The northern provinces (Klamath Mountains, Cascade Range, and Modoc Plateau) then are discussed, followed by the Basin

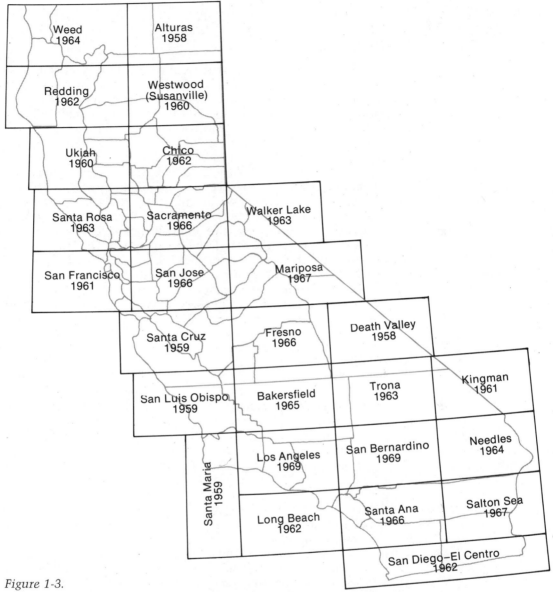

Figure 1-3.

Index to California geologic map sheets. (Source: California Division of Mines and Geology)

Ranges, the Mojave Desert, the Colorado Desert, the Peninsular Ranges, the Transverse Ranges, the Coast Ranges, the Great Valley, and the Offshore, with a special section on the San Andreas fault.

It is assumed that users of this book are familiar with fundamental principles of geology. For review, however, the appendices provide summaries of several basic concepts: Appendix 1—Glossary; Appendix 2—Common Minerals and Rocks; Appendix 3—Geologic Sequence and Time; Appendix 4—Some Theories Pertinent to California Geology. The standard geologic time scale has also been provided.

REFERENCES

Wherever possible, publications by the California Division of Mines and Geology and the U.S. Geological Survey have been cited, because they are generally more widely available in local libraries. Lists of available geologic publications may be obtained from: California Division of Mines and Geology, Publication Sales, P.O. Box 2980, Sacramento, California 95812; U.S. Geological Survey, 555 Battery Street, Room 504, San Francisco, California 94111; U.S. Geological Survey, 300 N. Los Angeles Street, Room 7638, Los Angeles, California 90012. Geologic maps published by both agencies and lists of the maps available may be obtained from: Geological Inquiries Group, U.S. Geological Survey, Washington, D.C., 20244. The authors recommend that readers acquire the U.S. Geological Survey's colored geologic map of California. This is available from: Branch of Distribution, U.S. Geological Survey, Federal Center, Denver, Colorado 80225. It also may be purchased over the counter (not by mail) at the U.S. Geological Survey, 345 Middlefield Road, Menlo Park, California. The official title of this map is Miscellaneous Geological Investigations Map 1-512.

The following four references are especially pertinent and are frequently suggested as collateral reading.

California Geology (formerly Mineral Information Service). Calif. Div. Mines and Geology. Monthly digest of new data on mining of metals and nonmetal mineral products, plus general geology and announcements of publications pertinent to California geology.

Geologic Maps of California. Calif. Div. Mines and Geology. Usually referred to by name of sheet and date of publication.

Bailey, Edgar H., ed., 1966. Geology of Northern California. Calif. Div. Mines and Geology Bull. 190. Series of papers by different authors, organized into 10 chapters, each with extensive bibliographies.

Jahns, Richard H., ed., 1954. Geology of Southern California. Calif. Div. Mines and Geology Bull. 170. Series of papers by different authors, organized into 10 chapters, with 34 map sheets and 5 geologic guides.

Other general references are:

Anderson, F. M., 1932. Pioneers in the Geology of California. Calif. Div. Mines and Geology Bull. 104.

GeoTimes. Published ten times yearly by the American Geological Institute, 5205 Leesburg Pike, Falls Church, Virginia 22041.

Harbaugh, John W., 1974. Geology: Field Guide to Northern California. William C. Brown Co.

Hinds, Norman E. A., 1952. Evolution of the California Landscape. Calif. Div. Mines and Geology Bull. 158.

Oakeshott, Gordon B., 1971. California's Changing Landscapes. McGraw-Hill Book Company.

Reed, R. D., 1933. Geology of California. Am. Assoc. Petroleum Geologists.

Sharp, Robert P., 1972. Geology: Field Guide to Southern California. William C. Brown Co.

Chapter Two

Sierra Nevada

*Go my sons, buy stout shoes, climb the
mountains, search the valleys . . . and the
deep recesses of the earth.*

Peter Severinus

The Sierra Nevada is California's topographic backbone. Consider
how different the state's geography and human resources would be if
the Sierra were lower or nonexistent. For instance, the Sierra
squeezes moisture from Pacific storms that must rise above its crest.
Another example is the presence of mineral resources near the surface because ancient rocks have been exposed by erosion in the
Sierra Nevada's deep canyons and on its slopes. Placer gold derived
from Sierran lodes was the main attraction for thousands of
pioneers, many of whom reached gold country only after an arduous
crossing of the range.

Sierran grandeur has evoked both wonder and inspiration. One
of its first and greatest interpreters was the famous naturalist John
Muir, who spent much of his life exploring and extolling this
stupendous landform. Muir recorded his observations meticulously,
and his journals provide probably unexcelled portrayals of Sierran
majesty. Furthermore, Muir's explorations resulted in corrections of
earlier, erroneous ideas regarding the origins of the natural features
in the Range of Light, as he called the Sierra.

Differences of opinion regarding the Sierra, exemplified by the
extended controversy about the origin of Yosemite Valley, resulted
in the initiation of a comprehensive study of the valley by the U.S.
Geological Survey. The project was undertaken by François E.
Matthes, who prepared the first large-scale topographic map of
Yosemite, and F. C. Calkins. The result was the classic publication
Geological History of the Yosemite Valley. Although it did not appear until 1930, this study established Matthes as a leader in Sierran
geology. Like Muir, Matthes was a persuasive and eloquent though
less prolific writer. Several posthumous papers, based on Matthes's
voluminous field notes, were prepared by Fritiof Fryxell. Primarily

for the layman, two of these papers were published by the University of California Press in 1950 on the occasion of the state's centennial. The U.S. Geological Survey published two more in 1960 and 1965.

GEOGRAPHY

The Sierra Nevada extends about 400 miles (640 km), terminating in the north at Lassen Peak in the Cascade Range province. The rock and structural patterns of the Sierra disappear beneath the lavas of the Cascades and the volcanic and sedimentary cover of the northern Sacramento Valley. They reappear in the Klamath Mountains, where rock units and faults have geologic histories similar to those of the Sierra. (The relationships between the Sierran and Klamath provinces are shown in the geologic map of Figure 1-1.) Most geologists think that the same basement patterns extend north into Oregon, curving northeast and disappearing again beneath the volcanics of southern Washington.

To the south, the Sierra grades into the Tehachapi Mountains, which enclose the southern end of the Great Valley. The Garlock fault, separating the Sierra-Tehachapi from the Mojave Desert, is the southern geologic boundary. The Great Valley's Cenozoic gravels (derived primarily by erosion of the Sierra) overlap and conceal the western extensions of the Sierra. The Sierran basement terminates near the western margin of the Great Valley, presumably in contact with the Franciscan formation. The exact relationship between these two major rock units, Sierran and Franciscan, is debated, but the contact is probably a major fault.

The Sierra is from 40 to 100 miles (64–160 km) wide. Its elevations vary from 400 feet (120 m) at the Great Valley boundary to summits of more than 14,000 feet (4250 m) adjacent to the Basin Ranges. Extensive vertical movement on the Sierra Nevada fault has produced an almost unbroken eastern wall for more than 100 miles (160 km), making descents to the adjacent valley extraordinarily precipitous. Important geographic names are given in Figures 2-1 and 2-2.

ROCKS: SUBJACENT SERIES

An important section of the earth's outer crust is exposed in the Sierra Nevada. The rocks are mainly igneous and metamorphic units of diverse composition and age, including volcanics and metasedimentary interlayered rocks. In the central and southern Sierra, plutonic igneous rocks, mostly silicic (granitic), form the multiple intrusions of the Sierra Nevada batholith and constitute 60 percent of the exposed rock. Early geologists called the metamorphic and igneous rocks (basement) the *subjacent series.* Overlying the basement are sedimentary and volcanic rocks, most prominent in the central and northern Sierra. These are known as the *superjacent series* and are considered later in the chapter.

Beginning in the middle Jurassic, the subjacent metasedimentary series became intensely deformed, primarily by folding and concomitant invasion of plutonic rocks (Figure 2-3). This tectonism, known as the Nevadan orogeny, produced the Nevadan Mountains—site of the modern Sierra Nevada. Today's Sierra Nevada represents rejuvenation by faulting of portions of this earlier range.

Basement Metasedimentary Rocks

The oldest known Sierran rocks are Ordovician metasediments, classified from fragmentary fossils found near Lake Crowley in the Mount Morrison roof pendant. Nearby, about 32,000 feet (9700 m) of hornfels, chert, marble, slate, and quartzite reflect nearly continuous Paleozoic deposition from the Ordovician through the Pennsylvanian and probably the Permian, although Silurian and Devonian beds have not been positively identified. In the west-central and northwestern Sierra, particularly north of the Mother Lode, the metamorphic rocks are often less deformed and metamorphosed. Slates, phyllites, and massive sandstones, with aggregate thickness possibly up to 50,000 feet (15,000 m), carry Silurian fossils in the uppermost part. Devonian sediments also have been identified, and presumably Devonian pyroclastic units up to 25,000 feet (7600 m) thick rest unconformably on the Silurian. Included also are two Mississippian and Permian fossil localities, suggesting depositional correlation with the Calaveras formation of the Mother Lode belt. Although no completely continuous Paleozoic section is known, all Paleozoic periods except the Cambrian are represented in the province. The Paleozoic shows higher percentages of volcanic and volcaniclastic rocks in the northern Sierra than in the central and eastern parts of the range.

In addition to the metasedimentary roof pendants definitely established as Paleozoic by fossils and radiometric chronology, extensive but apparently unfossiliferous pendants occur in the southern Sierra. These are composed of greenstones and other metamorphosed volcanics, with subordinate amounts of metasediments, and are usually designated Paleozoic. Although high percentages of volcanics are typical of pendants definitely established as Mesozoic (like those at Mineral King and in the Ritter Range), thick volcanic sections alone do not necessarily reflect Mesozoic age.

Mesozoic sedimentary rocks are especially widespread in the west-central and northwestern Sierra. Triassic sequences up to

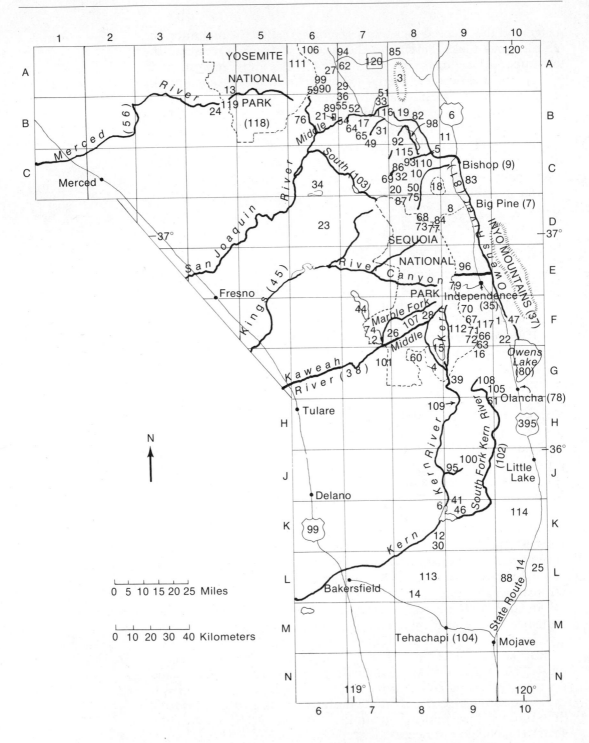

Figure 2-1.   Place names: southern Sierra Nevada.

1 Alabama Hills—F10
2 Ash Meadow (Sequoia)—F7
3 Benton Range—A8
4 Big Arroyo—G8
5 Birchim Canyon—C8
6 Big Blue Mine—K8
7 Big Pine—D9
8 Big Pine Creek—D9
9 Bishop—C9
10 Bishop Creek—C–D, 8–9
11 Bishop Volcanic Tableland—B–C, 8–9
12 Bodfish—K8
13 Cathedral Range—Yosemite—A4
14 Caliente—L8
15 Chagoopa Plateau—G8–9
16 Cottonwood Lakes—G9
17 Convict Lake—B7
18 Coyote Flat (Mountain)—C8–9
19 Crowley Lake—B8
20 Desolation Lake—C8
21 Devils Postpile—B6
22 Diaz Lake—F10
23 Dinkey Creek—D6
24 El Portal—B4
25 El Paso Mountains—L10
26 Giant Forest Village—F8
27 Grant Lake—A6
28 Great Western Divide—F–G, 8
29 Gull Lake—A6–7
30 Havilah—K8
31 Hilton Creek—B7
32 Horton Creek—C8
33 Hot Creek—B7
34 Huntington Lake—C6
35 Independence—E9
36 Inyo Craters—B7
37 Inyo Mountains—D–F, 10
38 Kaweah River—F–G, 6–8
39 Kern Lake—G9
40 Kern River—F–L, 6–10
41 Kernville—K9
42 Kings Canyon—E7–8
43 June Lake (see 29)—A6–7
44 Kings Canyon National Park—F7
45 Kings River—E–F, 5–8
46 Lake Isabella—K9
47 Lone Pine—F10
48 Marble Fork, Kaweah River—F8
49 McGee Creek (into Crowley Lake)—B7
50 McGee Creek (into Pine Creek)—C8
51 McGee Mountain—B7
52 Mammoth—B7
53 Mammoth Hot Creek (see 33)—B7
54 Mammoth Lakes—B7
55 Mammoth Mountain—B7
56 Merced River—B1–6
57 Middle Fork, Kaweah River—F8
58 Middle Fork, San Joaquin River—B6
59 Minarets—A6
60 Mineral King—G8
61 Monachee Mountain—H9
62 Mono Craters—A7
63 Mt. Langley—F9
64 Mt. Laurel—B7
65 Mt. Morrison—B7
66 Mt. Muir—F9
67 Mt. Russell—F9
68 Mt. Sill—D8
69 Mt. Tom—C8

70 Mt. Tyndall—F9
71 Mt. Whitney—F9
72 Mt. Williamson—F9
73 Mt. Winchell—D8
74 Moro Rock—F7
75 North Fork, Bishop Creek—C–D, 8
76 North Fork, San Joaquin River—B6
77 North Palisade—D8
78 Olancha—G10
79 Onion Valley—E9
80 Owens Lake—G10
81 Owens River—B8–F10
82 Owens River Gorge—B8
83 Owens Valley—A8–J10
84 Palisade Glacier—D8
85 Panum Crater—A8
86 Pine Creek—C8
87 Piute Pass—D8
88 Redrock Canyon—L10
89 Reds Meadow—B6
90 Reversed Creek—A6
91 Ritter Range (see 59)—A6
92 Rock Creek—B–C, 8
93 Round Valley—C8
94 Rush Creek—A6–7
95 Salmon Creek—J9
96 Sawmill Creek—E9
97 Sequoia National Park—D–G, 8–9
98 Sherwin Hill—B8
99 Silver Lake—A6
100 Siretta Peak—J9
101 South Fork, Kaweah River—G7
102 South Fork, Kern River—G–K, 9–10
103 South Fork, San Joaquin River—C6-7
104 Tehachapi—M8
105 Templeton Mountain—G9
106 Tioga Pass—A6
107 Tokopah Valley—F8
108 Toowa Valley—G9
109 Trout Meadows—H9
110 Tungsten Hills—C8
111 Tuolumne Meadows—A6
112 Upper Crabtree Meadows—F9
113 Walker Basin—L8
114 Walker Pass—K10
115 Wheeler Crest—B–C,8
116 Whitmore Hot Springs—B7
117 Whitney Portal—F9
118 Yosemite National Park—A–B, 4–6
119 Yosemite Valley—B4–5

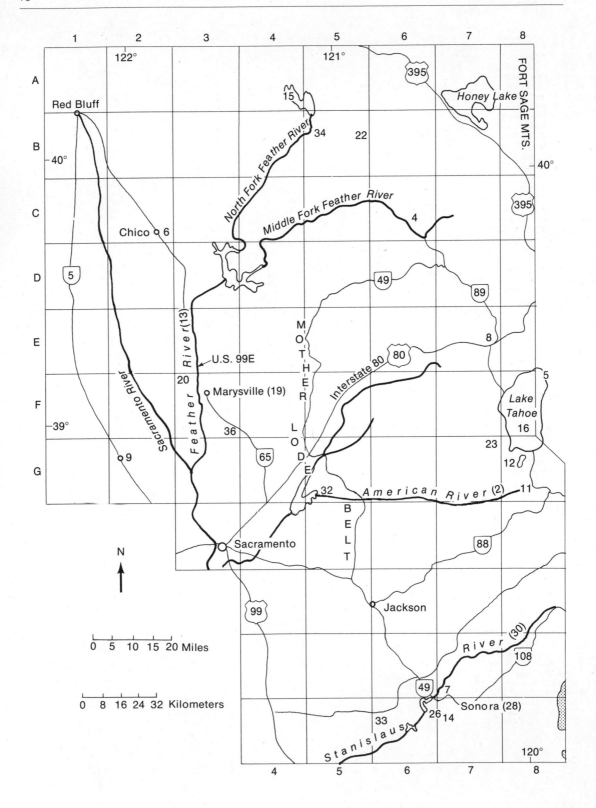

1   2   3   4   5   6   7   8

122°

121°

395

FORT SAGE MTS.

A

Red Bluff

15

Honey Lake

B

North Fork Feather River

34

22

40°

40°

395

C

Middle Fork Feather River

4

Chico 6

D

5

49

89

Feather River (13)

8

E

U.S. 99E

MOTHER

80

Interstate 80

5

20

Lake Tahoe

Marysville (19)

F

Sacramento River

16

39°

36

23

65

12

G

9

LODE

32

American River (2)

11

BELT

Sacramento

88

N

Jackson

River (30)

0 5 10 15 20 Miles

99

108

49

7

0 8 16 24 32 Kilometers

Sonora (28)

26 14

33

Stanislaus

120°

Figure 2-2.

Place names: northern Sierra Nevada.

 1 Adobe Lake Basin—L12
 2 American River—G–H, 3–8
 3 Black Point—K11
 4 Blairsden—C6
 5 Carson Range—F8
 6 Chico—C2
 7 Columbia—K7
 8 Donner Pass—E7
 9 Dunnigan—G2
10 Ebbetts Pass—H9
11 Echo Summit—G8
12 Fallen Leaf Lake—G8
13 Feather River—D–H, 3
14 Jamestown—L7
15 Lake Almanor—A4
16 Lake Tahoe—F8
17 Lundy Lake—K10
18 Lyell Fork, Tuolumne River—L10
19 Marysville—F3
20 Marysville (Sutter) Buttes—F3
21 Mono Lake—K–L, 11
22 Mt. Jura—B5
23 Mt. Tallac—G7
24 Negit Island—K–L, 11
25 Paoha Island—K11
26 Rawhide Flat—L6
27 Saddlebag Lake—L10
28 Sonora—L7
29 Sonora Pass—J9
30 Stanislaus River—J–L, 5–9
31 Sutter Buttes (see 20)—F3
32 Sutter's Mill—G5
33 Table Mountain—L6
34 Taylorsville—B5
35 Tuolumne River—L9–10
36 Wheatland—F3
37 White Mountain (Peak)—L13

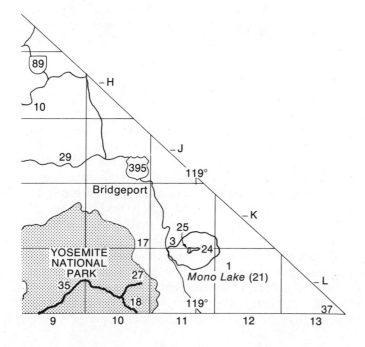

*Figure 2-3.
Split Mountain roof
pendant with grani-
tic rocks (light)
intruding
metasedimentary
units (dark). Note
other roof pendants
in the ridges be-
yond. A typical
view of Sierran
rocks, this picture
was taken looking
west across the
Sierran crest in the
vicinity of Big Pine.
Arcuate glacial
moraines resulting
from ice recession
are also shown.
(Photo by Spence
Air Photos, courtesy
of Department of
Geography, Univer-
sity of California,
Los Angeles)*

10,000 feet (3000 m) consist of metavolcanics and sedimentary rocks derived by weathering of volcanics. (Isotopic dates from the volcanics suggest Permian age for part of the volcanics.) Some limestones and marine shales carry Triassic fossils. In the eastern metasedimentary belt, Mesozoic units are found from the Ritter Range northwest about 150 miles (240 km). Here Jurassic rocks (largely metavolcanics and fossiliferous marine sediments) predominate and are more widespread and much thicker than the Triassic. The Mount Jura section near Taylorsville is nearly 15,000 feet (4550 m) thick.

It is currently acknowledged that early to middle Jurassic sediments were the youngest rocks involved in the Nevadan orogeny. Upper Cretaceous sediments lie by transgressive unconformity on basement in the northern and north-central parts of the range's west flank. This suggests that the Nevadan orogeny developed prominent mountains from the opening of the middle Jurassic through the middle Cretaceous.

Pre-Cenozoic sedimentation on the site of today's Sierra formerly was attributed to the presence of a (Cordilleran) geosyncline during much of the Paleozoic and most of the Mesozoic. This shallow trough was presumed to have extended from the Gulf of Mexico or the eastern Pacific north to the Arctic, periodically submerging much of western North America. This view has been superseded by the concept of a borderland with sedimentation occurring along continental margins in subsiding miogeosynclinal (shelf) and eugeosynclinal (slope) belts. The continental margin of North America in the Sierra-Klamath area is thought to have grown oceanward by tectonic accretion during the Phanerozoic, an idea consistent with continental drift and plate tectonic theory. The sedimentary prisms that accrue on the forward and trailing margins of plates are still commonly described as accruing in geosynclinal environments, however.

In summary, early Mesozoic volcanic and volcaniclastic deposits increased from southeast to northwest along the Sierran axis. Thick Paleozoic sediments in the south were followed by equally thick Mesozoic volcanics to the northwest, which suggests major shifts in regional deposition in the sedimentary basins. Only fragments of the cover survive today, as roof pendants. It appears that during the Paleozoic miogeosynclinal conditions dominated, but eugeosynclinal conditions gradually replaced miogeosynclinal—as shown by the extensive Mesozoic volcanic sedimentation. The relationship of the Permian and the Triassic is uncertain, however, since evidence throughout the Sierra is contradictory.

Basement Igneous Rocks

Plutonic igneous rocks form more than 100 plutons in the Sierra Nevada. The earliest intrusives are small, mafic, often coarse-textured with hornblende phenocrysts in a plagioclase matrix, and generally dioritic and gabbroic. Remnants of these intrusives became inclusions in later granitic rocks, and occasionally irregular intrusives and dikes occur in metamorphic sequences. The dark plutonic intrusives currently are considered forerunners of the widespread plutonic invasion during which most of the Sierran batholith was constructed. Some workers have emphasized the dioritic and gabbroic plutonic phases, suggesting that perhaps they represent remnants of a larger magmatic episode preceding the Jura-Cretaceous granitic invasion. Isotopic dates from gabbroic-dioritic bodies yield ages 10 to 15 million years older than those of the major granitic bodies. Such age differences are inconclusive for establishing an earlier mafic intrusive episode, although Devonian mafic intrusives have been recently identified in the northern Sierra.

Sierra Nevada

Sierra Nevada Batholith

The Sierra Nevada batholith is composed of granitic rocks variously described as granite, quartz-monzonite, granodiorite, and quartz-diorite. These rocks were intruded into pre-Jurassic units as a series of overlapping plutons and have ages of about 77 to 210 million years. Plutons of similar granitic rocks in the White Mountains east of the Sierra have been dated at 70 to 225 million years. Both sets of plutons are considered part of the general Cordilleran plutonic episode of western North America. As study of the batholithic rocks has progressed, the Sierra Nevada batholith itself has come to embrace the expanse of silicic plutonic rocks in the Inyo-White Mountains and western Nevada, plus the granitic rocks of southern California (Figure 2-4).

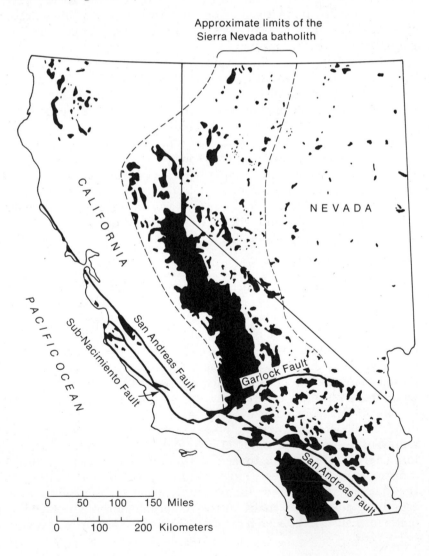

Figure 2-4. Distribution of granitic rocks in the Sierra Nevada batholith. (Source: Geological Society of America)

Sierran granitic units comprise an unknown number of individual plutons, of which several dozen have been mapped. Each pluton has specific temperature-pressure-composition characteristics that determine the final rock product. The span of ages currently established suggests that the plutons were emplaced from early Triassic in the east-central Sierra and White Mountains to late Cretaceous west of Whitney Crest. A study of about 250 of the isotopic dates available through 1973 indicates that the plutons were emplaced during a span of approximately 150 million years in bursts of plutonic activity with intervening periods of quiescence or readjustment. These plutonic episodes constitute the minimum age spread of the Nevadan orogeny, assuming, as is customary, that the orogeny began when the plutonic activity first affected the roots of the mountains and that it closed when plutonic activity ceased. Plutonic activity is summarized in Table 2-1.

The batholithic granites display varied grain size and mineral content. Dark-colored minerals occur in high proportions in some rocks, but are absent in others (alaskite). Consequently, the rate of weathering and the forms of the weathered rocks vary substantially.

A common granite is a porphyritic rock in which feldspar crystals, often an inch (2.5 cm) on a side, are embedded in a light-colored, finer-grained matrix. This granite may occupy many square miles and its crystals are often so abundant that a square foot (0.1 m²) of outcrop may show several dozen. Porphyritic plutons and plutons of even grain with typical "salt and pepper" appearance often are closely associated. These differences in granite have been described in detail in the Yosemite Valley region, where careful mapping has delineated many variations in the rocks.

Other differences distinguish the granitic plutons. For example, large areas frequently occur where thousands of fine-grained dark blebs are contained in the parent rock. These inclusions sometimes are derived by the fragmentation of rocks of the older metamorphic cover into which the plutons were intruded (xenoliths). In other

Chronologic Age (max–min in m.y.)	Geologic Age	Intrusive Epoch
85–70(?)	Earliest Tertiary	White Mountains
90–79	Late Cretaceous	Cathedral Range (south of Yosemite)
121–104	Early Cretaceous	Huntington Lake
148–132	Late Jurassic	Yosemite
180–160	Early and Middle Jurassic	Inyo Mountains
210–195	Middle and Late Jurassic	Lee Vining
225(?)	Earliest Triassic	White Mountains

Table 2-1
Sierran Plutonic Activity

Source: Geological Society of America and U.S. Geological Survey.

cases, the blebs represent segregations crystallized from the magma itself before final congealment (autoliths).

Dikes, sills, and veins—usually lighter in color than the enclosing rocks—may represent late-stage intrusions into the parent. This process is thought to occur near the final consolidation point of the magma. In other cases, some plutons show surfaces of foliation and lineation where minerals are dragged out before complete consolidation of the magma, producing gneissic patterns. Such rocks should not be confused with metamorphic gneisses.

How do plutonic bodies form? What combinations of energies are at work? Are all plutons of similar origin? Is there such a thing as a parent magma? Are some plutons formed in place by heating but not melting of older rocks, thus producing nonmagmatic (granitized) plutons? What theories of origin best explain Sierran rock bodies? Commonly offered models are as follows.

1. Magma is generated in the lower crust by melting of crustal basement. Added to this are sedimentary and volcanic materials that were deposited on the margins of continental boundaries as the continents were depressed by downfolding into a zone of melting.

2. Magma forms in subduction zones where oceanic plates are carried beneath continental plates. Subsequent magmatic differentiation occurs as magma rises through the crust toward final emplacement.

3. Magma is mantle-generated, pulsating upward from deep-seated "hot spots" into the crust during periods of increased temperatures. (This is the interpretation for magmatic origin in areas like the Hawaiian Islands.)

4. Various combinations of 1, 2, and 3.

The origin of Sierran intrusives is debated vigorously. In all consideration of granitic batholiths, however, it must be remembered that either they are restricted in worldwide distribution to areas underlain by material composing continental plates or they are associated with continental shield areas as these enlarge by accretion. All lines of evidence strongly suggest crustal sources for the magmas that produced Sierran plutons.

Most research on the origin of intrusives of western North America has been conducted by the U.S. Geological Survey. A major contributor to Sierran petrogenesis is Paul C. Bateman, who has mapped and interpreted Sierran geology for over a quarter of a century. His studies have been complemented by the work of such senior scientists as L. D. Clark, J. B. Eaton, N. K. Huber, Anna Hietanen, J. G. Moore, and C. D. Rinehart.

In 1967, Bateman proposed that magma was generated by depressing and downfolding of a geosynclinal accumulation of Precambrian and early Paleozoic sediments whose compositions favored reconstitution into granitic magma. Emplacement accompanied and followed compression in the tectonic episode initiating the Nevadan orogeny.

By 1974, new field data had prompted Bateman to present an alternate model. He proposed that magma was generated in an east-dipping subduction zone, where the Pacific plate slid beneath the North American plate. The magma then migrated upward to fuse lower crustal material. This process generated magmas with composition intermediate between those derived primarily from mantle sources and those derived from crustal sources.

Studies of the rubidium and strontium isotopic components of Sierran plutons reveal a high incidence of rubidium in the trace-element chemistry, with highest percentages in the range's eastern axis. This supports the view that granitic magmas were derived primarily from crustal materials, contaminated only minimally by increments of mantle. Studies in the northern Sierra reinforce the position that granitic magmas are crustally derived, whereas the older gabbroic and dioritic plutons are probably from mantle sources.

Structure

Faults Prebatholithic rocks are complexly faulted, especially in the northwest and north-central Sierra. Faults of the Mother Lode both predate and postdate the batholithic invasions. Prominent internal fault patterns were partially responsible for the idea advanced by earlier workers that an ancestral Nevadan orogeny occurred in the Permian or late Pennsylvanian. This view is being revived because investigators have reported evidence in the Klamath province for late Paleozoic age of thrust-fault sheets and for a second, Jurassic episode of less defined pattern. These episodes also seem to be recorded in the stratigraphy of the northwestern Sierra, where pre-Jurassic faults are regionally significant. Although correlation between prebatholithic faulting in the northwestern Sierra and the Klamath Mountains cannot be established unequivocally, circumstantial evidence from both structural relations and rock sequences definitely corroborates the similarity of Sierran and Klamath histories.

Significant postbatholithic faulting does occur in the basement complex, but it is older than the present Sierran block faulting. Prominent older faults are the Kern Canyon fault and the sheared and lineated margins of Dinkey Creek and other plutons. Additional older faults are the lineaments along subordinate crests of the Sierra like the Great Western Divide, where aligned scarps have been interpreted as fault-line features.

The Kern Canyon fault is one of the largest structures entirely within the Sierra. Located in the north-south trending upper Kern River drainage, the fault parallels the river for 75 miles (120 km)

Sierra Nevada

Figure 2-5. Kern Canyon: view to the south from approximately Golden Trout Creek. This segment of the canyon is cut along the Kern Canyon fault, seen in the west wall (right foreground). (Photo courtesy of U.S. Geological Survey)

(Figure 2-5). Latest movement was more than 3.5 million years ago, because the fault is truncated by an elevated erosional surface on which rests a basaltic flow of that age. Some geologists have suggested a connection between the Kern Canyon fault and the White Wolf fault of the Tehachapi Mountains—Great Valley boundary, but little evidence supports this view. Figure 2-5 shows how strongly the Kern Canyon fault has influenced drainage direction and present topography. The Kern River follows the fault for some of its drainage, producing the prominent fault-line valley, but Figure 2-6 shows that fault and present-day drainage may not always coincide.

Figure 2-6.
Kern Canyon fault:
view to the south
from south of Sal-
mon Creek. Kern
River follows ir-
regular canyon to
the right of the
fault-line valley.
(Photo courtesy of
U.S. Geological
Survey)

Folds   Folding has been continuous and complex since the late Paleozoic. At the culmination of the Nevadan orogeny, the thick Sierran sediments had been transformed by progressive waves of deformation into a giant overturned synclinorium with steep west-dipping beds on the western margin and equally steep east-dipping beds on the eastern margin. This gross structure included many lesser folds of local significance. How many times folding deformed the subjacent formations is not known, but evidence for at least several episodes has been reported.

Unconformities   Many local and regional interruptions occurred in the history of Cordilleran sedimentation. There were periods of deformation, erosional interludes, and intervals of no deposition at all—preceded and followed by major and minor folding. This pattern has been established by detailed studies of the Precambrian, Paleozoic, and Mesozoic beds exposed in ranges from the Wasatch of Utah to the coast of California and Oregon.

Joints   Sierran joint patterns, displayed mostly in plutonic rocks, are important in the modern landscape. Both vertical and horizontal directions can be seen. Of the vertical type, northwest-southeast and northeast-southwest systems are especially prominent. They are imposed on all basement rocks regardless of age and have been critical in developing giant domes like those in Yosemite Valley and Kings Canyon. The horizontal pattern is expressed by sheet joints that appear as horizontal or nearly horizontal structures in the plutonic rocks. These provide the "layer-cake" rock patterns typical of many high ridges in the Sierran interior.

Many faultlike structures occur in plutonic igneous rocks, but they are not mapped easily because there are no readily available planes of reference for establishing separation. Some of these structures are probably no more than joints, although others may be faults of large separation. High-altitude satellite photography has revealed hitherto unmapped lineaments in Sierran terrains. Figure 2-7 suggests rectilinear lineament patterns, prominent in the larger

Figure 2-7.   Lineaments in Sierran granitic rocks. (Photo courtesy of U.S. Geological Survey)

exposures of plutonic rock. The patterns probably reflect intersecting joints, but ancient fault traces like the Kern Canyon show just as clearly. Field work may eventually locate these linear expressions, but often vegetation, soil, and uniformity of rock type preclude field confirmation.

ROCKS: SUPERJACENT SERIES

Overlying the Sierran basement are late Mesozoic and Cenozoic sediments and volcanics deposited after the long erosional interval that followed the Nevadan orogeny. Although the region generally had low relief during this interval, local relief may have reached 3500 feet (1050 m). The Cenozoic volcanic flows and volcaniclastics were deposited discontinuously. Nonmarine basins received continental, lacustrine, and fluviatile deposits, and marine sediments overlapped the western margin of the former Nevadan Mountains. These disconnected deposits have been labeled the *superjacent series.*

North of the Tuolumne River, the superjacent series covers ancient river channels. South of the Tuolumne, the superjacent cover is extremely fragmented, irregular, and discontinuous. Apparently volcanism was restricted or else Cenozoic erosion has removed much of a once-extensive cover. The limited volcanic cover in the southern Sierra is often quite young, sometimes even interbedded with Pleistocene glacial deposits. Prominent and extensive erosional surfaces are preserved under remnants of volcanic flows in the southern Sierra west of Whitney Crest, and in the Kern, San Joaquin, and Merced river drainages.

Northern Sierra

Cretaceous marine sediments overlap Sierran basement rocks along the Sacramento Valley margin. The sediments crop out discontinuously and include fossiliferous sandstones, shales, and basal conglomerates. Maximum exposed thicknesses are about 2800 feet (850 m). Well records have shown that Cretaceous fossiliferous sandstones occur near the edge of Sierran basement as far south as the San Joaquin River.

Eocene sedimentation is represented by up to 1000 feet (300 m) of marine and fluviatile deposits resting on deeply weathered basement. The beds are composed of discontinuous, interbedded clay, sand, and shale with occasional lignite beds. This suggests a marginal marine and deltaic, nearly tropical environment. The upper Ione formation contains quarries that produce glass-sands and clays.

The post-Eocene gravels and conglomerates distributed along the western third of the Sierra north of Tuolumne River are younger than the Ione formation but older than the initial volcanism characteristic of the superjacent series. Study of the oldest gravels (extensively mined for gold) has permitted reconstruction of early Tertiary drainage patterns and gradients. Eocene and Oligocene streams drained west and southwest, forming a major network of five rivers. Subsequently, other gravels were deposited and interbedded with volcanics whose eruption greatly altered the older drainage by shifting and burying former channels. In some places these interbedded gravels are gold-bearing, but they are less extensive and have less consistent values than the prevolcanic sequences.

Volcanic rocks are abundant in the northern Sierra, where there have been three major Tertiary volcanic episodes: (1) late Oligocene to early Miocene—prominently rhyolitic; (2) middle Miocene to late Pliocene—thick and widely distributed andesitic and basaltic extrusives; and (3) late Pliocene–early Pleistocene to Recent—discontinuous but regionally important dominantly basaltic volcanism. These episodes are based primarily on potassium-argon radiometric dates. Episode 1 gives ranges of 19.9 to 33.2 million years, episode 2 ranges of 8.8 to 9 million years, and episode 3 ranges of 2.9 to 3.5 million years and younger.

Southern Sierra

Typically, southern Sierran volcanics have been extruded from isolated vents, are limited areally, and belong to later volcanic episodes than those in the northern Sierra. An excellent example of this volcanic pattern occurs in the Owens Valley, north of Independence. Here the steep Sierran escarpment shows spectacular, deep, V-shaped canyons from which streams are building huge alluvial fans that form aprons extending almost the width of the valley. In the canyon cut by Sawmill Creek, a basaltic lava flow emerged from vents high on the escarpment, flowed down Sawmill Canyon, and displaced the stream. Subsequently the stream recut its canyon through the lava. Field study has established that the lava erupted between glacial stages. Morainal deposits are preserved above and below the lava, which congealed no more than 90,000 years ago. Lava flows and eruptive centers in isolated areas of the southern Sierra correlate in age with the extensive extrusions of the last two volcanic episodes of the northern part of the range.

Of the three volcanic episodes of the northern superjacent series, only the younger two are recognized in the south. Basalts characteristic of the second episode have been found in Pliocene flows on the San Joaquin River and on Coyote Flat west of Bishop. Basalts of the third episode are represented by flows in the Kern River Canyon at Trout Meadows, in the South, Middle, and North forks of the San Joaquin, in the Owens River gorge, and at McGee Mountain. Pleistocene to Recent volcanism occurs at Mono Lake, Devils Postpile, Sherwin Hill, the volcanic fields of the Owens Valley, Toowa Valley volcanic field, Mammoth Mountain, and Mono and Inyo craters.

HISTORY OF NEVADAN MOUNTAINS

The story of the Sierra Nevada begins with the deposition of thousands of feet of sedimentary rock into the north-south oriented basin of the Cordilleran geosyncline. (As indicated previously, plate tectonic theory has raised doubts that the basin was a classical geosyncline, since the sediments may have accumulated over subduction zones as the North American plate rode west.) This nearly continuous depositional episode is recorded in the roof pendants and large inclusions described earlier. The primarily depositional Sierran regimen of clastic sedimentation and volcanism (much of it submarine) continued until the Nevadan orogeny. By the early Jurassic, the Nevadan orogeny had culminated in a large, dominantly folded mountain belt with extensive granitic batholiths. The Nevadan Mountains faced the sea along what is now the eastern edge of the Great Valley, and subsequent marine invasions have not extended any farther east.

The development of the Nevadan Mountains poses some intriguing geologic problems. What is the magmatic source for the plutonic episode? How much rock of the Cordilleran depositional basin was available for deformation originally? How much was used, if any, in making the granitic magmas? What percentage of the original sediment do today's outcrops and inferred distributions represent? How high were the Nevadan Mountains? How much erosion of cover (roof) rock occurred before the granitic bodies could be exposed?

How high were the Nevadan Mountains? Speculation is easy, but evidence regarding elevations of any former mountain range is usually obscure. Normally, at least three approaches are used: (1) thickness of deposits that were derived from the mountain mass; (2) slopes of reconstructed land surfaces preserved under later deposits; and (3) high mountain ridges, especially those with north-south trend, which block approaching weather systems and create climatic effects possibly reflected in nonmarine deposits in the lee of the range.

Evaluating the Nevadan Mountains under the first approach incorporates the fact that post-Jurassic marine sediments in the western Sacramento Valley are at least 25,000 feet (7600 m) thick. Sedimentary structures establish that the Cretaceous sediments were derived from sources to the east and northeast, presumably the Nevadan Mountains. If a 5.5-mile (8.8 km) thickness of sediment accumulated in an adjacent marine basin, did the Nevadan Mountains concomitantly rise 5.5 miles (8.8 km)? Probably not, although some unroofing of the mountain mass is likely. A cubic foot of consolidated rock fragmented by erosion occupies more volume when deposited than the original did, so a direct volume correlation between mountains and basin is unjustified.

The Cretaceous rocks involved are all shallow-water marine types, with fossils and conglomerates throughout the section. Consequently, the environment must have been primarily nearshore and relief of the source area reasonably great. Since the conglomerates contain cobbles of moderate (fist) size, it is plausible also that there existed either large streams with extensive drainage areas or short streams on steep slopes, or both. Furthermore, many of the cobbles are granitic and can be traced mineralogically to the granitic plutons of the Nevadan batholith.

The second approach, reconstructing ancient slopes, does not usually yield many clues because most buried profiles have been altered significantly by tilting. Nevertheless, interest in locating gold-bearing channels has produced more data about stream slopes of the Nevadan Mountains than exist for most fossil ranges, and substantial height for the ancient mountains is inferred.

Under the third approach, both pollen grain studies and strictly geologic criteria have been used. Marked aridity definitely occurred east of the Sierra, implying a high range in the path of rain-bearing storms from the Pacific. A persistent and prominent rain shadow thus is inferred.

Based on these data, speculation that the Nevadan Mountains were as high or higher than the modern Sierra seems justified, although estimates as low as 4500 feet (1370 m) have been given. Certainly the Nevadan Mountains were more extensive geographically than the Sierra Nevada. Moreover, it is not unreasonable, in view of the isoclinal fold axes in some roof pendants, to suggest a significantly high elevation for at least some parts of the Nevadan Mountains and minimal crests of 10,000 feet (3000 m).

In summary, Nevadan orogenesis followed deposition of dominantly marine sediments in the east and dominantly submarine volcanics in the west. Repeated disturbances both during and after deposition produced a major asymmetric synclinorium, characterized by extensive faulting on its west limb. Granitic magmas, generated below or within the synclinorium and probably heated by convection, penetrated the fractured synclinal rocks. The resulting plutons were concentrated centrally in the syncline. Assimilation and differentiation during magmatic emplacement produced the compositional variety within and between plutons. With impetus from depth and concomitant compression, the Nevadan orogeny built a substantial range that contributed sediment primarily to the west, forming the Great Valley sedimentary sequence.

The erosion that stripped 7 or more miles (11 km) of cover seems to have been largely completed by the close of the Eocene, leaving slowly moving streams draining from an eastern crest. The superjacent sequences of volcaniclastic and volcanic rocks were deposited upon the eroded roots of the Nevadan Mountains. Subsequent elevation by faulting rejuvenated the region and produced today's Sierra Nevada.

GEOMORPHOLOGY OF TODAY'S SIERRA

An impressive feature of today's Sierra is the even skyline that stretches for miles across the summit and near the summit (Figure 2-8). These subdued summit uplands are interpreted as evidence that erosional activity is destroying a fossil landscape, previously eroded across Nevadan roots and subsequently rejuvenated tectonically to present elevations. In this view, the undulating upland surface is the plain (peneplain) to which the Nevadan Mountains eventually were reduced. This landscape was a lowland, with elevations possibly as high as 3500 feet (1050 m), but with the crestline of the Nevadan roots near today's Great Western Divide, at an elevation of 2000 to 2500 feet (600–760 m). The present landscape results from erosion in the late Tertiary and Quaternary and rejuvenation primarily due to movement on the Sierra Nevada fault. A new crestline, the Whitney or Muir Crest, was formed. Rapid uplift aggregating more than 10,000 feet (3000 m) of vertical movement (mostly in the past 3 million years) produced the escarpment facing Owens Valley. The entire block was tilted west during uplift, renewing the erosional power of streams that rapidly cut new features into older topography (Figure 2-9). The Sierra Nevada's elevation augmented the cold period dominating the earth and glaciers developed, further modifying stream canyons.

Structure

Until the pioneer California geologist A. C. Lawson briefly studied the Kern River drainage in the early 1900s, the significance of Sierran high-level surfaces had not been noted. Lawson's classic papers on Kern River topography, followed by François Matthes's studies of Yosemite Valley, stimulated geologists to look more carefully.

Matthes systematically presented the generalized interpretation of the Sierra Nevada as a range: faulted extensively on the east and less so on the west, lifted about 10,000 feet (3000 m) on the Sierra Nevada fault, and containing rejuvenated streams that dissected older subdued landscape. Matthes's simplified concept is reproduced in Figure 2-10. Furthermore, Matthes recognized that streams incise their valleys in response to changes of baselevel. This produces canyons within valleys and canyons within canyons, features that are particularly helpful in interpreting geomorphic history.

Evolution of the Sierra and its rate of uplift reflect the fault

32

Figure 2-8. *Even skyline summit of Sierra Nevada. Looking southwest from latitude of Bishop, view includes the entire summit, gently sloping west and south to the San Joaquin Valley seen in the distance. Note erosional contrast between the summit Sierra and the lower western flank. Farmlands in the foreground (center) are in Round Valley. Steep ridge at lower right is Wheeler Crest. Pine Creek drains into Round Valley through paired moraines of Tahoe-Tioga glaciation. Mount Tom is in right center foreground, south of Pine Creek. Tungsten Hills lie between Round Valley and farmlands in left center. The lower gorge (bottom center) of the Owens River has been cut through Bishop tuff. Bishop Creek (center) drains into the Owens Valley south of the Tungsten Hills. In the hazy far left is the canyon of the south-draining Kern River. Note subordinate flats like Coyote Flat southeast of Bishop Creek and the flat immediately southwest of Mount Tom. The black spot on this flat is Desolation Lake, largest natural lake in the southern Sierra. (Photo courtesy of U.S. Air Force and U.S. Geological Survey)*

Sierra Nevada eroded fault scarp, as seen from the Owens Valley (east). *Figure 2-9.*
Fresh trace of the fault is indicated by the abrupt change of slope and the
fault facets terminating each ridge. Note that facets on the right do not
align with those on the left. Note also the upland flats to the west, beyond
the crest; the conical peak on the right is Monachee Mountain. (Photo
courtesy of U.S. Geological Survey)

systems on which uplift is occurring. Evidence for boundary faults is
prominent along the east side of the range, but is rarely seen along
the west side except in the Kern River–Tehachapi area.

Sierra Nevada Fault System Like other faults, the Sierra
Nevada is a zone of movement, not a line. It has a zigzag or en
echelon course and dominantly dip-slip offset (Figure 2-11). Vertical
displacement is up to 5000 feet (1500 m) in the north and 11,000 feet
(3350 m) in the south. Movement on the fault began in the middle
Miocene and accelerated in the Pleistocene, as demonstrated by the
overlap of glacial deposits, volcanics, and fault scarps. The Sierra
Nevada fault resembles Basin Range faults, of which it is probably
the westernmost expression. The Sierran block is the largest of the
north-south trending tilted fault blocks scattered across Nevada east
to the Wasatch block of central Utah.

Greenhorn Fault System Displacement from approximately
the San Joaquin River canyon to the Tehachapi Mountains has long
been attributed to uplift on parallel faults of the Greenhorn system.
Total uplift of 2000 to 7000 feet (600–2150 m) has been suggested
for these faults. Movement on the Greenhorn is thought to have
accelerated the deformation initiated by large-scale doming of the

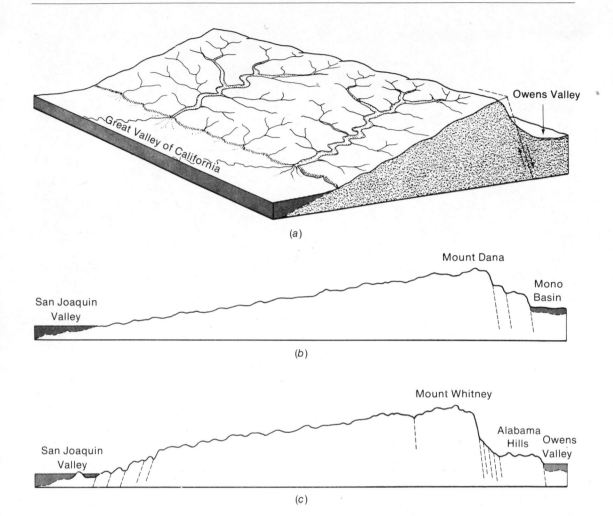

Figure 2-10.   (a) *Generalized diagram of part of tilted Sierran and Owens Valley blocks. Relative directions of movement are indicated by arrows. Height and slant of the Sierran block are much exaggerated. Streams are shown in characteristic arrangement, the main rivers flowing down the western slope with many of their tributaries at approximately right angles. No specific streams are represented. In front is a strip of the Great Valley, with thick layers of sand and silt derived from the elevated Sierran block. At the back is a strip of the Owens Valley, veneered with a thinner layer of sediment.*

   (b) *Idealized section across the Sierra Nevada, representing the simple "textbook conception" of its tilted-block structure. This representation approximates the facts known about the range in the latitude of Yosemite Valley.*

   (c) *Idealized section across the Sierra Nevada in the latitude of Sequoia National Park. Step faults exist at both western and eastern margins. Some distance from the western foothills, peaks of the down-faulted, buried fault block emerge as isolated hills above sediments of the San Joaquin Valley. (Source: U.S. Geological Survey)*

southern Sierra. Prominent scarps occur irregularly. The Greenhorn presumably contributed to the development of the plateaulike form of the southern Sierra Nevada, which contrasts with the tilted-block pattern north of the San Joaquin River.

The existence of a western fault system has been questioned, however, because the scarps of the supposed faults are both discontinuous and irregularly aligned. Since the faults (if they exist) are presumably high-angle normal faults, the irregularity is almost inexplicable even if erosion modified initial escarpments. In addition, basalts in the San Joaquin River drainage were extruded before the stepped topography was produced and apparently are not displaced.

Upland Surfaces

Sierran upland surfaces are more conspicuous in the south than in the north. Between 1904 and 1906, Lawson identified the erosional surfaces of the southern Sierra and suggested that the uppermost is two related surfaces, which he called the Summit Upland and the Subsummit Plateau. Since then, other designations have been suggested. At present, surfaces of the southern Sierra are named according to the terms given in Figure 2-12.

The upland surfaces are best developed in the Kern River drainage, probably because of drainage instability since initiation of up-

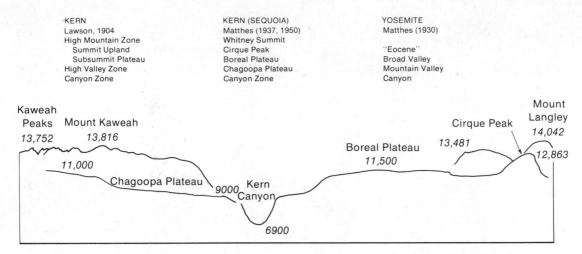

KERN
Lawson, 1904
  High Mountain Zone
    Summit Upland
      Subsummit Plateau
  High Valley Zone
  Canyon Zone

KERN (SEQUOIA)
Matthes (1937, 1950)
  Whitney Summit
  Cirque Peak
  Boreal Plateau
  Chagoopa Plateau
  Canyon Zone

YOSEMITE
Matthes (1930)

  "Eocene"
  Broad Valley
  Mountain Valley
  Canyon

Kaweah Peaks 13,752   Mount Kaweah 13,816   11,000   Chagoopa Plateau   9000   Kern Canyon   6900   Boreal Plateau 11,500   Cirque Peak 13,481   Mount Langley 14,042   12,863

Figure 2-12.   *Correlation of names applied to Sierran erosional surfaces and simplified profile across the upper Kern Basin, showing remnants of four ancient landscapes (erosional surfaces). (Source: U.S. Geological Survey)*

lift. The Kern is the only major north-south drainage among Sierran streams; all others flow west or southwest. As indicated previously, direction is controlled by the Kern Canyon fault, which has dominated the region since the Nevadan orogeny.

The highest surface is the summit flatland of Whitney Crest at nearly 14,000 feet (4250 m). Presumably it is the remnant of the oldest (Nevadan?) surface and so represents a peneplain formed during degradation of the Nevadan Mountains. (It should be noted here that we cannot assume that all benches, flat uplands, and associated mountain summits represent rejuvenated surfaces of former peneplains. According to some geologists, it is possible for so-called peneplains to develop at any altitude, though evidence for such alternatives is inconclusive.) Movement on the Sierra Nevada fault probably caused the uplift of the Whitney Crest subdued landscape. Quiescence followed, after which streams again dissected and partially leveled the region. This produced a Cirque Peak, sub-Whitney erosional surface about 1500 to 2000 feet (450–600 m) lower than the Whitney Crest. Before all evidence of the Whitney Crest was erased, however, there occurred rapid uplift and then another hiatus. During this quiescence the Boreal Plateau erosional surface appeared, 1000 to 1500 feet (300–450 m) below the Cirque Peak surface. Uplift occurred again, this time amounting to about 2500 feet (760 m), and subsequent erosion to the 9000-foot (2750 m) level produced the Chagoopa Plateau erosional surface. The Chagoopa stillstand was interrupted by uplift that continued into the present and produced the Kern River canyon. This canyon has an elevation of 6900 feet (2100 m) at the latitude of the profile of Figure 2-12.

Each of these surfaces is preserved southward, particularly the Boreal and the Chagoopa, in the Kern River canyon. Figures 2-13 and 2-14 illustrate the bases for establishing stages of uplift.

Debate regarding the number of recognizable upland surfaces is substantial, but it is exceeded in intensity by the debate concerning the ages of these features. Unfortunately, there are few posterosional surface rocks in the southern Sierra. Age assignment is therefore speculative, except where volcanic flows cover surfaces and radiometric dating can be applied.

The Kern Basin lava flows that partly cover the Chagoopa surface are 3.5 million years old, making the surface Pliocene. The lava has buried canyons that had 800 to 900 feet (240–275 m) of relief, proving that the Chagoopa was considerably dissected before the lava extruded. A date on the Coyote Flat lavas gives an age of 9.6 million years. Since Coyote Flat is presumed to be the somewhat warped correlative of the Chagoopa erosional surface, the surface must be early Pliocene at the youngest. Dates in other parts of the Sierra also imply early Pliocene age for the Chagoopa. Although the evidence is inconclusive as yet, this analysis does suggest the type of sequences still to be established for post-Nevadan geomorphic history.

Figure 2-13. Looking north across a glaciated section of the Chagoopa Plateau in the headwaters of the Big Arroyo, a tributary of the Kern River. (Photo courtesy of U.S. Geological Survey)

Figure 2-14. Looking north from Cottonwood Lakes across the Whitney (Muir) Crest (left foreground) and the Boreal Plateau (right center). Note cirque embayments of the Whitney Crest. Each is a different size, depending primarily on the extent of headward stream erosion that preceded glacial excavation. (Photo courtesy of U.S. Geological Survey)

GLACIATION

A wide array of glacial features exists in the Sierra Nevada. Almost everywhere in the higher parts of the range the landscape is the product of glacial and periglacial processes. Nevertheless, despite the ubiquitous evidence of past glaciers, only about 60 small glaciers remain today. These are probably best described as glacierets although the largest, Palisade Glacier, is about a mile (1.6 km) long. Similarly small patches of glacial ice persist on the upper slopes of Mount Shasta and in a few sites in the Klamath Mountains. There are no sizable modern glaciers anywhere in California. Moreover, the state's existing glaciers nearly disappeared during the warm interval that affected much of the world from about 1850 to 1950. Even a moderate climatic warming would destroy California's remaining glaciers.

During the last main glacial stage (the Wisconsin), the Sierra Nevada had a discontinuous ice cap that extended about 275 miles (450 km), from the Feather River to the upper Kern River. In several places, this ice cap was more than 30 miles (48 km) wide. Few, if any, areas outside the polar regions sustain such a large ice cap today. The longest glaciers from this summit ice cap extended about 60 miles (96 km) down the gentle western slope toward the San Joaquin Valley. Glaciers on the eastern slopes did not exceed 10 miles (16 km) in length.

Glaciers have sculptured some of the world's grandest scenery, whether it be the landscapes so well displayed in Glacier and Jasper national parks in the Rocky Mountains of the United States and Canada or the dramatic coastal scenery in the fjord country of Norway and New Zealand. Glacial landscapes were definitely the main reasons for establishing Yosemite Valley as a state park in 1866 and later as a national park and for establishing Kings Canyon and Sequoia national parks in the southern Sierra Nevada.

Evidence

Although the signs of former glaciers are everywhere in the high Sierra, it has been only about a hundred years since the influence of vanished ice sheets was accepted by the scientific community. The Sierra Nevada's numerous steep-sided, U-shaped valleys, including Yosemite (Figure 2-20) and the upper Kern River (Figure 2-5), all reflect glaciation. Cirques, often with nearly vertical walls, frequently contain lakes enclosed sometimes by bedrock and sometimes by moraines deposited during the last glacial recession. (Older moraines are usually destroyed by subsequent glacial activity.)

The varied effectiveness of the ice as it moved over bedrock produced many depressions now occupied by small lakes called *tarns.* Sierran glaciated valleys often show distinctly stepped profiles. Gentle or even reversed slopes on the treads of these stairs are interrupted by steep, abrupt risers whose upper edges may form rims just high enough to contain small lakes, called *beaded or paternoster lakes.*

Rocks excavated at higher elevations by glacial ice often have been carried some miles downslope and then abandoned when the ice melted. It is usually apparent that these rocks are derived from distant sources. The rocks commonly lie scattered over glacially planed, bare surfaces, from which they obviously were not plucked by the moving ice sheets. Most erratic boulders are composed of the same kind of rock as that on which they rest. If, however, an erratic is distinctive enough to establish its source, direction and amount of ice movement can be calculated.

Other glacial geomorphic features are the numerous moraines that may form ridges 400 feet (120 m) high. Generally, moraines are better developed on the steep eastern slope of the Sierra than on its western side. Lateral moraines particularly are conspicuous near the mouths of east-facing Sierran canyons between Tioga Pass and Bishop. In Figure 2-8, the two streams flowing into the cultivated areas are Bishop Creek (left) and Pine Creek (right). Both have prominent lateral moraines of Tahoe age. Those of Pine Creek extend from the base of the mountains onto the floor of Round Valley. Loop-shaped terminal moraines occur throughout the Sierra, but perhaps the best known are at Convict Lake north of Bishop. Here young moraines help form the basin that contains the lake. Another notable example is south of Lake Tahoe at Fallen Leaf Lake, where 10 to 15 arcuate moraines curve around the lake's lower end.

Sequence

Although the existence of several stages of Sierran glaciation had been suspected, François Matthes proposed the first set in 1930. Based on his studies in the western slopes, he recognized three stages: from oldest to youngest, Glacier Point, El Portal, and Wisconsin. Assigned to the Wisconsin were the youngest and least weathered moraines, both lateral and terminal. El Portal deposits, on the other hand, were more deeply weathered and generally incorporated lateral moraines without terminal loops. El Portal was believed to reflect a more extensive glaciation than did the Wisconsin. The Glacier Point stage was thought to be preserved only as scattered erratics on weathered bedrock that had escaped later glaciation. In 1931, as a result of his work on the eastern slopes, Eliot Blackwelder proposed a four-stage sequence: from oldest to youngest, McGee, Sherwin, Tahoe, and Tioga. Both sequences have been greatly augmented since Blackwelder and Matthes did their pioneer work. Table 2-2 gives the sequence of glacial tills now recognized, chiefly on the eastern slopes.

Most tills of 50,000 years and younger are dated by radiocarbon methods, but much glacial chronology remains relative. There are few radiometrically controlled dates older than 50,000 years, the

Table 2-2
Sierran Glacial Sequence: Eastern Slope

Epoch	Till	Approximate Time (Ages given are before A.D. 1950)
HOLOCENE	Matthes till	0–650 years (before existing glaciers)
	Unnamed till	1000 years
	Recess Peak till	2000–6000 years
	Unnamed till	6000–7000 years
		10,500 years
PLEISTOCENE	Hilgard till	11,000 years
	Tioga till	20,000 years
	Tenaya till	26,000 years
	Tahoe till	50,000 years
	Mono Basin till	87,000 years (Illinoian?)
	Donner Lake till	250,000 years
	Casa Diablo till	400,000 years
	Sherwin till	750,000 years (Kansan?)
	McGee till	1,500,000 years (Nebraskan?)
	Deadman Pass till	3,000,000 years
PLIOCENE		

Figure 2-15.
Bishop tuff resting on Sherwin till at Sherwin Grade summit on U.S. Highway 395. (Photo by Robert M. Norris)

limit of the radiocarbon method. Moreover, because of the uncertainties of the carbon-14 method, even the dates for younger glacial events do not necessarily equate with calendar years. The ages of older tills are even less well controlled, but there are some anchor points in the sequence.

One of these is the basaltic flow that nearly reaches the mouth of Sawmill Canyon near Independence. This flow has yielded a potassium-argon date of 60,000 to 90,000 years. It rests on an older till, presumably of Sherwin age, and is in turn overlain by a Tahoe till. The Bishop tuff has a potassium-argon age of 710,000 years and rests on Sherwin till (Figure 2-15), which in turn rests on a basalt dated at 3.2 million years. Hence the age of the Sherwin till is older than 710,000 years and much younger than 3.2 million years; the best guess today is 750,000 years. The Casa Diablo till is capped by one basaltic flow and rests on another. The overlying basalt has a potassium-argon age of 280,000 ±67,000 years and the underlying an age of 440,000 ±40,000 years; the age for the Casa Diablo till is estimated at 400,000 years. Basaltic flows dated at 2.6 million years lie beneath the McGee till. Both basalt and till have been elevated as much as 3000 feet (900 m) by faulting largely completed before the Bishop tuff was extruded. This substantial displacement emphasizes the accelerated movement on the Sierra Nevada fault system during the late Pleistocene. Unfortunately, this last example neither provides a tight age control for the McGee till nor allows us to distinguish it from the Sherwin till. Along Rock Creek, however, undoubted Sherwin till rests on an older, more weathered till, possibly of McGee age. Finally, the Deadman Pass till is assigned an age of about 3 million years, making it the oldest Pleistocene glacial deposit so far recognized in temperate latitudes. The basis for this age is a capping quartz latite flow dated at 2.7 million years and an underlying andesitic flow dated at 3.1 million years. This till is found along the crest of the Sierra in the Minarets region.

Of the glaciations listed in Table 2-2, the Tenaya and Tioga are best displayed on the western slopes and the Tahoe on the eastern slopes. Older glacial deposits seldom have any topographic expression and usually are buried under younger materials. On the east side of the Sierra, the largest and most prominent moraines that extend from canyons are generally of Tahoe age, and the less prominent ridges and morainal loops within the Tahoe embankments are usually Tioga. Tahoe and Tioga moraines with this association are well displayed at Convict Lake. Tenaya and Tioga moraines have similar patterns and often are indistinguishable, but the older Tenaya moraines are more apt to be modified by later erosion. Correlating the glacial deposits of the western and eastern slopes with those of the central and northern parts of the Sierra has been difficult. The reason is that most tills are confined to canyons and are not continuous for appreciable distances. Moreover, the older tills have been disrupted by subsequent glacial advances and stream erosion.

Modern Glaciers

The 60 or so glacierets remaining in the Sierra Nevada are restricted to cool, moist, north-facing or northeast-facing slopes that receive minimal direct sunshine even in summer. The glaciers are limited to these sites because no part of today's Sierra is high enough to permit snow to lie on south-facing slopes indefinitely. The hypothetical elevation above which some snow would persist on south-facing slopes is called the *climatic firn limit.*

It is still widely supposed that today's little glaciers are merely the shrunken remnants of grander Pleistocene forebears. François Matthes pointed out, however, that it is more likely that *all* present glaciers are less than 4000 years old, the opinion now sustained by most Sierran glaciologists. The relatively warm period that followed the Wisconsin glacial stage is known as the *climatic optimum* or *hypsithermal interval;* it probably began between 7500 and 9000 years ago and ended about 3000 years ago. It is unlikely that the Sierra contained any glacial ice at all during most of the warmer hypsithermal.

The first of three recent ice advances known collectively as the Neoglaciation or Little Ice Age began about 2800 years ago. The smallest advance reached a maximum about A.D. 1350, and the last and probably largest about A.D. 1750. Marked shrinkage of all Sierran glaciers took place between about 1850 and 1950, although even then some small ice advances occurred. If continued, the widely documented cooling tendency observed since about 1950 undoubtedly will induce another advance of Sierran glaciers.

SUBORDINATE FEATURES

Buried Topography of the Western Sierra

The Sierra Nevada contains a network of ancient stream channels carved during elevation of the Sierran block. Periodically the channels were filled with alluvial debris from upstream diversions or stream capture, and lava flows displaced some streams and buried the channelways. This pattern was repeated particularly in the Sierran foothills from the Merced River north to the Feather River.

Because many of the buried channels carry placer gold, their recognition constituted an important practical geologic problem, and many ancient channels have been plotted by geophysical surveys. Several channel cycles are known, and their slopes and directions are used to unravel Sierran landscape history from the close of the Cretaceous. Eocene channels have been identified, plus many of Miocene and Pliocene age.

Volcanics of Table Mountain

West of Jamestown and Sonora, in the Mother Lode foothills, is flat-topped, sinuous Table Mountain. This mesa is composed of lava flows up to 300 feet (90 m) thick and has a maximum width of 2000 feet (600 m) (Figure 2-16). The flows are 9 million years old, placing them in the intermediate volcanic sequence of the superjacent series. Superficially, Table Mountain resembles other mesalike landforms: a resistant lava cap burying weaker rock undercut by

Figure 2-16. *Topographic inversion at Table Mountain: view to the northeast, from Jamestown. Light bands on the flow are unrelated to its course and may be the surface expression of northeast-trending regional jointing, propagated upward from the basement. (Photo by Spence Air Photos, courtesy of Department of Geography, University of California, Los Angeles)*

erosion to maintain the steep cliffs on the more durable lava flow. Actually, however, Table Mountain is part of a larger lava flow that coursed down a river canyon, filling the channel and displacing the river. Subsequent erosion has inverted the topography so that the former stream valley is now a ridge—Table Mountain—whose lava flows can be mapped discontinuously for more than 40 miles (64 km). Such concise examples of reversed topography are rare.

Tehachapi Mountains

The Tehachapi Mountains separate the Great Valley from the Mojave Desert and rise to 6934 feet (2115 m) at Bear Mountain. The Great Valley–Tehachapi topographic boundary is primarily between sedimentary and crystalline rocks and is expressed by a fault-line scarp eroded along the White Wolf fault. Movement on the White Wolf fault presumably triggered the 1952 Tehachapi and Bakersfield earthquakes.

The rocks of the Tehachapi Mountains are compositionally like those of the Sierra Nevada. Moreover, they reveal particularly well the late Paleozoic or early Mesozoic sedimentary record and the early Mesozoic intrusive record of the Nevadan orogeny. Tehachapi roof pendants show the same northwest-southeast trend as those of the Sierra, but as a mountain block the Tehachapi trend is more nearly east-west. This appears to reflect left-slip movement on the Garlock fault, the master structure separating the Tehachapi Mountains from the Mojave Desert. The Garlock fault either cuts off the Sierra Nevada fault in the vicinity of the El Paso Mountains or the Sierra Nevada fault is sheared en echelon so that it curves west to join the Garlock. In any case, the result is the major barrier of the Tehachapi Mountains, which separate northern and southern California. Since early Tertiary time, this barrier has protected the Mojave desert from marine waters with the brief exception of a late Miocene invasion into the western Antelope Valley.

Within the Tehachapi Mountains are several flat-floored valleys described as grabens. Although unequivocal evidence of boundary faults is not available, the valleys either have internal drainage or show evidence of former internal drainage. Consequently, graben origin is essential.

Round Valley–Bishop Creek Embayment

Owens Valley is a narrow linear valley over 150 miles (240 km) long, usually less than 8 miles (13 km) wide, and bounded by major faults. Near Bishop, a wide reentrant increases the width to almost 20 miles (32 km). This reentrant is Round Valley, bounded on the west by an extremely steep Pleistocene scarp that is the east face of Wheeler Crest. (Wheeler Crest is sometimes called Wheeler Ridge and should not be confused with the oil-producing area of the same name in the western San Joaquin Valley.) To the south, Round Valley rises about 4200 feet (1280 m) across an alluvial fan to form a ramplike surface (the Coyote Warp) that culminates at the head of Bishop Creek at about 10,000 feet (3000 m). Wheeler escarpment dissipates to the south, merging into the chain of Sierran summits of which Mount Tom is the northern anchor. East of Bishop Creek, Coyote Flat drops steeply to the Owens Valley from 12,000 feet (3650 m). To the north, Round Valley is blocked by the Bishop volcanic tableland. Figure 2-8 places Round Valley in topographic perspective.

Movement and distribution of individual faults of the Sierra Nevada system are nowhere better demonstrated than in the Round Valley reentrant. The Sierra Nevada block faces east at Wheeler Crest. Here the major active Sierran fault is the Wheeler Crest fault, which emerges from beneath the volcaniclastic rocks and glacial debris of the Bishop volcanic tableland to become the major boundary between the Sierra Nevada and Round Valley for 10 to 15 miles (16–24 km). To the south, the main Sierran escarpment is offset eastward for about 10 miles (16 km) by another Sierran fault, which bounds Owens Valley in the Bishop–Big Pine area. This en echelon pattern typifies the Sierra Nevada fault system throughout most of its 400-mile (640 km) definable length.

Bishop Volcanic Tableland

Bounding Round Valley on the north is a major volcanic tableland that covers about 325 square miles (850 km²) (Figure 2-17). It is composed of the Bishop welded tuff, which is probably derived from the Long Valley–Crowley Lake area. The tableland extends east and south to within a few miles of Bishop, and its tuff forms bold escarpments and other prominent landscape features in northern Owens Valley. The tuff was deposited across an irregular ridge-valley topography cut by both glacial and preglacial streams. As mentioned before, its association with glacial tills has provided a means for dating some Sierran glaciations.

The Bishop tuff is not a single volcaniclastic unit, but is composed of many layers of tuff, ash, and pumice. Typical Bishop tuff is dark gray when fresh and weathers to pink, white, and brown. It includes glass and fragments of earlier-formed volcanic materials, plus granules of silicic minerals. The tuff's surface is irregular, with hummocks produced by gas that escaped during eruption, and shows

Figure 2-17. *Bishop volcanic tableland (upper right), with Wheeler Crest in the background. The Tungsten Hills are the low hills in the upper left. City of Bishop is in the foreground. The scarp at the edge of the tableland has been cut by the Owens River. (Photo by Spence Air Photos, courtesy of Department of Geography, University of California, Los Angeles)*

prominent columnar and radial joints. The rock is considered a welded tuff (ignimbrite), the product of hot, gaseous clouds that showered the countryside, ultimately producing a pasty flow of incandescent near-magma that moved downslope into the Owens Valley and also covered a large area toward Mono Lake.

Because of its significance in glacial chronology and as an unusual volcanic type, the Bishop tuff has received considerable attention. It has been mined for building stone and pumice, and the picturesque gorges of Owens River and Rock Creek are cut through the tuff into the basement rocks below.

Long Valley and Crowley Lake

Long Valley, a major feature between Bishop and Mammoth, was occupied during glacial times by a lake up to 300 feet (90 m) deep. The valley now is occupied by man-made Crowley Lake, formed by damming the Owens River at the valley's southeastern end. The valley covers about 75 square miles (195 km²) and is bounded on the west and southwest by the Sierra Nevada, on the north by Glass Mountain, on the east by the Benton Range, and on the south by a low plateau of volcaniclastic materials that include the Bishop tuff. It is drained by Owens River, whose major tributaries from the Sierra are Hilton, McGee, Convict, and Mammoth-Hot creeks. Geophysical studies indicate that Long Valley is a sinking block in which volcanic, volcaniclastic, and alluvial debris have accumulated to thicknesses of more than 10,000 feet (3000 m).

On the southwest side of Long Valley is an en echelon fault of the Sierra Nevada system. One segment extends south from Long Valley into the Sierra, nearly paralleling Hilton Creek. (U.S. Highway 395 follows the major escarpment of this fault for a mile or two south of the Convict Creek crossing.) Recent studies suggest that Long Valley is bounded on all sides by faults that collectively form a circular or semicircular trace. Although such faults are difficult to establish because of overlapping volcanic rocks, one hypothesis is that the circular fault pattern may define a caldera whose collapse followed emission of the fiery clouds that produced the Bishop tuff. The size of the caldera is suggested by the estimate that the Bishop tuff involves 30 to 40 cubic miles (126–168 km³) of erupted material.

Inyo and Mono Craters

Inyo and Mono craters are two separate but aligned volcanic centers often seen by visitors to the high Sierra. The alignment results from eruption along a major northwest-southeast fissure (probably a major fault in the Sierra Nevada system) that extends about 40 miles (64 km).

Figure 2-18. *Looking southwest over Mono Craters. The plug-in-crater pattern shows clearly in the largest crater (left center). The rounded mountain in front of the main Sierra in the upper right is Mammoth Mountain. (Photo by John S. Shelton)*

The southern sector contains the two explosion pits of Inyo Craters. Small lakes nestle in the pits, associated with flows of obsidian and pumice. These flows are the youngest volcanic features of the chain and yield radiocarbon dates between 500 and 850 years.

The northern domes of the chain are the Mono Craters (Figure 2-18), a group of primarily rhyolitic tuff rings and obsidian and plug domes that constitute at least a dozen eruptive centers. The steep-sided rhyolitic flows have oozed out of tuff rings and are visible from U.S. Highway 395. The plug-dome character is apparent because some younger effusions have pushed older materials up, "floating" them as it were with the younger material squeezing out underneath. On the other hand, some younger plugs have moved up through older material and overflowed the crater rims.

Ages of the Mono Craters are from 6000 to 6500 years. The northernmost eruptive center is Black Point on the northern shore of Mono Lake, where a basaltic cinder cone (most Mono centers are rhyolitic) appears to be as young as the last highstand of glacial Lake Russell. A minor eruption may have occurred beneath the waters of Mono Lake about 80 years ago. Some observers, however, consider this incident only another of the many fluctuating hot springs that periodically form and disappear in this very new volcanic area.

June Lake District

The June Lake area lies at the base of a particularly steep part of the Sierra Nevada's eastern face. The dominant topographic feature is a horseshoe-shaped trough whose toe abuts the steep escarpment. The depression contains four lakes: Grant in the northwest arm, Silver near the toe of the horseshoe, and Gull and June in the southeastern arm (Figure 2-19). It is the drainage rather than the lakes themselves, however, that makes this area of special interest.

In 1889, I. C. Russell was struck by the peculiar course of Reversed Creek, which drains Gull Lake and flows directly toward the Sierra instead of away from it as is usual for Sierran streams. Before reaching Silver Lake, Reversed Creek is joined by Rush Creek at the toe of the horseshoe, and the combined stream (called Rush Creek)

Figure 2-19. Drainage of Reversed Creek, June Lake area.

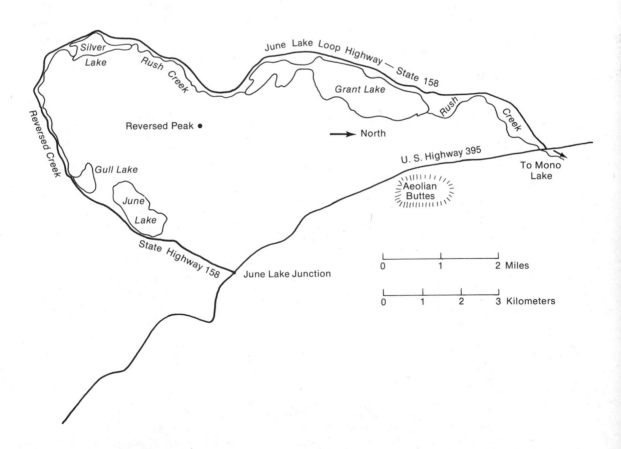

flows into Silver Lake and from there into Grant Lake, which drains into Mono Lake. Since the extension of the Los Angeles aqueduct system into the Mono area, Grant Lake has been used as a reservoir, so now much of Rush Creek's flow is intercepted before it can reach Mono Lake.

Russell was the first to offer an explanation for the anomalous course of Rush Creek. He suggested that the horseshoe-shaped trough was formed by a glacier forced to divide into two arms by Reversed Peak (9478–2891 m), which rises about 1800 feet (550 m) above June Lake. He contended that the lobe that excavated the June Lake arm left a terminal moraine across the valley just below June Lake's present position and that a large block of glacial ice remained in the basin while the parent glacier melted back into the Sierran uplands. It was thought that the ice mass and the moraine produced meltwaters that subsequently cut a channel toward the Sierran front.

Half a century later, J. E. Kesseli examined the area and pointed out the difficulty of explaining why some of the glacier obligingly remained in the June Lake depression while its bulk melted back into the higher mountains. He proposed that the horseshoe-shaped trough was an old, entrenched meander curve of a large river that flowed from the north through Grant, Silver, Gull, and June lakes. The subsequent reversal of flow was accounted for by tilting of the valley block to the northwest, probably by faulting. This explanation also necessitated vigorous stream erosion in the Grant Lake arm, which is almost 400 feet (120 m) deeper than the June Lake arm. The weakness of Kesseli's interpretation is that little independent evidence substantiates either the large meandering river or the tilting and deep erosion of the Grant Lake arm.

In 1949, W. C. Putnam concluded that the Rush Creek glacier did divide into two lobes as it moved around Reversed Peak, but excavation was more effective in the Grant Lake area because bedrock there is less resistant than in the June Lake arm. The Grant Lake lobe of the glacier also was narrower and much thicker. The reversed slope in the June Lake arm merely expresses increased rock resistance along the arm. Furthermore, glaciers frequently cut valley floors with reversed slopes because glacial erosion depends on the slope of the ice surface, not on the slope of the valley floor.

SPECIAL INTEREST FEATURES

Gold Mining

Most of California's gold came from the Mother Lode belt of the foothills of the western Sierra Nevada, but gold also was produced in quantity in the Klamath Mountains and to a lesser extent in the

Basin Ranges. Estimates of California's production vary, but it is probably a fair guess that about $1 billion in gold has been extracted, mostly before 1900 and predominately from the Mother Lode. Interest in the Mother Lode remains high, mainly on the part of tourists and those interested in California history.

California gold is found mainly in two occurrences, lode and placer. Thick, rich quartz veins and lodes (networks of veins) were discovered by prospectors soon after the publicized discovery of gold at Sutter's Mill in 1848. The veins occur primarily in the Jurassic Mariposa slate, into which Sierran granitic rocks were intruded during the Nevadan orogeny. It is believed that concentrations of gold-bearing liquids on the upper and outer margins of plutons during magmatic cooling impregnated wall rocks with quartz and gold. Why some plutons concentrated rich gold vapors and liquids while others did not is unresolved. Many Sierran localities have characteristics like those of rich-producing veins, but still lack significant gold content. The quartz veins of the Mother Lode belt are white and contain pockets, seams, and veinlets of native gold and pyrite (fool's gold) plus a few other precious and gangue (nonvaluable) minerals.

Initially gold was mined by placer methods. This involved working stream gravels and sands—the sediments derived from erosion of gold-bearing parent rock. Hydraulic mining (placer mining with high-pressure water) also was done, but the resulting destruction of the landscape and choking of stream channels prompted the outlaw of hydraulic mining in California late in the nineteenth century (Figure 3-2). Apart from hydraulic operations and gold dredging, most placer mining is a small-scale process employing sluice boxes, of which there are countless designs. A flow of water is directed across a confined channel (box) on which a series of cleats (riffles) has been placed. The sluice box is inclined, the debris is passed down the channel, and the gold (if it is not too fine) is caught in the riffles because of its high specific gravity.

Periodically, hunters, prospectors, and hikers turn up rich gold samples, but in the past 50 years none of these finds has been important. Nevertheless, gold prospecting continues, and increases in the price of gold have sparked renewed interest in commercial production, particularly in the Mother Lode belt. The lure of a "Mother Lode"—the geologic myth that there is a single parent for gold of unusual value—still persists.

Yosemite Valley

Because Yosemite National Park embraces a large part of the high Sierra already discussed, this section is confined to the characteristics and geologic evolution of the valley itself and its immediate surroundings. It is the valley, moreover, that is the park's commanding feature and is so renowned for its imposing scenery (Figure 2-20).

The verdant floor certainly contributes to Yosemite's beauty, but it is the sheer rock walls, the waterfalls, and the monolithic granitic peaks that make the valley special. Other Sierran valleys possess some of Yosemite's features, but only Yosemite unites them all in a single area. Figure 2-21 gives the locations of Yosemite's important topographic features.

The portion of the valley seen by most visitors is only about 12 miles (19 km) long. It lies between El Portal and the junction of the valley's two main branches, Little Yosemite Valley and Tenaya Canyon. Like Yosemite proper, these branches are fairly flat in their lower reaches, but are broken by abrupt risers above their junction. The south or Merced River branch is more spectacular, for its channel is a glacial stairway of grand dimensions. Vernal Fall plunges over the even, vertical cliff of the first riser as an unbroken curtain of water 317 feet (97 m) high. A short distance upstream is the second riser, lying between Vernal Fall and Little Yosemite Valley above. This riser, the site of Nevada Fall, is higher (495 ft or 181 m) but less regular than the first. Instead of the smooth curtain of Ver-

Figure 2-20.   Yosemite Valley. (Photo by Fairchild Aerial Surveys, courtesy of Department of Geography, University of California, Los Angeles)

1 Basket Dome
2 Bridalveil Fall
3 Bunnell Point
4 Cascade Cliffs
5 Cathedral Rocks
6 Cathedral Spires
7 Clouds Rest
8 Dewey Point
9 Eagle Peak
10 Echo Peaks
11 El Capitan
12 Glacier Point

13 Government Center
14 Half Dome
15 Indian Creek
16 Leaning Tower
17 Liberty Cap
18 Little Yosemite
19 Merced River
20 Mirror Lake
21 Mount Broderick
22 Mount Florence
23 Mount Lyell
24 Mount Maclure

25 Mount Watkins
26 North Dome
27 Quarter Domes
28 Ribbon Fall
29 Royal Arches
30 Sentinel Dome
31 Sentinel Rock
32 Sunrise Mountain
33 Taft Point
34 Tenaya Canyon
35 Washington Column
36 Yosemite Falls (top of)

Figure 2-21.

Main topographic features of Yosemite Valley. (Source: University of California Press)

nal Fall, Nevada is a complex of dartlike jets of water that unite near the fall's base into a foaming, continuous cascade that pours into the Merced River below. Upstream, Little Yosemite Valley has no significant tributaries and so lacks notable waterfalls. Nevertheless, during the spring, when snow melts rapidly in the high country, many cascades exist temporarily.

The park's highest falls plunge over the steep north wall of the main valley and are fed by tributary streams that occupy hanging valleys abandoned by glaciers. Of these, Ribbon Fall is the highest, dropping 1612 feet (492 m). This is not a single leap, however, because much of the drop is confined to a narrow crevice cut by Ribbon Creek. Since it drains only a small area, once the snow has melted this creek dwindles to a trickle, and by late summer Ribbon Fall often dries completely.

Almost directly across the valley from Ribbon Fall is Bridalveil Fall, fed by Bridalveil Creek, another small stream that normally diminishes after the snow melts. Early in the season Bridalveil Fall often has impressive volume. Its name reflects its early summer appearance, however, when the smaller volume of water breaks into a fine mist as it pours out of a deep notch in the valley wall and drops 850 feet (260 m). About 620 feet (189m) of this is a clear leap over a vertical precipice.

The valley's most spectacular waterfalls are the Yosemite Falls, two separate falls whose combined height is 2565 feet (782 m). The upper fall is one of the world's highest, 1430 feet (435 m). All but the upper 70 feet (21 m) form a clear, arc-shaped leap that carries the water so far from the cliff face that normally even the sloping base of the precipice is easily cleared. Lower Yosemite Fall drops 320 feet (98 m) over the lowest cliff into the valley, a clear leap when water volume is high.

The hanging valleys have floors high above Yosemite Valley itself. François Matthes believed these valleys to be of two origins. Some, typical of glaciated regions everywhere, were left hanging when the main glaciers widened and deepened their channels. The more modest efforts of the tributary glaciers resulted in far less erosion, producing the discordant junctions seen today. The other type of hanging valley arises because some resistant, massive, unjointed granitic rocks effectively withstand the erosion of small, often intermittent streams. Consequently, the streams have been unable to keep pace with the downcutting of the large, perennial Merced River into which they drain.

The sheer, almost unjointed cliffs on both sides of the main Yosemite Valley are imposing features. Yosemite wouldn't be Yosemite without El Capitan, Glacier Point, Half Dome, and the Washington Column. All were produced by widening of the valley as glaciers moved down the Merced River canyon. Actually, however, neither El Capitan, with its 2898-foot (884 m) high cliff, nor Half Dome, which rises 4352 feet (1327 m) above the valley floor, was ever fully overridden by ice. Their smooth sides away from the valley were sculptured by exfoliation rather than glacial scour. Furthermore, most of the other towering cliffs that rim Yosemite Valley are mainly the result of the El Portal glaciation and not the most recent Wisconsin glaciation. The last glacier occupying the valley, though certainly augmenting the widening done by earlier glaciers, was apparently no thicker than about 1500 feet (450 m). Nevertheless, it was this last glacier that produced the widespread, almost mirrorlike polished surfaces on the region's granitic rocks. Earlier glaciers certainly produced rock polish too, but none survives today. Indeed, though only about 10,000 years have passed since the last Wisconsin ice streams melted, weathering has roughened and destroyed considerable areas once highly polished. In such places, only the deeper gouges still persist.

Matthes divided the geologic history of Yosemite into three stages. These are still widely accepted, though the dates he assigned have been revised. During the first or Broad Valley stage, the Sierra was of modest height with few peaks over 3000 feet (900 m). At this time, the modern drainage direction to the southwest was established. Matthes suggested an early Tertiary age for this stage. The Mountain Valley stage followed, after some middle Tertiary uplift that steepened the stream gradients and enhanced erosion. The ancestral Merced River responded by cutting a steep-sided gorge about 1000 feet (300 m) deep in the Yosemite region. During the Pliocene and perhaps into the Quaternary, great uplift and tilting of the Sierran block occurred along eastern faults. Stream activity was greatly augmented, resulting in rapid deepening of the ancestral Merced River canyon to as much as 2000 feet (600 m). This period represents Matthes's Canyon (third) stage.

By the onset of glaciation 2 to 3 million years ago, the Merced River occupied a deep, V-shaped, stream-cut canyon down which the ice eventually flowed. As indicated chiefly by moraines, earlier glaciations (particularly the El Portal) were characterized by greater extent and thickness of ice than was the Wisconsin glaciation. Curiously, the evidence reveals that ice in the source areas was about as thick during the Wisconsin as it was previously. Why, then, were the earlier ice sheets in the Yosemite region so much larger and thicker? Matthes's answer was that the earlier glacial stages lasted longer than the Wisconsin and so promoted greater accumulation of ice in peripheral areas. For example, during the El Portal glaciation, ice in Yosemite Valley was probably about 6000 feet (1800 m) deep because it excavated bedrock at least 2000 feet (600 m) below the present valley floor. In contrast, the Wisconsin glacier was never more than about 1500 feet (450 m) deep anywhere in the valley.

The El Portal glacier had a maximum length of 37 miles (60 km) and extended 7 miles (11 km) beyond the edge of the Sierran ice cap. It ended just below El Portal, where its terminal moraine can be seen today. The smaller Wisconsin glacier ended in the main valley just above Bridalveil Meadow, where there is a well-preserved terminal moraine. This glacier received no ice from any of the canyons entering the main valley; it was solely dependent on the ice streams moving down Little Yosemite and Tenaya valleys. Maximum ice thickness of about 1500 feet (450 m) occurred near the junction of the two streams. At this time, Glacier Point towered more than 2000 feet (600 m) above the ice. In contrast, during the El Portal stage the ice covered Glacier Point to a depth of 500 feet (160 m). Furthermore, the earlier glaciers were over 2700 feet (820 m) thick at Bridalveil Meadow, where the Wisconsin glacier terminated.

Once the Wisconsin glacier had melted and the main valley was free of ice, Lake Yosemite filled the basin upstream from the terminal moraine. The lake was 5.5 miles (8.8 km) long and occupied the main part of the valley from wall to wall. The original depth of this lake is unknown, but the present valley floor is underlain by about 2000 feet (600 m) of lake deposits that rest on glacially scoured granite. Some of the lake beds may have been deposited in pre-Wisconsin lakes, however.

Lake Yosemite was destroyed by a combination of two processes. First, entering streams, mainly the Merced River and Tenaya Creek, quickly built large deltas into the lake. Smaller deltas were built by smaller streams entering the lake from the north. Second, the morainal dam at the lake's lower end was notched by the outflowing Merced River. This lowered the lake by at least a few tens of feet. The Merced River subsequently regraded and leveled the valley, leaving only small patches of the former valley floor as low benches about 15 feet (5 m) above present valley level.

Sequoia National Park

Sequoia National Park includes more than 600 square miles (1560 km²) of the highest part of the Sierra Nevada, from the Whitney Crest on the east to the western foothills. It embraces the greatest range of altitude of any American national park or monument outside Alaska—13,500 feet (4100 m).

Topographically the park is divided by the north-south Great Western Divide, which in any other setting would be a notable range in its own right. As it is, the Divide is really a mountain range within a range. Seen from Moro Rock on the west, the Great Western Divide may seem to be the Sierran crest proper and has the craggy, alpine appearance of the main ridge. It is almost as high as the actual crest, with several peaks of more than 13,000 feet (4000 m).

East of the Divide is the true high Sierra with its numerous lakes and relatively barren, heavily glaciated terrain. The area is drained by the Kern River and its gorge, which contrast strongly with the Kaweah River drainage west of the Great Western Divide. Not only does the Kern follow an almost straight north-south course across Sequoia park, but also the river's upper reaches occupy a pronounced U-shaped glaciated valley. In addition, the Kern River canyon has an openness of form that is unlike the deep gorges of the Kaweah system. This results mainly from the broad Chagoopa benchlands fringing the Kern and the heavily glaciated uplands above the plateau. Bare rock and glacial lakes characterize much of this high country. Glaciers have stripped away almost all soil, further limiting tree growth already inhibited by the high-altitude climate. On the other hand, glaciers have left dozens of sparkling lakes that fill nearly every ice-scoured depression.

Rocks in the eastern section of the park are almost exclusively granitic. The older metamorphic rocks so prominent elsewhere in the Sierra are conspicuously missing; their few exposures are confined to the Mineral King area and the western foothills below Giant Forest village. Volcanic rocks are likewise absent from the eastern region, though they do occur a few miles below the park's boundary.

Evidence of glacial scour prevails throughout the high eastern part of Sequoia, reflecting the existence of a long glacier that extended down the Kern River canyon to about 7 miles (11 km) beyond the park's southern boundary. This glacier terminated at an elevation of 5700 feet (1750 m), at latitude 36°14'N, and is the southernmost glaciation known in the Sierra Nevada. The southern 100 miles (160 km) of the range were too low and dry to sustain glaciers.

Unlike the Kern, the Kaweah River and its branches form a dendritic pattern. They occupy particularly deep canyons, some lying almost 7000 feet (2150 m) below the high peaks on either side. Usually, though, relief is between 4000 and 5000 feet (1200–1500 m). The landscape of the western slope differs markedly from that on the eastern slope: benchlands tend to be missing; the openness characteristic of the Kern River canyon is not apparent; and glaciation was more often confined to the canyons, leaving uplands untouched and covered with thicker soil and forest. The deep valleys do provide considerable evidence of glaciation, however. In Tokopah Valley on the Marble Fork of the Kaweah, the canyon walls are nearly perpendicular and are spectacularly grooved and polished by the latest Pleistocene glaciers. The polish seen here and other places in the high country is almost without exception the result of the final ice advance.

Although most glacial ice has melted, Sierran rocks continue to be shattered and reduced to smaller pieces by frost action—probably the dominant weathering process in the high country. Sometimes huge boulders look as if they had been split by a giant cleaver, with the discrete pieces still lying close to one another. Frost action causes this, attacking both large and small blocks with such indiscriminate vigor that whole valleys are choked with the resulting angular debris. The topography of the high country is thus gradually smoothed and rounded.

Limestone or more properly marble caverns are a minor, but interesting, feature of Sequoia. At least four are known, all in the metamorphic rocks of the Kaweah drainage. None is large, but all contain limestone cave features such as stalactites, stalagmites, and pillars. The most popular is Crystal Cave not far from Giant Forest. Others are Paradise Cave near Ash Meadow and Palmer Cave and Clough Cave on the walls of the South Fork of the Kaweah River.

Moro Rock, overlooking the Middle Fork of the Kaweah River, is typical of the massive, bulbous, granitic domes throughout the Sierra Nevada. Usually glaciation is not involved in formation of these domes. Instead they result from rock joints being more widely spaced than normal. This produces monolithic masses of rock with fewer avenues for weathering agents to enter. The normally jointed rocks surrounding the domes are more quickly reduced to small blocks and soil, leaving a projecting unjointed mass of rock that responds to weathering chiefly by exfoliation. In this process, thin shells are spalled off the exposed rock, gradually producing the rounded surfaces so characteristic of these domes.

Kern River—Lake Isabella Recreational Area

To control floodwaters, a dam was built across the Kern River at its confluence with the South Fork. As a result, Lake Isabella was formed, and the area has been developed into a major recreational region.

The Kern River drains south from its headwaters at the west base of Mount Whitney (Figures 2-5, 2-6), its course being somewhat determined by the Kern Canyon fault. The fault presumably poses little hazard since it apparently has been inactive for about 3 million years. Many geologic formations occur near the recreational area, including the chief rock units of the Nevadan orogeny. A typical Sierran limestone cavern, Packsaddle Cave, occurs about 15 miles (24 km) north of the lake. Below the dam is the deep gorge of the Kern, where giant potholes are spectacularly displayed.

At the south end of the lake is the old mining town of Bodfish, once a stage stop for travelers passing from the gold-mining areas of old Kernville (now beneath Lake Isabella) to the antimony mines of Havilah, to the Walker Basin, and to the railroad that crossed the Tehachapi Mountains. The Big Blue gold mine, productive until about 1940, may still be seen on the west side of the Kern River north of Lake Isabella.

Mount Whitney

Mount Whitney (14,495.881 ft or 4521.247 m) is the highest mountain in the coterminous United States (Figure 2-22). Its spire is unspectacular from lower elevations, however, because it is one of several high peaks forming the Whitney Crest (Figure 2-23). Six thousand feet (1800 m) below Mount Whitney is Whitney Portal, part of the steep Sierran scarp that is about 8000 feet (2440 m) high in this latitude.

Mount Whitney was named in 1864 by Clarence King for J. D. Whitney, director of California's first geologic survey. (King was a member of the party organized by Whitney to explore the Sierra Nevada.) In recent years the peak has become a mecca for hiking enthusiasts. One Fourth of July over 1000 people reportedly signed the register on the Whitney summit.

Mount Whitney patrol station near the timberline (elevation 10,720 ft or 3270 m) is one of the country's highest ranger stations and is only 8 miles (13 km) west of the summit by trail. The federal government originally reserved Mount Whitney for weather observations, but this plan did not materialize. It was later used as a base for solar radiation and Mars spectrum studies.

Figure 2-22. Mount Whitney. Iceberg Lake, elevation about 12,700 ft. — 3874 m, one of the higher lakes in the 48 coterminous states, is in the right foreground. This cirque lake does not thaw completely from one year to the next. The Great Western Divide is in the background. (Photo by Spence Air Photos, courtesy of Department of Geography, University of California, Los Angeles.)

Figure 2-23. *Whitney (Muir) Crest, with Mount Whitney in almost the exact center. Crest is 14,000 feet (4200 m) in average elevation, with many peaks almost equal to Mount Whitney. Alabama Hills and Diaz Lake are in the foreground. (Photo courtesy of U.S. Geological Survey)*

Alabama Hills

At the east base of the Sierra are the Alabama Hills, which extend about 10 miles (8 km) and rise 300 to 400 feet (90–120 m) immediately west of Lone Pine. The hills are a series of fault slivers raised in the Sierra Nevada fault zone (Figure 2-10). They are neither unique nor important in the total geology of the state, but they have achieved some prominence nevertheless. First, in the nineteenth century the persistent myth that the Alabama Hills are the oldest hills in the world was fabricated. Second, in 1872 an earthquake (possibly the strongest in recorded California history) destroyed Lone Pine, killed about 30 people, and produced significant surface ruptures at the base of the Alabama Hills and throughout the Owens Valley. Third, many motion pictures, particularly westerns, have been filmed here. The cavernous weathering along multiple and closely spaced joints in the Mesozoic granites makes an unusual, scenic backdrop.

The notion that the Alabama Hills are the oldest in the world results from two common misconceptions: that weathered features are of great age and that granite is always an old rock. As a matter of fact, Lone Pine and several other creeks flow across the hills (Figure 2-23, right center), indicating that the streams are antecedent or superimposed and that the hills are younger than the streams draining them. This means uplift of the region is very young indeed, probably late Pleistocene. The myth of old age also was fostered by the peculiar appearance of the weathered granites, in contrast to the granites of the Sierra Nevada. Geologists recognize, however, that different climatic regimens produce different weathering patterns in similar rocks and would expect the Whitney summit climate and the arid environment of the Alabama Hills to produce different topographies.

Arid weathering as the cause of the boulder-pile topography of the Alabama Hills has been questioned. It is suggested that instead the area once was buried under alluvial cover. Percolating groundwater then promoted chemical decomposition of the covered granite, with decomposition being accelerated along joints. Subsequent shifts of erosional baselevel or change to arid climate exhumed the granitic terrain, eroding the overlying alluvial outwash and removing the decomposed material along the joints. This is an attractive explanation for the Alabama Hills, since they slope north and south beneath the giant alluvial fans derived from the Sierra.

The 1872 earthquake corroborates the youth of the topography of the Alabama Hills. Surface displacements occurred along the base of the Sierra and the Alabama Hills for as much as 120 miles (192 km), producing many sags in which intermittent lakes form. Vertical displacement was as much as 17 feet (5 m), with a significant horizontal component. The resulting scarp shows little erosional effect in the 100 years that have elapsed.

Palisade and Lyell Glaciers

The two largest modern glaciers in the Sierra are Palisade Glacier on upper Big Pine Creek and Lyell Glacier at the head of the Lyell Fork of the Tuolumne River. Palisade Glacier is slightly larger, but neither glacier is more than about a mile (1.6 km) long (Figure 2-24).

Palisade Glacier lies below a precipitous north-facing rock wall capped by three high peaks. On the east is Mount Sill (14,162 ft or 4319 m), in the center is North Palisade (14,242 ft or 4344 m), and on the west is Mount Winchell (13,768 ft or 4199 m). The southernmost true glacier in the United States today (latitude 37°04′N), Palisade is only about half the size it was in 1850. Though the steep cirque wall to the south shields the glacier from direct sun, a renewal of the 1850–1950 warming trend would bring it close to extinction relatively soon.

Like most Sierran glaciers, Palisade is rimmed by a fresh and youthful-looking moraine. During the latter part of the summer, a crescentic crack (bergschrund) opens up near the head of the glacier as melting allows the ice to pull away from the cirque wall above. During the ensuing winter, the crack is filled with snow and ice that is added to the glacier during the following years.

Lyell Glacier is almost 2 miles (3.2 km) across, but it is shorter than Palisade Glacier. Lyell lies in a cirque on the north side of

Mount Lyell (13,095 ft or 3994 m) at the Sierran crest in eastern Yosemite. In most respects Lyell Glacier resembles Palisade Glacier 60 miles (96 km) to the south. They are about the same size, move at about the same rate, and lie at about the same elevation (12,000 ft or 3650 m) in comparable settings.

In 1933, two park naturalists were surprised to see a mountain sheep ram standing on the toe of Lyell Glacier. Their surprise was occasioned by the knowledge that mountain sheep had been unknown in the Yosemite region since about 1880. Closer inspection revealed that the naturalists were looking at the ram's mummified remains, exposed as the glacier melted. Apparently the ram had fallen into the bergschrund near the glacier's head about 250 years earlier. He had been frozen into the glacier as the bergschrund filled with snow and had been moved slowly through the advancing ice until melting released his body near the toe of the shrinking glacier. At the time of discovery, two of the ram's feet were still frozen in the ice.

Mammoth Mountain and "Earthquake Fault"

Mammoth Mountain, a ski center for thousands, is a young silicic volcanic cone in the chain that includes the cones and domes of Inyo and Mono craters. The volcanic rock of the mountain has been dated at 370,000 years.

In the same area is a linear crack a few feet wide and several hundred yards long that continues not as an open crack but as a fault trace. This feature is known locally as the "earthquake fault." Precise time of movement on the crack has not been confirmed, but the fault is definitely part of the Sierran series of north-south trending faults that cut the area's volcanic rocks. The view that the earth may open up in cracks like the earthquake fault is seldom supported by geologic observation. Such a feature is a rarity and is not the normal result of earth movement.

It has been suggested that the crack might have resulted from shrinkage of volume during cooling of the magma. The volcanic rock where the fissure is exposed is of Tertiary age, however. It is most improbable that a cooling crack would be preserved this long because weathering would long since have obliterated it.

Devils Postpile National Monument

Near the headwaters of the San Joaquin River is an example of columnar jointing known as the Devils Postpile. This feature developed in lava that was extruded from a fissure near Red's Meadow and subsequently flowed downriver for several miles. The flow was part of the volcanic episode that began 3 to 3.5 million years ago.

Figure 2-24.
Palisade Glacier. Note glacial moraines in the cirque below the glacier. Compare with Figure 2-3. (Photo by Spence Air Photos, courtesy of Department of Geography, University of California, Los Angeles)

The columns of the Postpile are long, regular, and numerous. Although it does not rank with Devils Tower (Wyoming) or Giants Causeway (Ireland), the Devils Postpile is California's best display of columnar jointing.

Hot Creek

Mammoth Lakes drain into Mammoth Creek, which hugs the base of a low escarpment cut in young volcanic rocks as it flows east past the town of Mammoth. At intervals hot springs emerge along the creek from the base of the escarpment. As Mammoth Creek crosses the meadow toward Owens River, it becomes Hot Creek, along which hot springs also occur (Figure 2-25).

These hot springs derive their heat from the still-cooling young volcanic rocks. Sometimes pressure even produces small geysers along the creek, and earlier this century geyser activity was substantial. The area has been investigated for geothermal power sources, but so far none seems adequate commercially. The clay near many of the hot springs has been mined commercially for the past 50 years.

Sierra Nevada

Figure 2-25.
Looking west at the gorge cut by Hot Creek. The dissected tableland is rhyolitic. (Photo by Mary Hill)

Mono Lake

Mono Lake occupies the bottom of a large valley at the east base of the steep Sierran escarpment near Lee Vining. It is a shallow water body, with no outlet and with a salinity exceeding that of sea water. Mono Lake is the remnant of larger, glacial Lake Russell, whose highstand is recorded by conspicuous shorelines around the basin (Figure 2-26). Lake Russell is presumed to have drained east, ultimately into Owens Valley, prior to eruptions of the Bishop tuff. Mono Lake was much larger before 1940, when streams entering the lake were diverted to augment the water supply of Los Angeles. Eventually, Mono Lake will dry completely because evaporation now usually exceeds inflow.

An unusual feature of Mono Lake is the display of tufa pinnacles that are forming from the briny waters. These occur along the eroded fault scarp at the lake's western edge and resemble those found in other Basin Range lakes. Several new minerals have been described from the brines of Mono Lake. Its volcanic islands, white Paoha and black Negit, are younger than the lake itself and were formed by outpourings of flows and the building of volcanic cones. They are as young as the last highstand of Lake Russell.

*Figure 2-26.
Mono Lake, from
the east. The
shorelines of Lake
Russell and dwin-
dling Mono Lake
are apparent. The
even skyline of the
Sierra Nevada is
prominent. (Photo
courtesy of U.S. Air
Force and U.S.
Geological Survey)*

Prospecting for geothermal power in the Mono Lake area has been discontinued. Several wells were drilled, but low temperatures were encountered and the bedrock complex was reached at shallow depth.

Lake Tahoe

California's most famous lake, Tahoe (1645 ft or 501 m deep), is cradled in a down-faulted block in the central Sierra Nevada (Figure 2-27). It is bounded on the east by the Carson Range and on the west by Mount Tallac ridge. The lake basin has been enlarged by the deposition of morainal material and volcanics that have clogged the outlet. Unfortunately, urban growth threatens the future of this spectacular lake, and stringent regulations probably must be enforced if Tahoe is to remain beautiful.

Figure 2-27. Looking east across Lake Tahoe, with South Tahoe (California) and Stateline (Nevada) clustered at the southern end. Washoe Lake is in the upper center. (Photo courtesy of U.S. Air Force and U.S. Geological Survey)

REFERENCES

General

Bateman, Paul C., 1967. Sierra Nevada Batholith. Science, v. 158, pp. 1407–1417.

———, 1969. Geology of the Sierra Nevada. Mineral Information Service (now California Geology): Part 1, v. 22, pp. 39–42; Part 2: v. 22, pp. 61–66.

———, 1974. Model for the Origin of Sierran Granites. California Geology, v. 27, pp. 3–5.

——— and Clyde Wahrhaftig, 1966. Geology of the Sierra Nevada. *In* Geology of Northern California. Calif. Div. Mines and Geology Bull. 190, p. 107–171.

Calkins, F. C., 1930. The Granitic Rocks of the Yosemite Region. *In* F. E. Matthes, Geological History of Yosemite Valley. U.S. Geological Survey Prof. Paper 160, pp. 120–129.

Hill, Mary, 1975. Geology of the Sierra Nevada. Univ. Calif. Press Natural History Guide no. 37.

Lindgren, Waldemar, 1911. The Tertiary Gravels of the Sierra Nevada of California. U.S. Geological Survey Prof. Paper 73.

Rinehart, C. D., and D. C. Ross, 1957. Geology of the Casa Diablo Mountain Quadrangle, California. U.S. Geological Survey Map GO-99.

———, 1964. Geology and Mineral Deposits of the Mt. Morrison Quadrangle, Sierra Nevada, California. U.S. Geological Survey Prof. Paper 385.

Wright, William H., III, 1975. The Stanislaus River—A Study in Sierra Nevada Geology. California Geology, v. 28, pp. 3–10.

Special

Anonymous, 1963. Basic Placer Mining. Mineral Information Service (now California Geology), v. 16, no. 12, pp. S1-S16.

Axtell, Lawrence H., 1972. Mono Lake Geothermal Wells Abandoned. California Geology, v. 25, pp. 66–67.

Burnett, John L., 1964. Glacier Trails of California. Mineral Information Service (now California Geology), v. 17, pp. 33–34, 44–51.

———, 1971. Geology of the Lake Tahoe Basin. California Geology, v. 24, pp. 119–127.

——— and Robert A. Mathews, 1971. Geologic Look at Lake Tahoe. California Geology, v. 24, pp. 128–129.

Buwalda, J. P., 1954. Geology of the Tehachapi Mountains, California. *In* Geology of Southern California. Calif. Div. Mines and Geology Bull. 170, pp. 131–142.

Clark, William B., 1965. Tertiary Channels. Mineral Information Service (now California Geology), v. 18, pp. 39–44.

———, 1970. Gold Districts of California. Calif. Div. Mines and Geology Bull. 193.

Crippen, J. R., and B. R. Pavelka, 1970. The Lake Tahoe Basin, California-Nevada. U.S. Geological Survey Water Supply Paper 1972.

Curry, Robert R., 1968. California's Deadman Pass Glacial Till Is Also Nearly 3,000,000 Years Old. Mineral Information Service (now California Geology), v. 21, pp. 143–145.

Hill, Mary, 1975. Living Glaciers of California. California Geology, v. 28, pp. 171–177.

———, ed., 1972. The Owens Valley Earthquake of 1872. California Geology, v. 25, pp. 51–64.

———, ed., 1975. Geologic Guide: Sierra Nevada—Basin and Range Boundary Zone. California Geology, v. 28, pp. 99–119.

Huber, N. K., and C. D. Rinehart, 1965. The Devils Postpile National Monument. Mineral Information Service (now California Geology), v. 18, pp. 109–118.

Jenkins, Olaf P., 1964. Geology of Placer Deposits. Mineral Information Service (now California Geology), v. 17 (passim). (Reprint of Calif. Div. Mines and Geology Bull. 135.)

———, ed., 1948 and 1963. Geologic Guidebook Along Highway 49: The Sierran Gold Belt—The Mother Lode Country. Calif. Div. Mines and Geology Bull. 141.

Kistler, R. W., 1966. Geologic Map of the Mono Craters Quadrangle, Mono and Tuolumne Counties, California. U.S. Geological Survey Map GO-462.

Matthes, F. E., 1930. Geologic History of the Yosemite Valley. U.S. Geological Survey Prof. Paper 160.

————, 1950a. Sequoia National Park: A Geological Album. Univ. Calif. Press.

————, 1950b. The Incomparable Valley—A Geological Interpretation of the Yosemite. Univ. Calif. Press.

————, 1960. Reconnaissance of the Geomorphology and Glacial Geology of the San Joaquin Basin, Sierra Nevada, California. U.S. Geological Survey Prof. Paper 329.

————, 1965. Glacial Reconnaissance of Sequoia National Park, California. U.S. Geological Survey Prof. Paper 504A.

Mathews, R. A., and Charles Schwartz, 1969. Preliminary Bibliography, Lake Tahoe Basin, California-Nevada. Calif. Div. Mines and Geology Spec. Pub. 36.

Oakeshott, Gordon B., ed., 1955. Earthquakes in Kern County, California, during 1952. Calif. Div. Mines and Geology Bull. 171.

————, ed., 1962. Geological Guide to the Merced Canyon and Yosemite Valley. Calif. Div. Mines and Geology Bull. 182.

Putnam, William C., 1938. Mono Craters, California. Geog. Rev., v. 28, pp. 68–82.

————, 1960. Origin of Rock Creek and Owens River Gorges, Mono County, California. Univ. Calif. Pub. Geol. Sci., v. 34, pp. 221–280.

Rapp, John S., 1974. Hammonton Dredge Field. California Geology, v. 27, pp. 201–202.

Redmond, John L., 1966. French Meadows: The Geomorphic Expression of an Eocene River Channel. Mineral Information Service (now California Geology), v. 19, pp. 3–4.

Rinehart, C. D., and N. K. Huber, 1965. The Inyo Crater Lakes—A Blast in the Past. Mineral Information Service (now California Geology), v. 18, pp. 169–172.

Romanowitz, Charles M., 1967. Floating Dredges Used for Mining Purposes. Mineral Information Service (now California Geology), v. 20, pp. 82–87.

————, 1970. California's Gold Dredges. Mineral Information Service (now California Geology), v. 23, p.. 155–168.

Schumacher, Genny, and others, 1959. The Mammoth Lakes Sierra—A Handbook for Roadside and Trail. Sierra Club.

Short, Harry W., 1975. The Geology of Moaning Cave, Calaveras County, California. California Geology, v. 28, pp. 195–201.

Twain, Mark, 1965. Islands of Mono Lake. Mineral Information Service (now California Geology), v. 18, pp. 173–180.

Chapter Three

Northern Provinces

> *No vestige of a beginning—no prospect of an end.*
>
> > *James Hutton*

KLAMATH MOUNTAINS

Drained by the tortuous, deep, and rock-filled Klamath River, the Klamath Mountains constitute the least accessible and least known geologically of California's geomorphic provinces. (Important place names are given in Figure 3-1.) In fact, most of the geologic study that has been done has emphasized mineral resources. Shortly after the 1848 announcement of discovery of gold in California, prospectors explored the Klamath and found rich placer and lode gold deposits. Hostile Indians resisted encroachment, however, especially in the few areas suitable for cultivation. The Klamath River's swiftness made navigation almost impossible—and extremely hazardous. (A dam at Copco now partially regulates the flow.) In spite of these deterrents, the gold miners diverted the waters of the Klamath, Trinity, Scott, and other rivers and mined the gravels of the stream bottoms.

Subsequently hydraulic mining was done in widely scattered areas, and huge piles of cobbles blighted the landscape and spoiled fertile stream bottoms. At the South Fork of the Scott River from Etna Mills to Callahan, for example, several miles of gravel piles persist as evidence of the environmental consequences of hydraulic mining (Figure 3-2).

The absence of adequate base maps for geologic work was remedied about 1960, when aerial photographs and modern topographic maps of useful scale became available. Although systematic mapping is underway, much remains to be completed before an integrated and reliable appraisal of the province's geologic history becomes available.

Geography

The Klamath province, which geographically extends north into Oregon, is mostly a topographic upland, with a general elevation of

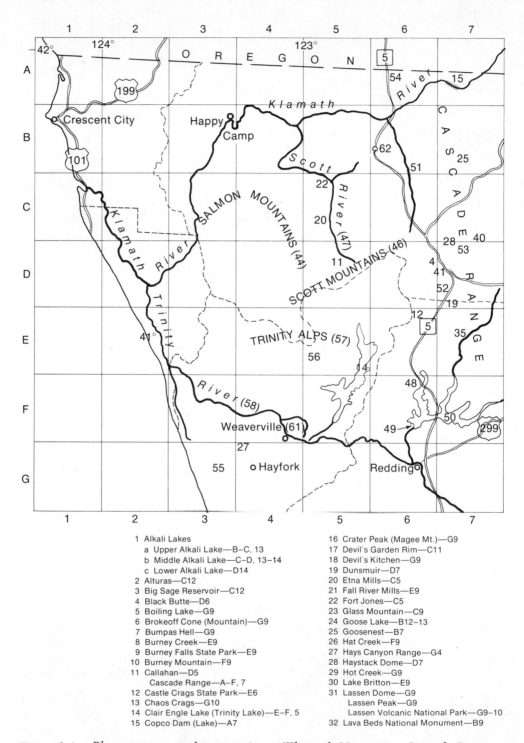

1 Alkali Lakes
  a Upper Alkali Lake—B–C, 13
  b Middle Alkali Lake—C–D, 13–14
  c Lower Alkali Lake—D14
2 Alturas—C12
3 Big Sage Reservoir—C12
4 Black Butte—D6
5 Boiling Lake—G9
6 Brokeoff Cone (Mountain)—G9
7 Bumpas Hell—G9
8 Burney Creek—E9
9 Burney Falls State Park—E9
10 Burney Mountain—F9
11 Callahan—D5
  Cascade Range—A–F, 7
12 Castle Crags State Park—E6
13 Chaos Crags—G10
14 Clair Engle Lake (Trinity Lake)—E–F, 5
15 Copco Dam (Lake)—A7

16 Crater Peak (Magee Mt.)—G9
17 Devil's Garden Rim—C11
18 Devil's Kitchen—G9
19 Dunsmuir—D7
20 Etna Mills—C5
21 Fall River Mills—E9
22 Fort Jones—C5
23 Glass Mountain—C9
24 Goose Lake—B12–13
25 Goosenest—B7
26 Hat Creek—F9
27 Hays Canyon Range—G4
28 Haystack Dome—D7
29 Hot Creek—G9
30 Lake Britton—E9
31 Lassen Dome—G9
  Lassen Peak—G9
  Lassen Volcanic National Park—G9–10
32 Lava Beds National Monument—B9

Figure 3-1.   *Place names: northern provinces (Klamath Mountains, Cascade Range, Modoc Plateau).*

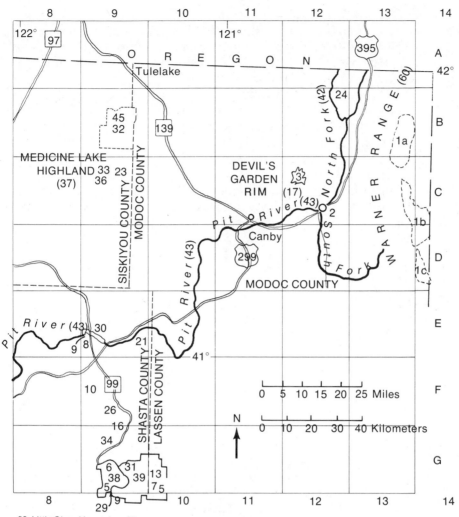

5000 to 7000 feet (1500–2150 m). The northern half of the California section is drained by the Klamath River, and the southern half by the Trinity River. From east to west, the province is about 70 miles (110 km) wide, yet the Klamath River travels nearly 150 miles (240 km) to drain the California section alone.

The term Klamath Mountains is not really precise geographically, since the province's rivers and their tributaries have cut the upland into several well-defined mountain ranges. Included are the Trinity Mountains with elevations over 8900 feet (2700 m) in the area known as the Trinity Alps. Other ranges that have regional names are the South Fork, Marble, Scott, and Salmon mountains. All are difficult to reach, although timber exploitation has forced the U.S. Forest Service to develop roads to assist in forest fire suppression. (These roads are not open to the public during much of the year.)

Rocks

Under the auspices of the U.S. Geological Survey, studies of the Klamath province were initiated by J. S. Diller in 1902 and continued until 1910. Diller's studies were confined primarily to the province's eastern edge. At about the same time, O. H. Hershey, also of the U.S. Geological Survey, penetrated farther into the province and published reconnaissance studies. After 1910, practically nothing was written on the geology of the California Klamath until 1950, when the U.S. Geological Survey began extensive studies. Principal workers have been W. P. Irwin (stratigraphy and structure) and G. A. Davis (plutonism, metamorphism, and structure).

Klamath rocks divide into subjacent and superjacent series. Rocks of the subjacent series were involved in the Mesozoic Nevadan orogeny. Superjacent formations are of early Cretaceous age and younger. The rock units of both series are often closely related to Sierran rocks. In particular, the sedimentary and volcanic sequences of the subjacent series are the equivalent of formations occurring on the western flank of the Sierra Nevada, including the Mother Lode belt. Gold deposits of the Klamath and the Mother Lode are geologically similar and apparently contemporary. They both originated in the period of quartz-gold mineralization associated with the intrusion of the late Jurassic and early Cretaceous granitic plutons.

Subjacent Series   The subjacent sequence ranges from Ordovician or Silurian to Jurassic. The lower portion consists of clastic sedimentary units, including marine limestones that carry diagnostic fossils like graptolites and brachiopods. Fossiliferous upper Paleozoic and Triassic marine volcanics, clastics, and limestones are present. The sedimentary-volcanic marine record was terminated by the granitic plutonic activity of the Nevadan orogeny. The Klamath granitic rocks are usually quartz diorites, representing less silicic plutons than the granodiorites common in the Sierra Nevada.

Figure 3-2.
Gold dredge on the South Fork of the Scott River, 1973. Gravels that previously made the valley a wasteland are now partly recycled as dredging for gold proceeds. (Courtesy of California Department of Water Resources)

The subjacent series occurs in two major belts in the Klamath. The western belt lies along the boundary of the northern Coast Ranges as a complex of Paleozoic and Triassic rocks, and the eastern Paleozoic belt abuts the Cascade Range. Total thickness of the sedimentary record is about 40,000 feet (12,200 m), but in the Oregon Klamath nearly 70,000 feet (21,300 m) of equivalent beds are known.

Ultramafic rocks of several ages compose prominent parts of the subjacent series. Intrusive sheets and sills of peridotite and plutons now altered to serpentinite are abundant. These serpentinite bodies are critical in defining structure, especially of the pre-Mesozoic. The interest in plate tectonics has conferred new importance on the ophiolite sequences and the ultramafic serpentinites. Plate tectonic theory postulates the existence of subduction zones of metamorphism where some igneous rocks are born and in which the serpentinization and metamorphism that develop blue and green schist facies play an important role. The relationship of the schists on the western margin of the Klamath province to the Franciscan rocks of the northern Coast Ranges remains uncertain.

Superjacent Series Superjacent sequences include Cretaceous marine sedimentary rocks that occur up to 5000 feet (1500 m) thick in the California Klamath. These sediments occur chiefly on the margins of the province. Though some fossiliferous, nonmarine sediments of Tertiary age do exist, they are restricted in thickness and area and reveal little about Cenozoic history.

Structure

The rocks of the Klamath have been intensely deformed by several periods of folding with intervening faulting. Two structural episodes predominate: the isoclinal folding that preceded and accompanied the batholithic emplacement of the Nevadan orogeny, and the middle and late Cenozoic normal and high-angle reverse faulting that produced the Klamath upland on which present topography is cut. In addition, significant overthrust faulting is recorded in the stratigraphy of the subjacent series, although number and correlation of episodes are uncertain. Continuing studies of the Klamath's stratigraphy and structure promise better understanding of the thrust faulting and eventually correlation with thrust faulting in the northwestern Sierra, where similar patterns appear.

Nevadan Folding and Overthrust Faulting Two periods of folding during the Nevadan orogeny are thought to have affected the Klamath province. One was a major episode of isoclinal deformation, as early as the Jurassic, in which complex deformation produced primarily eastward-dipping fold axes. The second, Cretaceous episode further deformed the region.

Several major thrust faults that dip east at low angles have been defined (Figure 3-3). The overriding thrust plates have moved from east to west with significant displacements. These faults are overlapped at the southern end of the province by the late Jurassic and Cretaceous Great Valley sequence, nearly 32,000 feet (9700 m) thick. The thrust faulting is of major importance in the evolution of the basement complex of the Klamath Mountains because the Klamath may be the northwest continuation of the Sierra Nevada. The thrust faulting preceded or accompanied the Nevadan orogeny and predates the granitic invasions of both the Sierra Nevada and Klamath provinces.

Cenozoic Faulting O. H. Hershey identified a western-bounding, high-angle reverse fault that prompted him to infer late Cenozoic uplift for the Klamath. The fault subsequently has been recognized as part of the South Fork Mountain thrust. The fault zone breaks Miocene terrace deposits, thus giving a minimal date for its movement. It appears inactive at present.

Within the province proper, there is little evidence of Cenozoic block faulting. Topographic control seems related primarily to

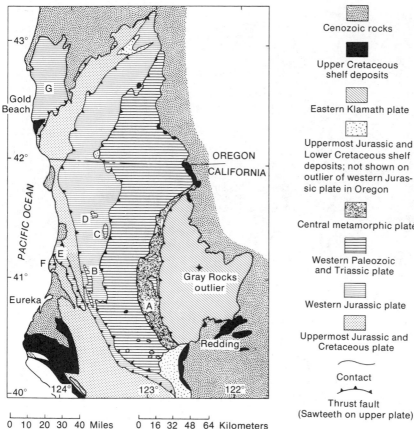

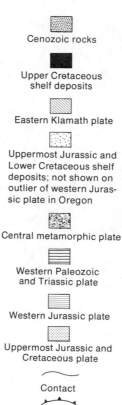

Cenozoic rocks

Upper Cretaceous
shelf deposits

Eastern Klamath plate

Uppermost Jurassic and
Lower Cretaceous shelf
deposits; not shown on
outlier of western Juras-
sic plate in Oregon

Central metamorphic plate

Western Paleozoic
and Triassic plate

Western Jurassic plate

Uppermost Jurassic and
Cretaceous plate

~~~ Contact

⊢⊣⊢ Thrust fault
(Sawteeth on upper plate)

Figure 3-3.
*Principal postulated
thrust plates of the
Klamath Mountains
and adjacent Coast
Ranges. Thrust out-
liers are indicated
by letter symbol: A,
Oregon Mountain;
B, Willow Creek; C,
Prospect Hill; D,
Flint Valley; E,
Redwood Moun-
tain; F, Patricks
Point; G, south-
western Oregon.
(Source: California
Division of Mines
and Geology)*

Nevadan folds and plutonic boundaries. In the eastern Klamath, adjacent to the Cascade province, some late Cenozoic normal fault- ing apparently occurs in valleys like those of the Shasta and Scott rivers. Positive fault locations have not been established, however.

Extent   Structure of the Klamath strongly suggests that Sier- ran basement rocks arc westward into the Klamath and then swing northeast beneath Cascade Range and Columbia Plateau volcanics. These volcanics cover most of the basement rocks of eastern Ore- gon. In the Blue Mountains, however, are rocks that project above the volcanics and are thought to correlate with the Klamath base- ment. Both units may be part of a continuous structure. If this in- terpretation is correct, the Klamath is merely the western bulge of a giant arcuate fold more than 400 miles (640 km) across.

Special Interest Features

Trinity Alps (Trinity Recreational Area)   North of Weaverville is a small but unusual mountain area known as the Trinity Alps. The Klamath Mountains' highest elevations occur here, culminating in Thompson Peak (8936 ft or 2725 m). Many glacial features, includ-

Figure 3-4. Shasta Lake and Shasta Dam. The lake has drowned major segments of the Sacramento, McCloud, and Pit rivers, thus emphasizing the ridge-valley topography of the Klamath province. Mount Shasta dominates the skyline. (Photo courtesy of California Department of Water Resources)

ing two glacial tarns, are prominent. The glacial sequences resemble Sierran glaciation, but cover a smaller area because of the lower elevations. Trinity (Clair Engle) Lake, a large man-made reservoir on the Trinity River, is in the center of the recreational area.

Shasta Lake Recreational Area   The Sacramento River canyon extends about 75 miles (120 m) from Redding north to the head of the Sacramento's drainage, near the base of Mount Shasta. As part of water-resource development and to reduce floods in the Sacramento Valley, Shasta Dam was constructed in the early 1940s (Figure 3-4). Ponding water headward to the junction of the McCloud and Pit rivers established a recreational area in which many geologic features are readily accessible by boat. The region includes the Shasta Caverns, an underground landscape typical of limestone regions. The caverns are formed in the McCloud limestone, which is mined commercially. All formations of the eastern sedimentary belt of the Klamath are well exposed in the Shasta Lake area.

Castle Crags State Park   West of the Sacramento River canyon, midway between Redding and Mount Shasta, is Castle Crags State Park. The park is noted for its series of monolithic peaks known as Castle Crags. The high relief of these features results from a small resistant granitic pluton that contrasts with the less resistant rocks it intruded.

CASCADE RANGE

The Cascade Range extends from southern British Columbia to Lassen Volcanic National Park. Major features (Figure 3-1) include the high volcanic peaks from northernmost Mount Baker (10,750 ft or 3279 m) to southernmost Lassen Peak (10,453 ft or 3188 m), nearly 500 miles (800 km) apart. Between lie 12 major and many minor cones built on or near the western edge of the Columbia Plateau escarpment. Some Cascade volcanos straddle the escarpment, others are built from orifices penetrating the Columbia Plateau lavas, and one major cone, Mount St. Helens (9671 ft or 2950 m), lies west of the Columbia Plateau. Geographically, the Cascade Range occupies only a small part of California. Nevertheless, because its most recent eruption was at Lassen Peak, many people identify the range primarily with California.

In its California section, the Cascade Range grades east into the Modoc Plateau, whose volcanism is typically dominated by lava flows rather than by the explosive, eruptive products characteristic of the Cascades. Medicine Lake Highland, a product mainly of effusion, is generally included in the Cascade province. On the other hand, Lava Beds National Monument, which abuts Medicine Lake Highland, is more properly part of the Modoc Plateau.

Geography

The Cascade Range is famous for Mount Shasta and Lassen Peak. Mount Shasta (14,161 ft or 4319 m) stands boldly above a generally undulating timbered surface with average elevation of 4000 feet (1200 m). The mountain is double-crested: the higher and older crest is Shasta, the lower and younger is Shastina. The pair make an imposing landmark visible more than 100 miles (160 km). Adjacent to Mount Shasta are smaller cones. To the north many cones rise to elevations of 7000 to 8500 feet (2150–2600 m), including older cones like Miller Cone and Eagle Rock and younger cones like Whaleback and Goosenest. All the cones are Quaternary, but Whaleback and Goosenest are so young that they seem scarcely touched by erosional processes. Black Butte is a young plug dome. Between Shasta and Lassen are several lesser cones, including Crater Peak (Magee Mountain), Burney Mountain, and Cinder Cone. These break the undulating volcanic platform that is a southwest extension of the Modoc Plateau.

Lassen Peak is a composite cone that dominates the skyline and is visible from as far away as Sacramento (Figure 3-5). Snow-capped in most years, it erupted sporadically from 1914 through 1917. Its most recent activity consisted of steam emissions that concluded the episode initiated in 1914. The upper portion of the mountain is still barren of vegetation and can be reached easily by road and trail.

Glacial features are found on and adjacent to both Shasta and Lassen. Glacial deposits interfingering with the volcanics permit isotopic dating of the Pleistocene and Recent volcanism.

Rocks

The exposed rocks of the California Cascades are predominately volcanics of great variety and form. Some lake-bed deposits including freshwater diatomite, some water-laid tuff and ash, and a few

Figure 3-5.
Lassen Peak. (Photo
by Fairchild Aerial
Surveys, courtesy of
Department of
Geography, Univer-
sity of California,
Los Angeles)

glacial morainal deposits occur near Mount Shasta, but they make up less than 10 percent of the province's rocks.

The oldest volcanic unit related to formation of the Cascade Range is the Tuscan (maximum thickness 1500 ft or 450 m), composed primarily of mudflow, ash, and breccias. The Tuscan forms a ramplike transition more than 60 miles (96 km) long between Lassen Peak and the Sacramento Valley. It is of Pliocene age and is tentatively considered the equivalent of the oldest exposed volcanics of the Modoc Plateau, the Cedarville series.

Stratigraphically above the Tuscan formation are thick, extensive andesitic flows extruded from vents throughout the California Cascades. After deep, prolonged erosion, late Pleistocene basaltic lavas were erupted, burying most of the andesites that had escaped erosion. Some basaltic flows cover as much as 50 square miles (130 km²), with the total volume of all erupted basalt being up to 2 cubic miles (8.3 km³). Cones of variable composition representing several episodes followed and overlapped the basaltic flow eruptions, culminating in Mount Shasta and Lassen Peak. Activity probably has not ceased altogether, but Lassen has been dormant since 1917.

Many cinder cones occur, usually composed of black or red clinkery fragments, sometimes blown from local vents as hardened chunks of rock. Cinders frequently developed in place from vesicular lavas fragmented by gas and liquid discharges during eruption. Hills are often domal, pushed upward by trapped gas. Sometimes basaltic scoria, pumice, and obsidian intermingle. On occasion such rocks "float" upward in the rising magma because the porous cindery materials are lighter than the new enclosing magma. Cinders are mined to make concrete blocks for lightweight construction and for road metal. Shades of red, brown, gray, and black can be obtained by using appropriate cinder sources, and color change from one highway jurisdiction to another is a striking feature of northeastern California roads.

Structure

Faulting has been important throughout the development of the California Cascades. In the Tuscan formation, the lowest beds are folded and eroded but the highest are horizontal, indicating significant local deformation between earlier depositional stages. During andesitic (earlier) volcanism, block faulting occurred; magma emerged along some of the faults, and cones and domes developed. Before and during basaltic (later) flow eruptions, there was vertical faulting along which cinder cones aligned. Moreover, important young faults cut the basaltic sequences. These continuing fault movements substantiate the importance of north-south and northwest-southeast fault patterns reflected in the geologic history of the Basin Range and Sierra Nevada provinces. Abundant evidence of recent movement on similar faults is found in the Nevada section of the Columbia Plateau, the Sierra Nevada of eastern California, and the Warner and adjacent ranges of the Modoc Plateau.

The pattern of Cascade volcanos is not adequately explained by assuming that faulting has occurred along parallel linear fractures. The volcanos have a nearly north-south alignment that transgresses the arcuate structure of the basement platform of Sierran-Klamath-Cascade rocks. Furthermore, the Cascade peaks of Oregon and Washington are built on thick marine and nonmarine interbedded clastics and volcanics that include flows as old as Eocene. The chain of volcanos is primarily a Quaternary event, however, and is probably still in progress. It is possible that such "hot spots" are related at depth to increased temperatures along the boundary of the North American and Pacific plates.

The block faulting that was more or less continuous during early Tuscan deposition produced enclosed drainages in which water collected. Alluvial fan, delta, and lake deposits, including water-laid ash, tuff, and diatomite accumulated in these lake basins. Sometimes deposition of the lake sediments was interrupted by lava flows.

Many basaltic flows throughout the province show well-developed columnar jointing and possess weathered zones and fossil soils. These occasionally contain charred remains of vegetation that grew on the flows between eruptive periods. In rare instances, woody material in the soil zones has petrified.

Volcanic Geomorphology of Mount Shasta

Mount Shasta is unusual for its double summit—Shasta and Shastina (Figure 3-6). Shasta, the older, is scoured by numerous glacial

*Figure 3-6.
Mount Shasta and
Shastina. Note the
flow ramparts, al-
most like natural
levees on a river
flood plain. (Photo
by Burt Amundson)*

valleys containing interfingering Pleistocene glacial deposits and
volcanics. Shastina lacks glacial valleys and is clearly postglacial in
origin, making it extremely young. Along the basal circumference,
especially on Shastina's northwest side, are domal upwellings of
magma and overlapping, steeply sloping mudflows and breccia flows
of nearly historic date, almost untouched by erosion. The most
prominent extrusive dome is Haystack Dome, so new its formation
is documented with unusual topographic clarity. South of Haystack

Dome is an avalanchelike protrusion of overlapping lava flows with an expanded bulbous base developed in late stages of magma congealment. Furrows and ridges are visible behind the escarpment at the base of the flows. The ramparts (levees) of earlier flows confine later magmatic effusions. Such features are unusual in most eruptions, but may develop as eruptions subside.

Medicine Lake Highland

Medicine Lake Highland is an area of youthful volcanism east and north of Mount Shasta. With an average elevation of nearly 5000 feet (1500 m), the Highland is astride the Cascade Range–Modoc Plateau boundary and contains volcanic features built on a broad volcanic platform (Figure 3-7). This platform was a Mio-Pliocene shield volcano about 10 miles (16 km) in diameter, whose original summit area collapsed and formed an elliptical caldera covering about 25 square miles (65 km²). The caldera was buried by younger lavas extruded from cones that developed along the caldera's rim. In the Pleistocene and Recent, rhyolitic, andesitic, and basaltic eruptions yielded over 100 cinder and lava cones whose explosive action produced pumice and tuff. The youngest of these rocks are basaltic flows and additional cinder cones, some erupted as recently as 500 years ago. Glaciation produced minor morainal deposits and small amounts of scouring during the volcanism. Faulting occurred during both volcanism and glaciation.

Medicine Lake Highland is bounded by volcanic escarpments that enclose a high central tableland. Medicine Lake itself is a recre-

Figure 3-7.
Mount Dome, with Mount Shasta in the distance. (Photo by Burt Amundson)

ational area located near the center of the old caldera, but the lake basin is the result of lava dams and not the caldera.

Located near the northeastern edge of Medicine Lake Highland, Glass Mountain and Little Glass Mountain are the most recent major volcanic features of the Cascade Range–Modoc Plateau and are composed of rhyolitic pumice and obsidian, extruded about 1100 years ago. The eruption began with pumice showers that built pumice cones up to 500 feet (160 m) in diameter and 300 feet (90 m) high. Then came extrusion of silicic lavas that formed obsidian domes generally larger than the earlier pumice cones. Obsidian flows at Glass Mountain moved as far as 3.5 miles (5.6 km) from vent sources. Little Glass Mountain to the west and other obsidian flows of the area have origins similar to those of Glass Mountain.

Is Eruption Imminent?

Volcanologists are aware that volcanos, though they may appear "dead," are part of the continuous progression of earth history. Recently, concern has been expressed regarding future Cascade volcanism and possible hazards to the population. What are the facts and potentials of Mount Shasta and Lassen Peak, California's two Cascade giants?

In the Cascade chain, Mount Baker appears to have erupted intermittently during the past 10,000 years. Mount Rainier and Mount St. Helens (Washington) and Mount Hood (Oregon) all show evidence of extensive activity as recently as 2000 years ago. Mount Mazama (Crater Lake) was destroyed as a mountain peak within the past 7000 years. Mount Shasta or Lassen Peak exuded steam and ash in 1786. Important pumice and obsidian flows were formed 1100 years ago at Glass Mountain, and Lassen Peak erupted in 1914 with sufficient violence to be extremely hazardous had the area been populated. By any standard, these are historic eruptions.

Geologists recognize active and dormant volcanos. ("Extinct" is a dubious usage.) This recognition is a value judgment, however, and some apparently dormant volcanos have erupted quite unexpectedly. In spite of sophisticated techniques for recording preeruption symptoms, objective data on which to base reliable predictions are still not available. Cascade volcanism is far from extinct, and even considering Shasta and Lassen dormant (with their considerable hot spring and steam vent activity) may be intellectually and practically hazardous. Future eruptions must be anticipated, because another eruption could occur within the next 50 or 100 years—or tomorrow.

Stratigraphic Record

Geologic studies of the Cascade Range–Modoc Plateau are fragmentary and primarily involve problems relating to Mount Shasta and Lassen Peak. Correlating volcanic sequences is usually extremely difficult. Many "simultaneous eruptions" are localized over small areas and must be correlated not only with each other but also with earlier and later episodes that may be part of the same, or different (older or younger), volcanic pulsation. Often large areas are involved, sometimes complicated further because the rocks may be deeply weathered and covered by vegetation. Avalanches of gas, water, and comminuted ash often are strewn by explosive eruption simultaneously with mudflows, cinders, and occasional lavas. Recurrence of all these events through time complicates the problem of unraveling geologic history. Yet the mineralogic character of ash showers may be so distinctive that layers only a few inches thick can be traced for hundreds of miles from source vents. Solving correlation problems usually depends on availability of complete geologic maps of the province in question. Less than 10 percent of the Cascade-Modoc provinces of California has been mapped on a detailed geologic basis.

Special Interest Features

Mount Shasta Recreational Area   Mount Shasta covers about 230 square miles (600 km²), and its cone is estimated to involve 80 cubic miles (330 km³) of rock. The area has been developed into a major recreational location. High annual snowfall and upper slopes barren of timber and vegetation have encouraged skiing. Many graded county roads encircle Mount Shasta, enabling visitors to see countless volcanic features some of which are unstudied. An improved road climbs nearly 4000 feet (1200 m) up the mountain's southwest flank and is a dramatic route for travelers.

Volcanic regions where rainfall is abundant usually contain springs, some quite large owing to the channelways and openings that occur in volcanic terrains. Volcanic rock also is more porous than most rocks because of the vesicles formed when gas is given off during eruption; thus surface water is minimal. Springs emerge: on contacts between flows of different source and age; from soil zones and gravels that are buried by overlying rock layers (such soil zones often carry water trapped in earlier geologic episodes, representing connate as well as meteoric water); from local drainage lines as new streams develop where volcanic flows have displaced earlier drainages; and from lava tubes and tunnels.

Shasta Springs is a famous old spa located in the canyon of the Sacramento River, between the towns of Dunsmuir and Mount Shasta. Here a series of andesitic lavas with prominent columnar joints overlies a porous, gravelly conglomerate from which huge springs emerge. The lava flow that caps these water-bearing gravels can be traced for more than 40 miles (64 km). Some of the springs are charged with carbon dioxide, and their water was marketed in the San Francisco area for many years.

Lassen Volcanic National Park   This national park was established by Congress in 1916, to preserve the volcanic features of the area, particularly those formed during the eruption begun in 1914. Lassen Peak is the only accurately recorded volcanic eruption in the coterminous United States. Though the Lassen eruption was minor geologically, the region is as rich in volcanic features as any in the world. Fumaroles, steam vents, sulfurous hot springs, and hundreds of other attractions can be seen.

Special volcanic features abound—for instance, Geysers, Boiling Lake, and the sulfurous areas of Bumpas Hell and Devil's Kitchen. Brokeoff Cone is thought to be the remains of the predecessor of modern Lassen Peak. The pre-Lassen mountain, Mount Tehama, was effectively destroyed, forming the base from which Lassen Peak grew. The mudflows of the Devastation Area of 1915 in the valleys of Hat and Lost creeks make spectacular scenery. Cinder Cone is dated at about A.D. 500, and several cones are even more recent, possibly less than 200 years old. Chaos Crags and Lassen Dome are among other interesting volcanic features, well marked for visitors. Many large volcanic bombs and spatter (driblet) cones are found throughout the park. Lava caves and tunnels, some large and some with underground streams, have been mapped. Recent faulting has produced significant scarps, especially in the valley of Hat Creek, although it is often difficult to separate such fault scarps from the scarps produced by flow fronts where pasty magma has congealed and a flow-front escarpment has formed.

MODOC PLATEAU

The Modoc Plateau is an undulating platform 4000 to 5000 feet (1200–1500 m) high, composed of assorted volcanic materials, principally Miocene to Recent basaltic lava flows. The region covers 10,000 square miles (25,900 km²) of the southwest corner of the Columbia Plateau. As such, it is merely a small part of the larger unit that embraces about 200,000 square miles (518,000 km²) of eastern Oregon, eastern Washington, southern Idaho, northern Nevada, northern Utah, and western Wyoming. Locations and geography are shown in Figure 3-1.

The Modoc Plateau grades west into the Cascade Range and south into the Sierra Nevada. The lava flows dissipate to the south, overlapping the older formations of the Sierra Nevada. The plateau also abuts the Basin Range province to the east where volcanic ramparts form escarpments with the valleys of the Basin Ranges. Basin Range faults either displace the volcanics or disappear beneath them; some of these faults are still active. Much of the plateau itself is broken by faults that have produced block mountains. Except for the Warner Range, these are generally low, with small escarpments, exposing only lava flows and none of the older rocks below. Because of its structure, some workers assign the Warner Range to the Basin Range province.

Despite its small size, the Modoc is important as a transitional area between the Cascades and the Sierra Nevada, providing details of geologic history not available elsewhere in the Columbia Plateau. There are impressive records of the extensive freshwater lakes that occupied parts of the Modoc Plateau during the middle and late Tertiary. Fragmental and nonbasaltic rocks that are exceptions to the usual Columbia sheetflood basalts also occur in the Modoc. Nevertheless, basaltic magmas poured from fissures and flowed like water for vast distances. These records are studied best in the Columbia province proper, where only restricted areas of recent volcanism are found. Much of the Modoc Plateau is veneered with recent cones, glass flows, pumice, and fragmental materials, interspersed with soil layers, lake-bed sediments, and volcanic plugs and domes.

Drainage

The Modoc Plateau is a rolling upland drained by a master stream, the Pit River, which crosses the plateau from northeast to southwest and joins the Sacramento River to drain to the Pacific (Figure 3-8). For much of its course, the Pit River flows lazily across meadows

Figure 3-8.

Looking across the Modoc Plateau over the city of Alturas and the North Fork of the Pit River. (Photo courtesy of California Department Water Resources)

developed on lava surfaces and around and along escarpments. Finally, as it leaves the plateau, it enters a gorge before joining the Sacramento River. Most of the gorge topography has been flooded by Lakes Britton and Shasta, so the Pit now flows into Shasta Lake instead of the Sacramento River.

Rocks and Structure

The oldest rocks of the Modoc Plateau are the Cedarville series, lava flows and fragmental materials that are best exposed in the Warner Range. The series has been assigned an Oligocene or early Miocene age on the basis of a few vertebrate and floral fossils. The lower sequence is rich in fragmental volcanics, with subordinate flows. The older lavas are andesitic, and the younger are basaltic. Rocks similar to and possibly correlative with the Cedarville crop out on the southern and eastern margins of the Modoc Plateau proper.

Late Miocene Basin Range faulting caused drainage disruption in the ensuing Pliocene. Shallow lakes developed in the basins produced by the faulting, and freshwater sediments, mudflows, ash, diatomite, and interbedded volcanics all were deposited intermittently. This period also was marked by widespread volcaniclastic deposition along the Modoc-Cascade boundary southwestward into the Sierra Nevada.

The basaltic lavas that occupy so much of the present Modoc surface were erupted into the sedimentary basins between the block-faulted ranges. Extensive lava floods occurred throughout the Pliocene and have continued sporadically, although Recent volcanic events tend to be concentrated on the Modoc borderlands. Large continuous sheets of Warner basalt occupy Devil's Garden Rim and cover about 700 square miles (1800 km²) there. The total thickness of the Warner basalt is unknown, but estimates are up to 600 feet (180 m). The gross thickness of materials of the Modoc Plateau is also unknown. It is assumed that the Cedarville series underlies the Warner basalt as far west as Medicine Lake Highland. The Cedarville's southern counterpart may be the Tuscan formation of the southern Cascades and western Sierra Nevada. Pleistocene and Recent volcanic features include obsidian flows, domes, plugs, and cinder cones, particularly those of Lava Beds National Monument.

Special Interest Features

Fall River Springs   The Modoc Plateau has generally low rainfall and sparse vegetation, except for the conifer forests at higher elevations on the western edge. One of the largest springs in the United States, however, is Fall River Springs near Fall River Mills. With a daily flow of over 1250 million gallons (4720 million l), the springs discharge water collected in underground channels from as far away as 50 miles (80 km). The water moves through weathered soil zones buried beneath the Warner basaltic flows, porous fragmental volcanics, and open lava tunnels. The springs are unusual because from Goose Lake to Fall River Mills the Pit River actually loses water to percolation, since the water table in the drainage basin lies below the Pit's channel through most of the upper half of the river's drainage.

Ice Caves   Lava tubes and tunnels are typical features of most volcanic areas. Where magmas that fed flows were highly fluid, lava surged forward after crusting over by initial cooling. Uncongealed magma drained from beneath the hardened crust, leaving open passages and forming extensive lava tunnels. In some cases, as in Craters of the Moon National Monument (Idaho) and Lassen Volcanic National Park, caves or tunnels nearly a half-mile (0.8 km) long have been mapped. Nearly 300 lava tubes are known in Lava Beds National Monument, and many others are unexplored. Only where a roof has collapsed is a tube accessible. Often these entrances establish air movements through the tubes, allowing rain and snow to enter. If cold, heavy air is trapped, ice accumulates and is preserved in so-called ice caves.

Burney Falls   McArthur Burney Falls State Park is a recreational area that preserves an outstanding example of a waterfall and stream fed by groundwater trapped between lava flows. Part of the flow's volume is provided by the surface water of Burney Creek, which was formerly a small tributary of the Pit River and now flows into man-made Lake Britton. Burney Creek drops 120 feet (37 m) over the lip of a durable basaltic flow. Beneath this flow lie tuffs and other porous materials from which large springs augment Burney Creek. These springs discharge an estimated 1.2 million gallons (4.6 million l) per day. The gorge of Burney Creek has been developed below the falls by rapid downcutting of weak rocks that underlie the durable basaltic cap rock.

Features like Burney Falls are common elsewhere in the Modoc Plateau, but on a smaller scale. Between volcanic eruptions soils are developed and vegetation grows, only to be obliterated by the next episode of eruption. Groundwater enters these soil zones, often assisting in the petrifaction of plants, and the zones act as aquifers for water storage.

Lava Beds National Monument   This small, little-known section of the Modoc contains examples of most volcanic features characteristic of the plateau. The area is adjacent to and north of Medicine Lake Highland and covers about 73 square miles (189 km²).

From Medicine Lake Highland, the surface slopes gently north with local irregularities where cinder cones, explosion craters, chimneys, fault scarps, and individual lava flows appear (Figure 3-9). The surface basalt is generally a dark-colored flow rock that produces a jagged topography on which minimal soil or vegetation can form. Some flows are less than 1000 years old. The monument has at

Figure 3-9. *Lava Beds National Monument: view from Schonchin Butte, looking west toward the Cascade Range. Note the scarp-step topography of the middle distance. (Photo by Mary Hill)*

least 12 prominent cinder cones from 50 to 700 feet (15–210 m) high. Mammoth Crater, the most spectacular of several funnel-shaped explosion craters, lies 375 feet (114 m) below the surrounding surface.

Among the most recent features are several groups of spatter cones and chimneys formed by gradual accumulation around vents of semimolten lumps and clots of frothy lava. These chimneys range from 2 to more than 50 feet (0.7–15 m) in height and from 1 to more than 100 feet (0.3–30 m) in diameter. Sinuous rope ridges, which extend several hundred feet from the bases of some chimneys, have hollow interiors and so are miniature lava tubes formed at the surface. Small domes of lava, which probably formed as blisterlike gas bubbles, occur at many places on the surface.

Perhaps the monument's most spectacular features are its caverns or lava tubes. Of the 293 tubes known, only 130 have been explored. There are usually two or more open caverns per tube; each shows a rich diversity of dimension, form, wall, and floor detail. Fossil remains of a mastodon and a prehistoric camel have been found, and various stalactitic forms (lavacicles) occur in many of the caves. Pictographs, stone artifacts, obsidian weapons, bones, and human skeletons are evidence of Indian occupation of some of the caves, and during the Modoc Indian War (1872–1873) several caves served as Indian retreats.

Geothermal Prospects   Geothermal activity has been reported from many areas of the Modoc Plateau and the Cascade Range. To date, however, none in the California Cascades has excited commercial interest. In Lassen Volcanic National Park, geothermal activity has long been a favorite tourist attraction, but under present regulations the park prohibits commercial ventures. In the Modoc Plateau, there is notable geothermal activity in the Surprise Valley east of the Warner Range. Though this area is on the border of the Modoc province, it derives its hot spring water from Columbia volcanism. Drilling for prospective geothermal power was conducted in 1959, but more exploration is required before actual potential can be predicted. Results so far do not suggest large-scale energy availability.

REFERENCES

General

Anderson, Charles A., 1941. Volcanoes of the Medicine Lake Highland, California. Calif. Univ. Dept. Geol. Sci. Bull., v. 25, pp. 347–422.

Geologic Map Sheets of California. Alturas (1958) and Redding (1962). Calif. Div. Mines and Geology.

Irwin, William P., 1960. Geologic Reconnaissance of the Northern Coast Ranges and Klamath Mountains, California. Calif. Div. Mines and Geology Bull. 179.

———, 1970. Geology of the Klamath Mountains. Mineral Information Service (now California Geology), v. 23, pp. 135–137.

Macdonald, Gordon A., 1966. Geology of the Cascade Range and Modoc Plateau. In Geology of Northern California. Calif. Div. Mines and Geology Bull. 190, pp. 65–95.

——— and T. E. Gay, Jr., 1968. Geology of the Southern Cascade Range, Modoc Plateau, and Great Basin Areas in Northeastern California. Mineral Information Service (now California Geology), v. 21, pp. 108–111.

McKee, Bates, 1972. The Klamath Mountains of Oregon (pp. 139–152), The Cascade Volcanos (pp. 193–217), and The Columbia Plateau (pp. 271–299). In Cascadia. McGraw-Hill Book Company.

Williams, Howel, 1932. Geology of the Lassen Volcanic National Park, California. Calif. Univ. Dept. Geol. Sci. Bull., v. 21, pp. 195–385.

Special

Anonymous, 1974. Geologic Hazards at Lassen National Park. California Geology, v. 27, p. 254.

Northern Provinces

Aune, Quentin, 1964. A Trip to Burney Falls. Mineral Information Service (now California Geology), v. 17, pp. 183–191.

———, 1970a. A Trip to Castle Crags. Mineral Information Service (now California Geology), v. 23, pp. 139–144.

———, 1970b. Glaciation in Mt. Shasta–Castle Crags. Mineral Information Service (now California Geology), v. 23, pp. 145–148.

Chesterman, Charles W., 1971. Volcanism in California. California Geology, v. 24, pp. 139–147.

Evans, James R., 1963. Geology of Some Lava Tubes. Mineral Information Service (now California Geology), v. 16, no. 3, pp. 1–7.

Hill, Mary R., 1970. Mt. Lassen Is in Eruption and There Is No Mistake about That! Mineral Information Service (now California Geology), v. 23, pp. 211–224.

Sharp, Robert P., 1960. Pleistocene Glaciation in the Trinity Alps of Northern California. Am. Jour. Sci., v. 258, pp. 305–340.

Slosson, James E., 1974. Surprise Valley Fault. California Geology, v. 27, pp. 267–270.

Stearns, H. T., 1928. Lava Beds National Monument, California. Geog. Soc. Philadelphia Bull., v. 26, pp. 239–253.

Woods, Mary C., 1974. Geothermal Activity in Surprise Valley. California Geology, v. 27, pp. 271–273.

Chapter Four

Basin Ranges

*Rivers do not rise with the first rainfall; the
thirsty ground absorbs it all.*

Seneca

Although the Basin Range province is more extensive in neighboring
Nevada, its most dramatic development is in California between the
Sierra Nevada crest and the Nevada state line. The province's west-
ern boundary is the Sierran crest itself, and some think that the
Sierra should be considered the highest and grandest of the Basin
Ranges. The Garlock fault separates the province from the Mojave
Desert to the south. The California part of the Basin Ranges includes
nearly all of Inyo and Mono counties, plus northeastern Kern and
northern San Bernardino counties (Figure 4-1).

The province also extends into California in the Susanville–
Honey Lake area and extreme northeastern California. Although
underlain by rocks like those of the Modoc Plateau, the northeastern
segment is characterized by high, north-south trending fault-block
mountains like the Warner Range and elongate, narrow depressions
like those occupied by Upper, Middle, and Lower Alkali lakes east of
the Warner Range. The Honey Lake Valley southeast has interior
drainage, a common characteristic of the province. The area is usually included in the Basin Ranges because it is a triangular wedge
dropped down between the granitic rocks of the Sierra Nevada and
the volcanics of the Modoc Plateau. Recent faulting is reflected by a
6-inch (15 cm) scarplet produced at the base of the Fort Sage Mountains during an earthquake in 1951.

A certain amount of confusion exists regarding the terms Basin
Ranges and Great Basin. The authors prefer to interpret the Basin
Range province as that part of the western United States characterized by interior drainage and elongate, approximately north-south
trending mountains and valleys. They interpret the Great Basin as
the more extensive area of internal drainage lying between the Sierra
and the Transverse and Peninsular ranges on the west and the
Wasatch Range of Utah on the east. Included in the Great Basin are
the Mojave Desert, much of the Colorado Desert, and the Basin
Range province itself.

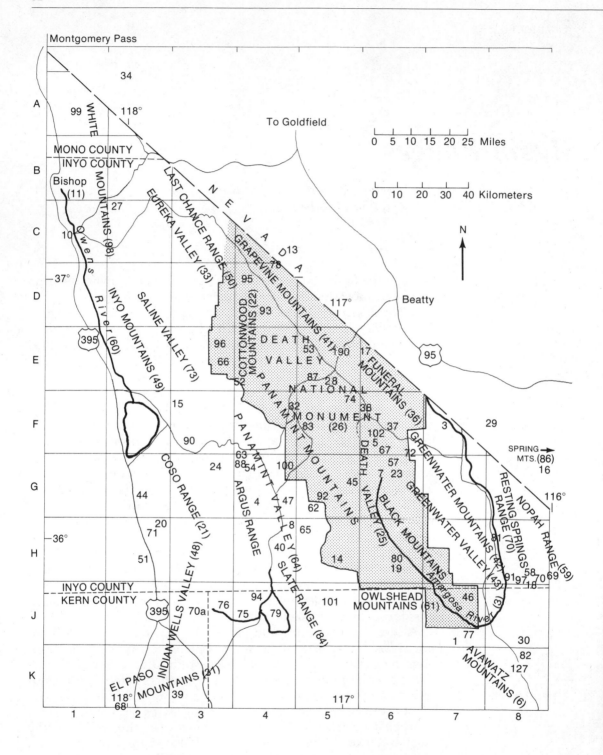

Figure 4-1.   Place names: Basin Ranges.

Adobe Lake Valley (Basin) (see Figure 2-2)
1 Alexander Hills—J7
2 Argus Range—G–H, 4
3 Amargosa River—F–J, 6–8
4 Ash Hill—G4
5 Artist's Drive—F6
6 Avawatz Mountains—K7–8
7 Badwater—G6
8 Ballarat—H4
9 Beatty (Nevada)—D6
10 Big Pine—C1
11 Bishop—B1
12 Black Mountains—G–H, 6–7
13 Bonnie Clare Playa (Nevada)—C4
14 Butte Valley—H5
15 Cerro Gordo Mine—F3
16 Charleston Peak (Nevada)
17 Chloride City (Chloride Cliff)—E6
18 Columbia Mine—H8
19 Confidence Hills—H6
20 Coso Hot Springs—H2
21 Coso Range—G3
22 Cottonwood Mountains—D–E, 4
Crowley Lake (see Figure 2-1)
23 Dante's View—G6
24 Darwin—G3
25 Death Valley—F–H, 6
26 Death Valley National Monument—D–J, 3–7
27 Deep Spring Valley—C2
28 Devil's Cornfield—E5
29 Devil's Hole Spring (Nevada)—F7
30 Dumont Dunes—J8
31 El Paso Mountains—K2–3
32 Emigrant Wash—F4–5
33 Eureka Valley—C2–3
34 Fish Lake Valley (Nevada)—A2
35 Fort Sage Mountains (see Figure 2-2)
36 Funeral Mountains—E–F, 6
37 Furnace Creek—F6
38 Furnace Creek Ranch—F6
39 Garlock (Station)—K3
40 Goler Gulch—H4
41 Grapevine Mountains—D–E, 4–5
42 Greenwater Mountains—G–H, 7
43 Greenwater Valley—G–H, 7
44 Haiwee—G2
45 Hanaupah Fan—G5–6
Honey Lake (see Figure 2-2)
Hot Creek (see Figure 2-1)
46 Ibex Mountains—J7
47 Indian Ranch—G4
48 Indian Wells Valley—H–J, 2–3
49 Inyo Mountains—D–F, 1–3
50 Last Chance Range—C–D, 3
51 Little Lake—H2
52 Lost Burro Gap—E4
53 Midway Well—E5
54 Minietta and Modoc Mines—G4
55 Mono Lake (see Figure 2-2)
56 Montgomery Pass (Nevada)—A1
57 Mormon Point—G6
58 Noonday Mine—H8
59 Nopah Range—G–H, 8
60 Owens River—B–F, 1–2
61 Owlshead Mountains—J6
62 Panamint City—G5
Panamint Mountains—E–G, 4–5
63 Panamint Springs—F4
64 Panamint Valley—F–H, 4
65 Pleasant Canyon—H5

66 Racetrack Playa—E3
67 Rainbow Canyon—F6
68 Redrock Canyon—K2
69 Resting Springs—H8
70 Resting Springs Range—G–H, 8
70a Ridgecrest—J3
71 Rose Valley—H2
72 Ryan—F6
73 Saline Valley—D–E, 2–3
74 Salt Creek—F5
75 Salt Wells Canyon—J4
76 Salt Wells Valley—J3
77 Saratoga Springs—J7
78 Scotty's Castle—D4
79 Searles Lake—J4
80 Shoreline Butte—H6
81 Shoshone—H8
82 Silurian Lake—K8
83 Skidoo—F5
84 Slate Range—H–J, 4–5
85 Soda Lake (see Figure 5-1)
86 Spring Mountains (Nevada)
87 Stovepipe Wells—E5
88 Surprise Canyon—G4
89 Susanville (Lassen County)
90 Talc City Hills—F3
91 Tecopa—H8
92 Telescope Peak—G5
93 Tin Mountain—D4
94 Trona—J4
95 Ubehebe Crater—D4
96 Ubehebe Mine—E3
97 War Eagle Mine—H8
98 White Mountains—A–C, 1–2
99 White Mountain (Peak)—A1
100 Wildrose Canyon—G4
101 Wingate Pass—J5
102 Zabriskie Point—F6

Basin Ranges

The Basin Range province includes some of the greatest relief in North America. For example, there are only 80 airline miles (130 km) from Mount Whitney (14,495 ft or 4421 m), the highest peak in the 48 contiguous states, to the lowest spot of dry land in the western hemisphere near Badwater in Death Valley (−283 ft or −86 m). In Inyo County alone, at least four spectacularly deep valleys occur between towering mountains. The best known is Owens Valley lying between the Sierra Nevada and the Inyo-White Mountains, each range containing peaks more than 14,000 feet (4250 m) high (Figure 4-2). The floor of Owens Valley rises gradually from 3500 feet (1050 m) at its south end to 5300 feet (1600 m) at the foot of Montgomery Pass in the north.

Immediately east of the Inyo Mountains is deep, almost circular Saline Valley, whose floor is nearly 2000 feet (600 m) lower than that of Owens Valley. Although the western side of Saline Valley is a 9000-foot (2750 m) escarpment, perhaps the valley's most interesting aspect of relief is its closure: if water filled Saline Valley to overflowing, a lake more than 3000 feet (900 m) deep would result—deeper than any lake in the western hemisphere.

Southeast of Saline Valley is Panamint Valley, about 55 miles (90 km) long. From the valley floor with elevation about 1000 feet (300 m), the Panamint Mountains rise to just over 11,000 feet (3350 m) at Telescope Peak. From this peak there is an impressive view into Panamint Valley to the west, about 10,000 feet (3000 m)

Figure 4-2.

Looking south from near Bishop over the Owens Valley and the town of Big Pine (center). Coyote Flat is the warped frontal block west and north of Big Pine. The Inyo–White Mountains parallel Owens Valley on the east. Saline Valley is in the upper left. Note the dark cone of Crater Mountain south of Big Pine. (Photo courtesy of U.S. Air Force and U.S. Geological Survey)

Figure 4-3.

Looking northwest across Death Valley from Dante's View in the Black Mountains. Devil's Golf Course is the large flat central area. The extensive alluvial fans at the east base of the Panamint Mountains are characteristic. Note the channel of Salt Creek, the major stream of the valley, leading into Devil's Golf Course from the north. (Photo by Mary Hill)

below, or Death Valley to the east, more than 11,000 feet (3350 m) below.

Death Valley, the easternmost of California's Basin Range depressions, also contains some spectacular relief. About 2 miles (3.2 km) east of Badwater is Dante's View (5475 ft or 1670 m), atop the Black Mountains. The slope from Dante's View to Badwater is so steep that a person standing at Dante's View cannot see either the base of the slope or Badwater itself (Figure 4-3). Death Valley possesses superb examples of alluvial fans and some remarkably recent fault features. In addition, surrounding ranges contain Precambrian to Recent rocks that reflect nearly all the main subdivisions of geologic time.

DRAINAGE

Although internal drainage characterizes today's Basin Ranges, parts of the region probably had external drainage as recently as the Pleistocene. Normally, at least some closed basins form in areas of uplifted and depressed blocks. Internal drainage then results if rain and snowfall produce less water than is lost by evaporation. Basin Range lakes commonly lose 7 or 8 feet (2–2.5 m) of water to evaporation annually. When this water is not replaced, the lakes become dry lakes or playas. Most of the province's lakes are ephemeral, holding water only after heavy rains. The few permanent lakes, such as Mono Lake, are fed by streams rising in high, well-watered ranges like the Sierra Nevada. Mono Lake has no outlet and is saline. Until about 1915, Owens Lake was another saline water body. Owing to construction of the Los Angeles aqueduct, however, the lake's main supply, the Owens River, was largely withdrawn, shrinking the lake. By the middle 1920s, the lake had become the vast, salty playa it remains today. Unusually heavy runoff from the Sierra sometimes fills the lake, but only for a time; the lake disappears again during dry years.

In the California part of the Basin Range province there are few permanent streams. The most prominent is the Owens River, which owes its existence to the Sierra Nevada. Actually, few permanent streams entering the Owens Valley south of Bishop manage to reach the Owens River as live streams. Huge fans built out onto the valley floor by Sierran streams during flooding provide permeable gravel cones that soak up the stream flow of normal years. Water moves toward the Owens River beneath the surface, emerging at the toe of the fans near the river, the lowest place in the valley.

Apart from a few short streams confined to mountain canyons, the only permanent stream in California's southern Basin Ranges is the Amargosa River, which rises in western Nevada. It flows south in a broad valley between the Greenwater and Resting Spring ranges, across the eroded bed of former Lake Tecopa, through a narrow canyon cut across the southern Ibex Mountains, and finally north into Death Valley. The river surfaces where bedrock is near ground level, but disappears where alluvium is deep. High evaporation has produced a white saline crust along much of the river's course. In wet years, the Amargosa may persist all the way to Badwater.

ROCKS AND GEOLOGIC HISTORY

Because of the prevailingly dry climate and the consequently sparse vegetation, Basin Range rocks are unusually well exposed. Furthermore, California's finest Precambrian and lower Paleozoic sedimentary sequences appear in the Basin Ranges.

Precambrian

Oldest Basin Range rocks are a complex of early Precambrian schists and gneisses, probably of sedimentary origin, and associated granitic rocks much like those exposed in the inner gorge of the Grand Canyon. Some of these ancient rocks have been dated as 1800 million years old. Good exposures are found in the western Black Mountains, which form the steep eastern wall of southern Death Valley.

Only slightly metamorphosed, the younger Precambrian is several thousand feet thick and composed of regularly bedded sandstones, shales, and conglomerates. The rocks are well displayed in the Funeral and southern Panamint mountains. Outcrops continue southeast for about 75 miles (120 km) to the Kingston Mountains and into Nevada. The sequence constitutes the Pahrump group: from oldest to youngest, the Crystal Spring conglomerate and sandstone (4000 ft or 1200 m), the Beck Spring dolomite and limestone (1000 ft or 300 m), and the Kingston Peak sandstone (200 ft or 60 m). These sedimentary rocks are probably nearshore materials representing earliest deposits in the Cordilleran geosyncline. Some of the Kingston Peak beds appear to be glacial marine deposits containing striated drop-stones rafted offshore by ice.

Precambrian—Paleozoic Transition

Cambrian strata are well represented in the Basin Ranges. The thick, fossiliferous lower Cambrian rocks of the Inyo Mountains are the North American type section for the lowermost Cambrian, or Waucoban. Farther east, Cambrian beds are well exposed in the Panamint Mountains and aggregate a thickness of about 17,000 feet (5200 m) southeast of Death Valley. Moreover, here the Cambrian is depositionally continuous with the Precambrian Pahrump beds, providing an unusually clear record of the transition. Some geologists assign the transitional beds to a discrete geologic period on the basis that much of the sequence was deposited when the earliest multicellular (metazoan) organisms first appeared, but before hard-shelled animals had evolved.

In the Death Valley area, the transition strata include three conformable units, the Noonday, Johnnie, and Stirling formations. The lowest, the Noonday, rests unconformably on the Precambrian Pahrump group. Previously, the Precambrian-Cambrian boundary was placed at the unconformity, although the youngest definitely Cambrian fossils were found as much as 10,000 feet (3000 m) higher.

Diagrammatically, the column is as follows.

Definite Cambrian	Wood Canyon	
	Stirling	
	Johnnie	Transition Beds
	Noonday	
	UNCONFORMITY	
Precambrian	Kingston Peak	
Pahrump Group	Beck Spring	
	Crystal Spring	

Although it is reasonable to set the base of the Cambrian at the unconformity, it is equally reasonable to place it at the base of the Wood Canyon formation because this unit contains the earliest undoubtedly Cambrian fossils. Another complication is that the Johnnie and Noonday formations resemble the underlying Pahrump group and might be included with the Pahrump lithologically. Finally, the Stirling, Johnnie, and Noonday formations could all be assigned to a transitional or "Eocambrian" period.

Paleozoic

Although in the Death Valley area Cambrian strata dominate the Paleozoic, most later Paleozoic periods also are represented—by extensive, but usually thinner limestones and dolomites. The Paleozoic section in the Inyo Mountains represents nearly all periods and is also extremely thick, implying more or less continuous deposition. Aggregate thickness is about 23,000 feet (7000 m), and nearly half of this is Cambrian.

Post-Cambrian Paleozoic rocks reach maximum thickness in the Inyo Mountains, thinning toward the east. Most are carbonate rocks (limestones and dolomites), indicating shallow warm waters with little incoming detritus. Included is the thickest section of Paleozoic carbonate rocks anywhere in North America. Although the extensive development of carbonates suggests a generally tranquil depositional environment, some interruptions are evident. The Ordovician Eureka quartzite represents an influx of abundant, clean, washed sand from adjacent land, and upper Permian rocks that include conglomerates rest by well-defined angular unconformity on Pennsylvanian deposits. Both the angular unconformity and the conglomerates imply initiation of orogenic activity in the Cordilleran eugeosyncline, from which the Sierra Nevada was eventually raised.

These same post-Cambrian Paleozoic rocks are well displayed in the Cottonwood and northern Panamint mountains near Ubehebe Crater. Richly fossiliferous Devonian rocks of the Lost Burro formation are exposed in Lost Burro Gap, about 20 miles (32 km) southwest of the crater.

Mesozoic

Geosynclinal conditions persisted through the Triassic. Some of the rocks deposited then are now well exposed in the Inyo Mountains, the Panamint Mountains near Butte Valley, and the Alexander Hills, where a xenolith of Triassic volcanic rock is surrounded by a younger intrusive porphyry. Lower Triassic rocks are primarily sedimentary, but by the close of the period huge amounts of volcanics were being erupted in the area now represented by the Sierra Nevada, Owens Valley, and Inyo Mountains.

Numerous granitic intrusives of probable Mesozoic age are found in California's Basin Ranges. The largest of these occurs in the Owlshead Mountains. Smaller exposures are present in the Greenwater, Black, Argus, Inyo, White, Last Chance, and southern Panamint ranges.

Cenozoic

The Basin Range area probably underwent an interval of erosion or at least nondeposition from the middle Mesozoic until the Oligocene, because the oldest Cenozoic rocks yet recognized are the Oligocene Titus Canyon beds northeast of Stovepipe Wells. These rocks are known for the discovery of titanothere remains in the 1930s. Related to the horse, the titanothere was a gigantic herbivore that lived in association with such animals as the rhinoceros, tapir, and camel. The fossil remains of these animals suggest an Oligocene environment of savanna, with moist climate and succulent vegetation. Later Tertiary rocks are abundant in the Basin Ranges, but few are fossiliferous.

Aridity increased during Tertiary time, and by early Pliocene rainfall was probably about 15 inches (585 mm) annually. Grasslands were widespread, and hot summers and mild winters the rule. The climatic pattern suggests to some geologists that erosion had reduced the ancestral Sierra Nevada to a range of low hills, allowing the moderating effects of the Pacific to extend farther inland than is now possible. The late Pliocene Coso Mountains formation has yielded impressive mammalian fossils, including hyenid dogs, horses, short-jawed mastodons, camelids, and peccaries. Other Pliocene rocks, found at Furnace Creek, have been dated by fossil leaves and animal footprints.

In the Greenwater and northern Black mountains, along the Furnace Creek and Artist's Drive fault zones, is a down-dropped wedge of rock partly occupied by Furnace Creek wash. This wedge contains a Tertiary section totaling almost 13,000 feet (4000 m). The Artist's Drive formation is nearly a mile (1.6 km) thick and has both sedimentary and volcanic members. This is the sequence's oldest unit and is at least partly contemporary with the Titus Canyon beds to the north. The overlying Furnace Creek formation of probable Pliocene age is composed of volcanic rocks, fanglomerates, and playa sediments that include borates and gypsum. These sediments indicate not only an arid or semiarid climate, but also the presence of undrained topographic depressions in which waters saturated with sulfates and borates could accumulate. The fanglomerates suggest an environment of vigorous erosion, flash floods, steep mountain faces, and deep basins.

Late Pliocene to Recent volcanic rocks are abundant in the Basin Ranges. Prominent Pliocene basaltic flows are exposed near Ryan in the Greenwater Mountains. Pleistocene and Recent cinder cones and flows are well displayed throughout the southern Owens Valley, particularly near Little Lake where basaltic flows and several red cinder cones dominate the landscape. Youthful cinder cones and flows are scattered northward in the Owens Valley and on the slopes of the Sierra Nevada and Inyo Mountains, especially near Big Pine. Just south of Big Pine is a group of bold volcanic peaks, the highest rising about 2000 feet (600 m) above the valley floor.

A center of recent activity is the southern Coso Range, where numerous basaltic flows and cinder cones have developed. Some eruptions discharged dark flows that poured over low divides and down modern canyons, appearing almost as fresh today as if they had just congealed. Obsidian and pumice cones and domes like those of the Mono area are also present (Figure 4-4), associated with boiling springs and steam vents.

The presence of Plio-Pleistocence saline beds and fanglomerates in the Basin Ranges implies that basins resembling today's were taking shape during these epochs. Saline deposits associated with clay-rich playa beds in Death Valley, Saline Valley, Searles Lake, and numerous valleys in the Mojave Desert province are chiefly of Pleistocene age. The saline layers in these basins are often separated by layers of silt and clay deposited during wetter periods that may have corresponded to glacial epochs in the Sierra Nevada.

The chemistry of these basins differed. Owens Lake, for example, accumulated deposits rich in sodium carbonates and borates that have been mined intermittently since 1904. Searles Lake has yielded the greatest array of saline minerals of any Basin Range valley; end products include potassium salts, borax, boric acid, salt, soda ash, bromine, and lithium carbonate. The lake has two main saline deposits. The upper, porous one is 70 to 80 feet (21–25 m) thick with many voids filled with saturated brines. It covers about 30 square miles (78 km^2). Below this deposit are about 12 feet (3.6 m) of mud, and below the mud is another saline body with brine about 35 feet (11 m) thick. The intervening mud layer contains organic material dated at 16,000 years, or late Pleistocene.

Death Valley was an early historic source of borate minerals, which had accumulated in the valley's muds during the Pleistocene.

Figure 4-4.
Obsidian and pumice cones and domes of the southern Coso Range. Compare with Mono Craters, Figure 2-18. (Photo by James Babcock)

The deposits were first worked in 1882; the ore was hauled to the railhead at Mojave, 165 miles (266 km) to the southwest, by the famed 20-mule teams.

STRUCTURE

The high mountain blocks of the Basin Ranges have a north-northwest trend and complex internal structure. Tight folds and faults occur, with flat thrusts, high-angle normal, and some strike-slip faults.

Garlock Fault System

The Garlock fault, which demarcates the Basin Ranges and the Mojave Desert, is a nearly vertical fault with about 40 miles (64 km) of left-lateral separation (Figure 4-5). It is no older than early Tertiary and may be much younger. Although the Garlock is extensive, only its eastern half is considered here: from Red Rock Canyon in

Figure 4-5.
View to the east
showing the trace of
the Garlock fault.
Town of Ridgecrest
is in the lower left.
Searles Lake is in
the left center, with
the Slate Range al-
most enclosing the
lake on the east.
The El Paso Moun-
tains are in the
center foreground.
The intersection of
the Death Valley,
Garlock, and Fur-
nace Creek fault
zones is evident in
the upper center.
(Photo courtesy of
U.S. Air Force and
U.S. Geological
Survey)

the El Paso Mountains to its junction with the Death Valley fault system at the eastern base of the Avawatz Mountains.

The Garlock system is an interlacing series of shorter faults enclosing narrow, elongate slices and wedges of rock. On the south face of the El Paso Mountains, for example, the El Paso fault is more prominent than the Garlock fault proper, which tends to be south of the range and buried beneath desert gravels. The El Paso fault is clearly exposed at the south entrance to Red Rock Canyon, where Pleistocene fanglomerates are in contact with Mesozoic granite. Here the two faults are separated by less than a mile (1.6 km). They then continue east for 10 miles (16 km) until the El Paso joins the Garlock. In this segment, the El Paso fault shows a strong component of vertical movement whereas the Garlock shows equally strong left-slip displacement. The slice between the two faults contains several small depressed blocks, springs, and offset streams. About 3 miles (4.8 km) east of Garlock Station, the faults enclose a well-developed graben, 50 feet (15 m) deep and nearly a mile (1.6 km) long. This feature is just out of sight from, but parallel to, the Garlock road. For 60 miles (96 km), from its juncture with the El Paso fault to its intersection with the Furnace Creek fault zone, the Garlock is marked by a fault-line valley until it becomes the northern boundary of the Avawatz Mountains.

Evidence of left slip on the Garlock is abundant, but amount of movement is not clear. Recent movements up to a half-mile (0.8 km) are demonstrated by offset streams and other topographic disruptions. In 1925, left-lateral displacement up to 6 miles (9.6 km) was suggested, based on inferred displacement of Paleozoic formations in the Randsburg area. Displacement up to 40 miles (64 km) was suggested subsequently, based on the presumed equivalence on each side of the fault of a network of basic dikes. The latest figures indicate from 16 to 30 miles (27–48 km) of displacement. Whatever the precise amount, geologists agree that displacement is substantial and that it is continuing, at least in the central section.

Death Valley–Furnace Creek Fault System

Another predominantly strike-slip fault zone is the Death Valley–Furnace Creek system, which has about 50 miles (80 km) of right-lateral separation (Figure 4-6). The Furnace Creek portion extends from northern Death Valley south to Furnace Creek, where it strikes east into the Amargosa River valley. Although the Furnace Creek system often is concealed by alluvial gravels, it is well exposed in low escarpments south and east of Stovepipe Wells. These escarpments and the sharp boundaries between fan deposits and the older rocks of the Funeral Mountains clearly delineate the Furnace Creek fault system between Furnace Creek Ranch and the Amargosa River valley. The fault system cannot be definitely traced all the way into the Amargosa River valley, but it probably does extend south and eventually may merge with the frontal fault of the Resting Springs Range.

The Death Valley fault system is best exposed at Shoreline Butte, in the Confidence Hills, and along the northeastern Avawatz Mountains where the system joins the Garlock fault zone. The system also includes the well-exposed frontal faults along the western Black Mountains, although specific names have been assigned to the large vertical separations that form the precipitous west-facing escarpment near the deepest part of the valley.

The junction of the left-slip Garlock and right-slip Death Valley fault systems poses some interesting problems. Evidence suggests that about 40 miles (64 km) of left-lateral separation occur along the Garlock only 8 miles (13 km) west of its junction with the Death Valley fault. At least three interpretations of this unusual situation have been offered.

Basin Ranges

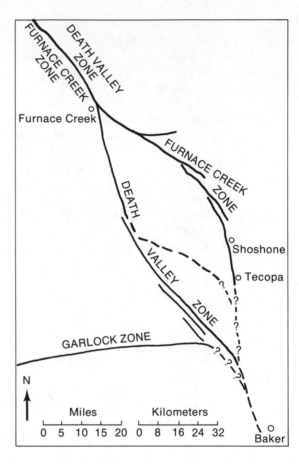

Figure 4-6.
Intersection of the
Death Valley, Gar-
lock, and Furnace
Creek fault zones.

1. Some believe that the Garlock terminates at the Death Valley fault zone and that the 40 miles (64 km) of displacement must be accommodated in the junction area.

2. Others suggest that the Garlock has been offset by a younger Death Valley fault and that an eastward continuation of the Garlock exists to the south. So far, no evidence for such a continuation has been found.

3. Another view is that the Garlock crosses the Death Valley fault and continues directly east. The differing structures of the Basin Ranges north of this extension and in the Mojave Desert to the south may substantiate this, but again there is no conclusive evidence of such a fault.

As yet no generally acceptable explanation is available. All investigators would agree, however, that the junction of these two large fault systems warrants further study.

Special Fault Features

Wine glass or goblet valleys are canyons whose lower walls are nearly vertical in contrast to their upper walls, which are more open and broadly V-shaped. Such valleys indicate rapid and recent uplift, which caused valley deepening to outpace valley widening and subsequently form the stem of the goblet. In the Basin Ranges, the volume of material visible in alluvial fans typically associated with these valleys is insufficient to account for the total material excavated from the drainage basins above. The explanation is that rapid, relatively downward movement of blocks like Death Valley depress old fan deposits and force the streams emerging from the steep valleys to construct new cones on top of older, buried fans.

Like the Garlock, the Death Valley and Furnace Creek fault zones are composed of interlacing faults that incorporate elongate, narrow slices of crust. Some of these slices have dropped downward throughout their lengths; some have been depressed at one end but raised at the other; and some have been raised primarily relative to adjacent blocks. Such movements produce varied topography and, in the relatively down-dropped blocks, preserve rocks that elsewhere are largely destroyed by erosion. Colorful examples occur at Zabriskie Point and Artist's Drive, where thick sections of the mostly yellow, late Tertiary nonmarine Furnace Creek formation are well exposed.

In the northeastern Avawatz Mountains are the colorful, predominantly red, Salt Basin beds. These probably correlate with the Furnace Creek formation, although the characteristic colors of each formation differ. The Salt Basin beds have been preserved because they are part of a block depressed between the merging Garlock and Death Valley fault zones.

Turtleback Faults

For many years there has been debate regarding the complex structure of the Black Mountains. The frontal fault system is not at issue; it is considered a set of high-angle normal faults separating the uplifted range from the depressed Death Valley block. Within the range, however, are features thought to be folded, broken, and disordered thrust sheets resting on a core of Precambrian crystalline rocks. In several places, the younger rocks of the thrust sheets have been stripped away, exposing smooth but folded fault surfaces. Some of these folded surfaces form anticlines up to 13 miles (20 km) long, called *turtlebacks* (Figure 4-7). Prominent turtlebacks occur at Badwater and at Mormon Point (the Desert Hound anticline).

Levi F. Noble spent years studying the Black Mountains, where rock exposures are almost too good and too complete. (The broad picture often can elude the geologist presented with an abundance of well-exposed detail.) Despite this, Noble established that the range's core is overlain in at least three places by disordered and exceedingly complex thrust sheets that he termed *chaos*. The lowermost is the Virgin Spring chaos, consisting chiefly of Precambrian to middle

Figure 4-7. Turtlebacks in Death Valley, south of Badwater. Note the alluvial cones, which reflect recent movement along the Death Valley fault. (Photo by Spence Air Photos, courtesy of Department of Geography, University of California, Los Angeles)

Cambrian brecciated and shattered blocks 200 to more than 2500 feet long (60–760 m). Atop the Virgin Spring is the Calico chaos, composed of colorful Tertiary volcanic rocks. Highest is the Jubilee chaos, which consists mainly of brecciated granitic blocks and Tertiary volcanic and sedimentary rock. The Jubilee also includes a megabreccia that seems to contain every rock type known from the Black Mountains.

Geomorphic Evidence of Faulting

Evidence of recent faulting in Death and Panamint valleys is provided by well-preserved scarps in unconsolidated alluvial gravels. The very existence of well-preserved scarps in such weak, recent deposits corroborates their youth, because in arid climates stream erosion quickly erases such features. Many of the scarps are probably only a few hundred to a few thousand years old. Apart from similar features in the Owens Valley known to be connected with the 1872 earthquake, none of the young scarps in the province's basins is associated with a recorded earthquake.

The best known of these youthful fault features is the Wildrose graben near the mouth of Wildrose Canyon on the fan near the western base of the Panamint Mountains (Figure 4-8). Here a block of fan material about 4 miles (6.4 km) long and nearly a mile (1.6 km) wide has dropped down about 200 feet (60 m). This has diverted some water courses, forcing them to follow along the base of the western scarp instead of flowing directly across it. The old courses of these desert streams form beheaded valleys notching the uplifted western wall of the graben.

Across the Panamint Valley at the base of the Argus Range is Ash Hill, capped with Quaternary basaltic flows. These lava flows dip to the east and match similar flows that cap the Argus Range and that disappear beneath the alluvial gravel at the range's base. The reappearance of these flows not more than a mile (1.6 km) east of the base of the range shows that both Ash Hill and the Argus Range have been faulted and tilted. This is an example of *louderback caps,*

Figure 4-8.

View to the north of the graben at the mouth of Wildrose Canyon. Note the streams beheaded by recent downfaulting and the subsequent readjustment. Northern Death Valley is in the far distance, with the Grapevine Mountains as backdrop. (Photo by John S. Shelton)

named by William Morris Davis after George Davis Louderback who cited such features as evidence for tilting of fault blocks in the Basin Ranges of Nevada.

Many of the Basin Ranges tend to be symmetrical, elongate blocks with bounding faults on their flanks. On the other hand, some ranges such as the Panamint, Funeral, and Grapevine mountains are tilted fault blocks. Each has a steep western face bounded by young normal faults and a gentle eastern slope that passes gradually from foothills into pediment and fan.

SUBORDINATE FEATURES

Pleistocene Lakes and Streams

Basin Range events of the past 2 million years are reflected everywhere in today's landscape. Because of this, it is considerably easier to assess the Quaternary environment than Tertiary or earlier environmental settings.

Geologic time distinctions within the Basin Ranges are continuously in dispute. As mentioned previously, for instance, the Precambrian-Cambrian boundary is still unconfirmed. Likewise there has been much debate about the Tertiary-Quaternary boundary. Although recognizing that this matter is not yet settled, the authors arbitrarily equate the opening of the Quaternary with the earliest recorded glaciation in the Sierra Nevada—about 3 million years ago.

The glacial epochs produced little permanent ice in the Basin Ranges. In fact, only the highest of the Basin Ranges had any glaciers at all (excluding the Sierra Nevada). Small valley glaciers probably existed near the summit of the White Mountains and on Charleston Peak in Nevada's Spring Mountains.

The glacial epochs profoundly affected Basin Range climates, which were cooler, wetter, and characterized by less evaporation than today. In addition, apparently climatic and associated vegetative zones shifted to lower altitudes. Thus the boundary between desert shrub vegetation and piñon-juniper woodland, which today occurs high on Basin Range slopes, lay at much lower elevation. Evidence from the desert Southwest confirms this lowering of vegetative belts and indicates that the climatic contrast between upland and valley was stronger than it is today. The deep basins like Death Valley seem to have remained dry during the glacial epochs, although they were probably much cooler and experienced appreciably less evaporation than at present.

Subsequently there developed permanent drainages and lakes in basins that today contain only salt flats, playas, and ephemeral streams. The seeming paradox of permanent lakes and streams existing in cool but arid valleys perhaps can be accepted more easily if modern parallels are considered. For example, alkaline Mono Lake lies in a cool, arid valley about 6400 feet (1950 m) high. Rainfall is low, but the lake and entering streams are fed by the much better watered Sierra Nevada to the west. Another example is Owens Lake. Until construction of the Los Angeles aqueduct and subsequent interception of the Owens River, Owens Lake was a large, rather shallow, saline body of water occupying the extremely arid southern end of Owens Valley. Both these lakes have lacked outlets since the end of the Pleistocene and both have become mineralized, further confirming the aridity of their settings.

Extensive lakes and connecting streams probably occupied many of the province's shallower basins in the glacial epochs, but most of the evidence was erased during the intervening drier periods. For the most part, only the record of the latest chain of lakes and connecting streams can be seen today (Figure 4-9).

Alkaline Mono Lake is appreciably smaller than its freshwater Pleistocene ancestor, Lake Russell, which was probably about 750 feet (225 m) deeper than the modern lake. Lake Russell spilled east into the valley of Adobe Lake and from there into upper Owens Valley.

Water from Lake Russell and the greatly augmented streams draining the glaciated Sierra undoubtedly made the ancient Owens River a formidable stream. Owens Lake, now only a few stagnant saline ponds in a vast, shimmering salt flat, was at least 250 feet (80 m) deep and extended south to the locality of Haiwee. The lake stayed fresh because it maintained an outlet channel across Rose Valley and a well-defined gorge was cut in the young volcanic rocks near Little Lake. This gorge, its falls, and the associated potholes developed during the final glacial stage, because the lava flows cut are no older than the last interglacial stage.

The now-vanished river that produced the falls continued south to Indian Wells and Salt Wells valleys, where a broad shallow lake formed. During this period, lime-secreting algae promoted development of tufa pinnacles around the lake's edge. The shallow lake occupying Indian Wells Valley spilled east across the southern Argus Range, cutting the winding gorge of Salt Wells Canyon through which the highway to Trona now passes.

The stream that cut this channel fed glacial Lake Searles, which was once about 640 feet (195 m) deep. As noted earlier, the floor of this basin contains two thick beds of saline minerals separated by a clay layer that accumulated during the final wet period, when the lake probably spilled east into Panamint Valley. These saline beds show that Lake Searles was sometimes too deep and arid for the entering streams to maintain a freshwater lake more than briefly. The Lake Searles depression was thus the final basin in the Owens–Death Valley chain of lakes for much of the Quaternary. On

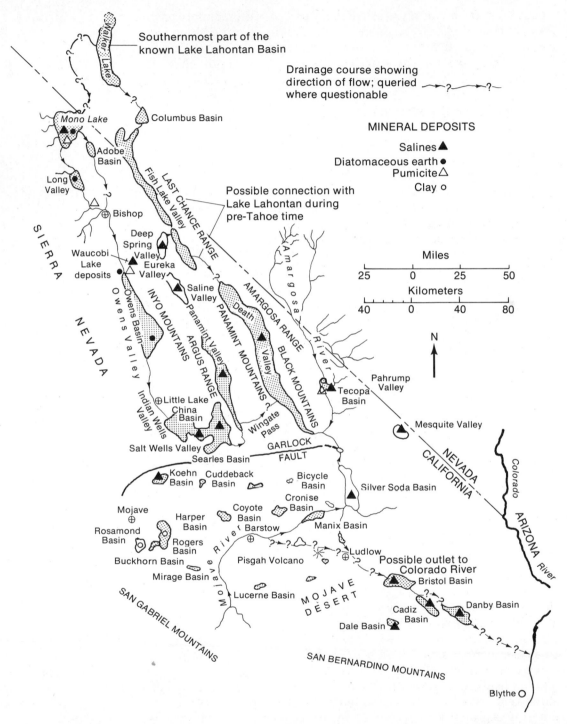

Figure 4-9.   Inferred Pleistocene drainage of the Basin Ranges. (Source: California Division of Mines and Geology)

the other hand, the lakes of Indian Wells Valley accumulated little saline material and were so shallow that they overflowed with even minimal inflow.

Both Mono and Searles basins are rimmed with many miles of well-preserved shorelines cut into surrounding ranges. Where the shorelines cross desert washes and alluvial fans, they form fossil beaches and sweeping, curved sand and gravel bars. In the Searles region, the most clearly developed shorelines are the lowest, further proof that the high-level, freshwater Lake Searles was more transitory than its lower-level saline counterparts.

California's finest tufa domes and pinnacles occur in the southern Lake Searles region, where there are more than 500 pinnacles (Figure 4-10). Some are 140 feet (43 m) high and more than 500 feet (160 m) in diameter, but most are 10 to 40 feet (3–12 m) high and 20 to 30 feet (6–11 m) in diameter. Like those of Indian Wells Valley, these pinnacles were formed by lime-secreting algae living in the lake and perhaps localized by the flow of springs rich in calcium carbonate.

During the wettest parts of the Pleistocene, Lake Searles drained east around the southern end of the Slate Range into Panamint Valley. Evidently Lake Panamint received limited local runoff and during drier periods was quickly isolated from Lake Searles, because the lake bed is chiefly clay rather than saline minerals. Nevertheless, at its maximum, Lake Panamint was more than 900 feet (275 m) deep, about 55 miles (90 km) long, and 10 miles (16 km) wide. When the lake was nearly full, several large streams entered it from the Panamint Mountains on the east. One of these streams deposited a large delta at the mouth of Pleasant Canyon, which can be seen just south of Ballarat. As the lake shrank, recessional shorelines were cut into this delta and elsewhere along the steep western face of the Panamint Mountains.

At its highest level, Lake Panamint's depth was limited by the level of Wingate Pass, which leads into the southern end of Death Valley. Death Valley was occupied by Lake Manly, about 90 miles (144 km) long and 600 feet (180 m) deep. Even during wet cycles, Lake Manly did not drain, but was the sink for minerals from Lake Panamint and from streams entering directly from the surrounding regions. Substantiating this are the thick accumulations of saline minerals in the valley bottom. Shorelines left by Lake Manly are preserved at several places in Death Valley, most notably at Shoreline Butte, a hill of young volcanic rock and Quaternary nonmarine beds located at the foot of Jubilee Pass. Other prominent shorelines occur along the west face of the Black Mountains near Badwater.

It seems probable that Lake Manly was connected southward with Silurian and Soda lakes in the Mojave Desert and from there with the Colorado River. Another probable connection existed with

Basin Ranges

Figure 4-10. (top) Tufa pinnacles at the southwestern shore of Searles Lake. Note old shorelines in lower left. (Photo by Fairchild Aerial Surveys, courtesy of Department of Geography, University of California, Los Angeles) (bottom) Detail of tufa pinnacles. (Photo by Robert M. Norris)

lakes in Fish Lake Valley and Walker Lake to the north, in Nevada. These intricate links have been proposed because of some unusual fish that live in Salt Creek and Saratoga Springs in Death Valley and in Devil's Hole Spring northeast of Death Valley Junction. The fish are small minnows, known as Desert Pup Fish. They belong to two genera, one closely related to fish of the Colorado River and the other related to fish in the Lahontan lakes and streams of west-central Nevada. It is thought that these genera of Desert Pup Fish once had free movement through the drainage system just described, but as the climate became drier the fish were isolated in permanent springs and streams.

The most recent flooding of modern Basin Range valleys probably occurred about 2000 to 5000 years ago, when small shallow lakes briefly occupied many basins. For instance, the pond in Death Valley was about 30 feet (10 m) deep, and its water persisted long enough to form faint shorelines. As the pond evaporated, a surface salt crust accumulated. In addition, faulting has deformed these young shorelines, making the eastern shore about 20 feet (6 m) lower than the western. This reflects the pattern of Quaternary fault movement also indicated by goblet valleys, steep frontal faults, and the eastward tilt of the valley floor.

Late Cenozoic Volcanism

Near Little Lake and Coso Hot Springs are many volcanic features, including about 15 late Quaternary basaltic cinder cones and about 30 rhyolitic tuff cones and domes (Figure 4-4). Most of these rest on the typical Sierran granite that composes the core of the Coso Range. Some of the lava flows are so young that they rest on Quaternary lake beds or form congealed lava streams in the canyons cut into the range. The canyons are probably Quaternary themselves.

The Coso region contains a variety of igneous rocks. Apparently, dark, rather fluid basalts were the first volcanics erupted. These were followed by explosive rhyolitic eruptions that produced tuff cones of light-colored, pumiceous frothy rocks often underlain by glassy obsidian. The basaltic eruptions continued, and some may be contemporary with the rhyolitic eruptions. The most recent eruptive activity has been basaltic, producing rather fluid lava flows and red and black cinder cones. Field evidence suggests that the basaltic magma rose from the upper mantle through the crust and the granitic rocks, melting some of the granites to produce the rhyolitic magmas. The chemical composition of the rhyolites resembles that of the underlying granite.

The Coso–Little Lake volcanic episode began during the Pleistocene, and there is nothing to suggest that it has yet concluded. The basaltic flows that poured down essentially modern canyons are one proof of continued activity. Another is the fresh, scarcely eroded shape of the cones themselves. Even eroded but still intact cones reflect youth, for there are few recognizable cones whose last eruptions predate late Tertiary.

Probably contemporary with the Coso field is the volcanic center in the northern Argus and Coso ranges on either side of State Highway 190, from Lone Pine and Olancha to Panamint Valley. South of Keeler are several distinct cones and a thick pile of sheet-like basaltic flows cut by north-south trending faults. Most of these volcanics are late Pleistocene, though a few are Pliocene. Additional flows occur just east of the junction with the Darwin road. Beyond the crest, a spectacular cross-sectional view is provided by Rainbow Canyon, which cuts the Argus Range north of the highway. The highway then slices across three cinder cones as it descends into Panamint Valley, giving the traveler an excellent worm's eye view of the inside structure of a cinder cone. Towards Panamint Springs, another roadcut reveals a thin basaltic dike cutting across extremely young unconsolidated alluvial gravels. This little feature is not unusual in itself, but it does demonstrate that volcanic activity has continued into essentially present time.

Ubehebe Crater, north of Tin Mountain near Death Valley, displays an unusual kind of volcanic activity (Figure 4-11). Some geologists consider the crater distinctive enough to be the type example of features designated *ubehebes*. Most investigators, however, call such wide, flat-floored, low-rimmed vents *maars* or *tuff rings*. It is thought that magma coming into contact with groundwater generates steam, with subsequent ejection of fragmental material and escaping steam. Usually a ring-shaped basal cloud is formed, composed of steam, ash, cinders, and rock fragments torn from the throat of the vent. At Ubehebe Crater, the erupted material is basaltic, and no lava flows are present.

Sand Dunes

Despite the popular association of dunes and deserts, dunes are not widespread in any of California's deserts. In the Basin Ranges, for example, rugged topography has precluded development of large tracts of dune sand. Certainly nothing in the American Basin Ranges compares to the sand seas that cover about a third of the Sahara or the seemingly endless linear dune ridges of the flat Australian desert. Basin Range dune systems are typically localized sand accumulations. They have developed where wind patterns allow the sand to collect and move to and fro, but not to move in any one direction for long.

Many of the desert basins have more than one group of dunes. Owens Valley has only a few, but they include several small areas southwest of Owens Lake that are blanketed with low, partially vegetated dunes probably derived from old lake beaches. Saline Valley also has a cluster of low dunes, none much higher than 50 feet

(15 m). They are northwest of the salt lake and indicate a wind pattern different from that prevailing in Owens Valley. Panamint Valley has a dune field at its far northern end, suggesting that winds transport sand north until they are slowed by being forced up the steep southern face of Hunter Mountain. These dunes are about 300 feet (90 m) high.

Although not more than 80 feet (25 m) above the valley floor, the dunes in Death Valley are the best known in the province. They have been seen by countless visitors and have provided the desert settings for many motion pictures. Associated with these dunes is the Devil's Cornfield, where deflation of sand and silt has left clumps of arrow-weed atop hummocks of sand and silt anchored by the plant roots. North of Stovepipe Wells and near Midway Well are some low dunes anchored by mesquite. The mesquite thrives in sandy areas where water is available at shallow depth, and its tangled branches trap sand until mounds about 20 feet (6 m) high develop. Many dunes of this sort are found in northern Death Valley.

Figure 4-11.

Ubehebe and Little Hebe craters, northern end of Tin Mountain on west side of northern Death Valley. Note that the material in the walls of Ubehebe is bedded sediment. (Photo by Spence Air Photos, courtesy of Department of Geography, University of California, Los Angeles).

Both the Panamint and Death Valley dunes show little persistent orientation of slope and slip face, although at any given moment the pattern of slip faces will be internally consistent. In addition, since both sets of dunes have relatively little vegetation, they can respond rapidly to wind changes.

The dunes of southern Eureka Valley are the highest and most striking dunes in the Basin Ranges, rising approximately 700 feet (210 m) above the valley floor. These dunes are so high, in fact, that some geologists think they may cover a rocky outcrop. Like the Panamint dunes, the Eureka dunes apparently formed where sand-laden winds were forced up against a mountain front.

The Dumont dunes have formed on a flat gravelly surface trenched by the Amargosa River near the southernmost part of its course. The center dune rises 420 feet (128 m) above the desert floor and resembles the dunes of Death, Eureka, and Panamint valleys. The margins, however, include many small barchans, star dunes, and seif ridges that extend away from the main dune.

Alluvial Fans

Although alluvial fans occur throughout the Mojave and Colorado deserts and in the semiarid parts of the Transverse and Coast ranges, their most spectacular California development is in the Basin Ranges. Perhaps the main reason for this is the extreme relief that characterizes the province, locally aided by vigorous streams. For example, along the steep eastern front of the Sierra is a series of large coalescing alluvial fans (Figure 2-11). These are partly the result of more than 10,000 feet (3000 m) of relief, but they are also due to the vigorous streams fed by snow at the Sierran crest. In contrast, the Inyo–White Mountains on the east of Owens Valley lie almost wholly in the rain shadow of the Sierra. Though they tower above the valley floor, the Inyo–White Mountains have few major streams and far fewer conspicuous fans than the Sierra.

The largest fans are those built out from the eastern Panamint Mountains into Death Valley. Of these, the Emigrant Wash and Hanaupah fans are the best examples (Figure 4-12). Starting at the fan's base near sea level at Stovepipe Wells, State Highway 190 climbs Emigrant Wash for 12 miles (19 km), reaching an elevation of 4300 feet (1300 m) at the top of the fan. Hanaupah fan is more than 2300 feet (700 m) high.

Desert Pavements

Basin Range desert pavements usually occur on the flat-topped divide surfaces between modern desert washes. There are at least two plausible explanations for these mosaics of shiny pebbles typically encrusted with desert varnish. They may be produced by selective

Figure 4-12.
Hanaupah alluvial
fan in the Panamint
Mountains. The
giant fans encroach-
ing high into the
Panamint Moun-
tains are in contrast
to the small alluvial
fans and cones at
the base of the
Black Mountains on
the east side of
Death Valley. Tele-
scope Peak, often
snow-capped until
midsummer, over-
looks Death Valley.
(Photo by Spence
Air Photos, courtesy
of Department of
Geography, Univer-
sity of California,
Los Angeles)

erosional removal of finer fractions of a sand-clay-gravel mixture, leaving the coarsest materials behind as surface armor protecting the undisturbed mixed materials below. Similar armors also may develop on surfaces of expansive clay layers that rest on gravel. As the clays expand and contract (owing to alternate wetting and dry-ing), loose pebbles from the gravel move up and eventually accumu-late on the surface. Especially fine examples of desert pavements are found in the southern Greenwater Valley.

Springs

For such an arid region, the Basin Ranges contain a surprising number of large springs. Several distinct types are known, but the precise origins of many of the springs are still undetermined.

Most of the streams draining into the Owens Valley from the Sierra carry substantial flows initially, but dwindle away as they cross the alluvial fans. Some stream water is lost by evaporation, but most of it sinks into the porous fan gravels. The residue emerges at the base of the fans as cienaga springs. Los Angeles taps much of this underflow for its water supply, so the cienaga springs are less pro-ductive than formerly.

Most of the large springs are controlled by faults, which often are concealed by superficial gravels. The faults provide extensive fracture systems along which water may collect, move, or emerge.

Springs of probable fault origin include Saratoga Springs in southern Death Valley, Tecopa and Resting Springs near Shoshone, and springs at the east base of the Argus Range.

Sometimes water moves to the surface along the frontal faults of ranges and supports conspicuous, linear patches of bright green vegetation that contrast sharply with the prevailingly olive drab desert vegetation. Such linear springs are seen along the west base of the Grapevine Mountains near Scotty's Castle, at the base of the Panamint Mountains near Indian Ranch, and along the Garlock fault at the base of the El Paso Mountains.

Hot and warm springs occur throughout the Basin Ranges. The distinction among cold, warm, and hot springs is somewhat arbitrary, but warm springs would probably range from about 80° to 110° Fahrenheit (27°–43° C). True hot springs are the least common in the Basin Ranges and are usually associated with recent volcanic activity. Two notable occurrences are the Coso Hot Springs just east of Rose Valley and the boiling springs along Hot Creek in the Crowley Lake area. These and other hot springs have provoked considerable interest recently as possible sources of geothermal power.

SPECIAL INTEREST FEATURES

Minerals and Mining

Many mineral species have been found in the Basin Ranges, and some of the metal mines have been worked since the middle of the nineteenth century. Most mines are no longer active, however, except for small operations that chiefly produce nonmetallic minerals.

Among the more famous metal mines is the Cerro Gordo, near the crest of the Inyo Mountains. First operated about 1860, considerable silver and lead were mined initially, but in the early twentieth century zinc became the main product. Mineralization was localized in Paleozoic carbonate rocks as replacements and fissure fillings adjacent to an intrusive granite. The same pattern of rock type and mineralization is typical of many mining districts in the region, although few have yielded the array of unusual minerals mined at Cerro Gordo.

One of the famous silver camps was at Panamint City, in the upper part of Surprise Canyon in the Panamint Mountains. Rich ore was produced from 1873 to 1876, and the town grew to a population of more than 1500. Theft of silver bullion from wagons creaking slowly down the steep canyon became a severe problem, however, and so the mine owners devised the following novel solution. Instead of casting silver in the standard bars of 20 or 30 pounds (9–13 kg), they cast it into cannonballs weighing 500 pounds (225 kg). Thieves were utterly frustrated, and the cannonballs reached San

Francisco without so much as a guard present en route. Panamint City ended about as spectacularly as it had begun. In 1876, when the ore was nearly exhausted and only the optimists were still in town, a cloudburst and flash flood descended on the canyon. Much of the town and the smelter were swept down to Panamint Valley in a torrent of rocks and water.

Other famous lead-zinc-silver mines are the Minietta and Modoc in the northern Argus Range, the Ubehebe in the northern Panamint Mountains, and the War Eagle, Columbia, and Noonday in the southern Nopah Range. Gold was mined at Skiddoo, Chloride City, and Ballarat about 1900, and at Bodie in the Mono Lake area during the 1870s and 1880s.

Apart from saline minerals, mining today is predominantly for talc. Talc mines operate intermittently in the Cottonwood and Ibex mountains, the Talc City Hills, the Argus Range, and particularly around the southern end of Death Valley in the Panamint and Black mountains. The talc is concentrated in contact metamorphic deposits, chiefly in magnesian limestones near intrusive igneous bodies.

Moving Stones at Racetrack Playa

North of Panamint Valley are several small upland valleys lying between the Cottonwood Mountains and Saline Valley. These valleys are undrained, and each contains a small playa. One of these playas is known as the Racetrack and lies about 3700 feet (1130 m) above sea level. The Racetrack was so named because of its oval shape and the presence of a syenite island that resembles and is called the Grandstand. In recent years people have driven on the playa when the surface was damp, leaving long-lasting wheel tracks, and the National Park Service has now closed the playa to automobiles.

The Racetrack is unusual not only for its smooth surface, but also for its peculiar "moving stones" (Figure 4-13). Years ago, visitors noticed that curving grooves an inch (2.5 cm) or so deep and up to a foot (30 cm) wide were etched onto the playa's surface. When followed, the grooves often led from the shore toward the center of the playa where they ended abruptly. At the ends of the grooves were rocks that weighed up to 100 pounds (45 kg). The shapes of the grooves and the rocks showed that the grooves had been formed as the rocks were dragged or shoved across the surface when it was wet and soft. Although there was agreement that the rocks had indeed formed the grooves, much difference of opinion existed concerning the method.

It was suggested that, following winter rains, strong winds had blown the rocks over the slick surface when the clays were moist and plastic. No one had ever observed this, however. Another view

Figure 4-13.
Racetrack playa,
showing moving
stones and tracks.
(Photo by Derek
Rust)

was that the stones were swept onto the lake by mudflows following rainstorms, but there was little associated debris as confirmation. Still another interpretation was that Indians had been involved, but no footprints could be found.

The most satisfactory explanation offered thus far was proposed by George Stanley. He envisioned that the Racetrack occasionally would be covered with a foot (30 cm) or so of water following heavy winter rains. This certainly happens every few years. Furthermore, temperatures may drop well below freezing in the clear and windy weather that typically follows a winter storm, and the lake easily could freeze to a depth of 6 inches (15 cm) or more. According to Stanley, the ice along the shore of the Racetrack would freeze around some of the pebbles and boulders, littering the playa's margin. Several stones might be included in a slab of ice that strong winds then might blow away from shore. If large enough, an ice slab certainly could support the incorporated rocks. In shallow water, the lower parts of the rocks would scribe the soft clay bottom. When the ice melted and the water evaporated, the stones would be left at the ends of the grooves they had cut. Stanley found some instances in which three or four rocks had cut parallel trails into the clay until the supporting ice had been rotated by the wind, causing the trails to make loops and turns in unison. This showed that the rocks had remained in the same position relative to one another, even though their supporting ice slab had twisted and turned.

Moving stones have been found on a few other playas in the Basin Ranges, for example, Bonnie Clare playa (in Nevada) east of

Scotty's Castle. It is clear that their occurrence requires conditions that seldom appear together on most Basin Range playas.

Redrock Canyon

Located near the western end of the El Paso Mountains, Redrock Canyon is a short, narrow gorge noted for rock outcrops of striking color and shape. The canyon has been the setting for many film and television productions.

Ricardo Creek and its tributaries cut the gorge of Redrock Canyon in response to uplift along the El Paso fault that probably involved one of two methods. On one hand, uplift may have been slow enough for Ricardo Creek to maintain its course and cut the channel, making the stream antecedent. On the other, the creek may have established its channel across an alluvial cover on the Ricardo beds and was then able to maintain its position as uplift accelerated. In this interpretation, Ricardo Creek is superimposed.

In the Redrock area, the Paleozoic-Mesozoic basement of the El Paso Mountains is overlain unconformably by the Ricardo series of nonmarine sands and water-laid ash beds and tuffs. Interbedded lava flows near the base of this section often contain zeolites and opal. The color of the canyon's red and pink rocks results from oxidation of iron-bearing minerals and from baking of underlying rocks where lavas flowed across older layers. Blacks and browns are from the outpourings of basaltic lavas. Grays and whites reflect the sandstone layers that are rich in quartz and light-colored feldspars.

Vertebrate fossils have been found in the younger layers of the canyon. Included are camel, primitive horse, and many rodent remains. In the late Cenozoic, when these beds were deposited, savanna grasslands apparently prevailed.

REFERENCES

Anonymous, 1968. Geology of the Basin Ranges. Mineral Information Service (now California Geology), v. 21, pp. 131–133, 137.

Berkstresser, C. F., Jr., 1974. Tallest(?) Sand Dune in California. California Geology, v. 27, p. 187.

Blackwelder, E., and others, 1948. The Great Basin. Bull. Univ. Utah, v. 38, no. 20, Biol. Series, v. x.

Cooke, Ronald, 1965. Desert Pavement. Mineral Information Service (now California Geology), v. 18, pp. 197–200.

Hazzard, John, 1937. Paleozoic Section in the Nopah and Resting Springs Mountains, Inyo County, California. Calif. Jour. Mines and Geology, v. 33, pp. 273–339.

Hill, Mary R., 1972. A Centennial . . . the Great Owens Valley Earthquake of 1872. California Geology, v. 25, pp. 51–54.

Basin Ranges

Knopf, Adolph, 1918. A Geologic Reconnaissance of the Inyo Range and the Eastern Slope of the Southern Sierra Nevada, California. U.S. Geological Survey Prof. Paper 110.

Noble, Levi F., and L. A. Wright, 1954. Geology of Central and Southern Death Valley Region, California. *In* Geology of Southern California. Calif. Div. Mines and Geology Bull. 170, pp. 143–160.

Oakeshott, Gordon B., and others, 1972. One Hundred Years Later. California Geology, v. 25, pp. 55–61.

Slosson, James E., 1974. Surprise Valley Fault. California Geology, v. 27, pp. 267–270.

Troxel, Bennie W., 1963. Mineral Resources and Geologic Features of the Trona Sheet, Geologic Map of California. Mineral Information Service (now California Geology), v. 16, no. 11, pp. 1–7.

———, 1974. Man-Made Diversion of Furnace Creek Wash, Zabriskie Point, Death Valley, California. California Geology, v. 27, pp. 219–223.

———, ed., 1974. Guidebook: Death Valley Region, California. Death Valley Publishing Co.

Chapter Five

Mojave Desert

*Truth generally lies in the coordination of
antagonistic opinions.*

Will Durant

The Mojave Desert and the Colorado Desert are so closely related geologically that considering them separately is almost more a matter of convenience than geologic unity. On the other hand, there is sufficient difference in late Cenozoic events that treating each area as a discrete unit does tend to result in a clearer overall picture.

The Mojave Desert occupies about 25000 miles² (65,000 km²) of southeastern California (Figures 5-1, 5-2). It is landlocked, enclosed on the southwest by the San Andreas fault and the Transverse Ranges and on the north and northeast by the Garlock fault, the Tehachapi Mountains, and the Basin Ranges. The Nevada state line and the Colorado River form the arbitrary eastern boundary, although the province actually extends into southern Nevada. The San Bernardino–Riverside county line is designated as the southern boundary.

The Mojave area contains Paleozoic and lower Mesozoic rocks, although Triassic and Jurassic marine sediments are scarce. The marine sediments may have been eroded away, or the Mojave may have been an early Mesozoic upland on which no such sediments were deposited. Jurassic and Cretaceous granitic rocks of the Nevadan batholith are widespread throughout the region's mountain blocks.

The desert itself is a Cenozoic feature, formed as early as the Oligocene presumably from movements related to the San Andreas and Garlock faults. Prior to the development of the Garlock, the Mojave was part of the Basin Ranges and shares Basin Range geologic history possibly through the Miocene.

Today the region is dominated by broad alluviated basins that are mostly aggrading surfaces receiving nonmarine continental deposits from adjacent uplands. The deposits are burying the old topography, which was previously more mountainous. In the late Tertiary, these mountains shed debris to the Pacific, but with the elevation of coastal ranges drainage began entering interior basins.

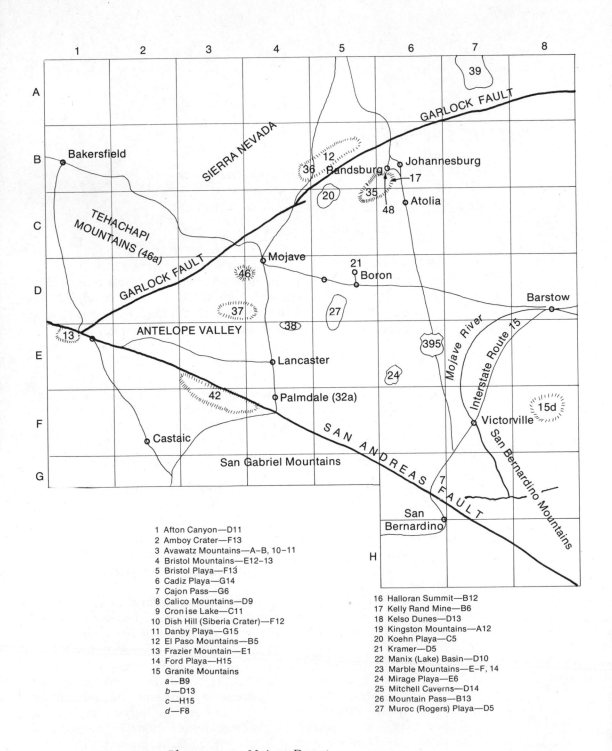

Figure 5-1. Place names: Mojave Desert.

1 Afton Canyon—D11
2 Amboy Crater—F13
3 Avawatz Mountains—A–B, 10–11
4 Bristol Mountains—E12–13
5 Bristol Playa—F13
6 Cadiz Playa—G14
7 Cajon Pass—G6
8 Calico Mountains—D9
9 Cronise Lake—C11
10 Dish Hill (Siberia Crater)—F12
11 Danby Playa—G15
12 El Paso Mountains—B5
13 Frazier Mountain—E1
14 Ford Playa—H15
15 Granite Mountains
    a—B9
    b—D13
    c—H15
    d—F8

16 Halloran Summit—B12
17 Kelly Rand Mine—B6
18 Kelso Dunes—D13
19 Kingston Mountains—A12
20 Koehn Playa—C5
21 Kramer—D5
22 Manix (Lake) Basin—D10
23 Marble Mountains—E–F, 14
24 Mirage Playa—E6
25 Mitchell Caverns—D14
26 Mountain Pass—B13
27 Muroc (Rogers) Playa—D5

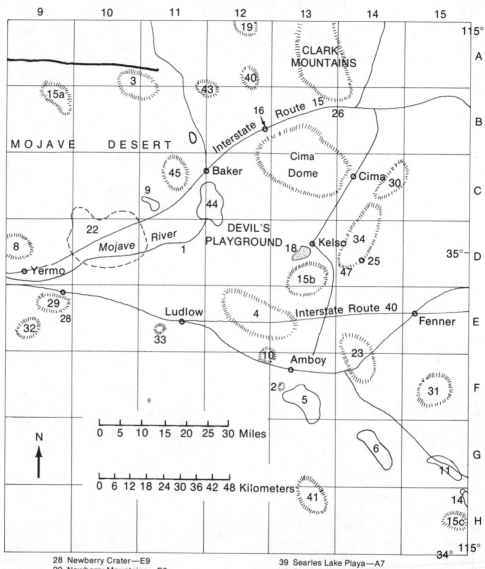

28 Newberry Crater—E9
29 Newberry Mountains—E9
30 New York Mountains—C14
31 Old Woman Mountains—F15
32 Ord Mountains—E9
32a Palmdale—F4
33 Pisgah Crater—E11
34 Providence Mountains—C–D, 14
35 Rand Mountains—B–C, 5–6
36 Redrock Canyon—B4
37 Rosamond Hills—D3–4
38 Rosamond Playa—E4

39 Searles Lake Playa—A7
40 Shadow Mountains—A12
41 Sheep Hole Mountains—G–H, 13
42 Sierra Pelona—E–F, 3
43 Silurian Hills—A–B, 11–12
44 Soda Lake Playa—C11–12
45 Soda Lake Mountains—C11
46 Soledad Mountain—D3 - 4
46a Tehachapi Mountains—C–D, 1–2
47 Vulcan Mine—D14
48 Yellow Aster Mine—B6

126

Mojave Desert

The highest general elevation of the Mojave Desert approaches 4000 feet (1200 m) along a northeastern axis from Cajon Pass to Barstow. Alluvial cover thins to the east, and pediment—often with thick regolith—occupies much of the surface. Westward into Antelope Valley, the cover of poorly consolidated nonmarine alluvial sediments thickens rapidly to a maximum of 4000 feet (1200 m). East from the Cajon-Barstow axis, the pediment is nearly constant in elevation, with undulations up to 300 feet (90 m), and finally slopes into the Colorado Desert where the alluvial cover again increases. Nowhere, however, is the alluvial thickness comparable to that in the Antelope Valley.

North of Barstow, extending to the Garlock fault, are several middle Tertiary depositional basins that carry thick sections of continental rocks. Eastward, along the Mojave River valley and at progressively lower elevations, are several Quaternary depressions.

Figure 5-2.   View of the western Mojave Desert, looking northwest from Palmdale. The general monotony of nearly zero relief is apparent. On the left is the San Andreas fault. The Tehachapi Mountains form the skyline. (Photo by Spence Air Photos, courtesy of Department of Geography, University of California, Los Angeles)

Northeast from Baker, the Mojave surface again becomes erosional and rises rapidly to nearly 4000 feet (1200 m) at Halloran Summit. Granitic pediment (Cima Dome) is exposed for many square miles in this area, interspersed by thin, discontinuous veneers of lava. West of Cima Dome are the prominent volcanic cones and flows of the Cima volcanic field. The Clark Mountains, almost 8000 feet (2440 m) high, form the northeastern corner of the province.

Throughout the Mojave, small hills rise above the alluvial fill, islandlike in seas of gravel. These are remnants of the mountainous topography that is now almost erased by erosion and buried by debris. Other prominent features of today's surface are the many playas, including Rosamond, Rogers (Muroc), Mirage, Bristol, Cadiz, and Danby. Every local internal drainage contains at least one playa, and the linear valleys of the eastern Mojave often have several since drainage usually is blocked by almost imperceptible rises between playas.

GEOGRAPHY

Like most deserts, the Mojave has highly variable rainfall—from 2.23 to 6.5 inches (56–165 mm) annually. A single stream, the Mojave River, drains most of the region. This river rises in the northern San Bernardino Mountains and flows northeast about 100 miles (160 km). Flow is primarily underground, except where bedrock at shallow depth forces the water to the surface. Exceptional flow occurs only during flash floods or unusually high rainfall.

Average elevation of the Mojave Desert is about 2500 feet (760 m). Relief is extreme, however, especially where mountain blocks rise above alluvial slopes or pediments. Examples are the Rand, Shadow, Calico, Ord, and Newberry mountains in the western half of the desert and the Bristol, Sheep Hole, Marble, Providence, New York, Clark, Avawatz, Kingston, and Granite mountains in the eastern half. (Some confusion exists regarding the Granite Mountains of the Mojave and Colorado deserts, since there are four "Granite Mountains" in four separate areas. In this text, designations are Granite Mountains A, B, C, and D; their locations are given in Figure 5-1.) Highest elevations are in the New York (7445 ft or 2290 m), Providence (6900 ft or 2100 m), Clark (7929 ft or 2440 m), Kingston (7323 ft or 2253 m), and Avawatz (6154 ft or 1898 m) mountains. Lowest elevations occur in Bristol, Cadiz, and Danby playas (all about 600 ft or 180 m) and Ford playa (500 ft or 160 m). Other playas are as high as 2300 feet (700 m).

The Mojave province displays an unusually large variety of landforms produced by intricate erosional and depositional processes. Volcanic features like the late Cenozoic basaltic flows and cinder cones near Pisgah, Amboy, and on Cima Dome and the plugs, domes, sheets, sills, and dikes of older rhyolitic volcanic episodes (Rosamond Hills and Soledad Mountain) are prominent. Many strike-slip and dip-slip fault features are clearly exposed. Scarps, slivers, offset streams, sags and sag ponds, grabens, and horsts are associated with many of the erosional and depositional landforms. The common landforms of pediment, bajada, bolson, panfan, alluvial fan, alluvial cone, goblet valley, badlands, and mudflow were first defined or are particularly well demonstrated in the Mojave. There are abundant examples of such weathering features as desert varnish, exfoliation, caliche horizons, and efflorescences. Desert pavements, sheetwash and sheetflood forms, and many features produced by wind action occur repeatedly. Most of these features result from the complex interactions of climate, rock type, and geologic history, although the origins of some are not yet fully understood.

ROCKS

Precambrian

The Precambrian is well represented in the Mojave Desert. The age is established in some instances because igneous and metamorphic complexes lie unconformably below sedimentary beds carrying lowermost Cambrian fossils. In other cases, radiometric dates from basement crystalline rocks are Precambrian. Oldest Precambrian rocks are highly schistose and gneissic and often are dynamically metamorphosed. Such rock units are exposed in the Newberry-Ord Mountains (granitic gneiss and marble with intruding porphyry), the Old Woman Mountains (gneiss, marble, quartzite, schist), the Marble Mountains (granitic basement), and near Kelso (granitic gneiss). Some age assignments are still unconfirmed, and their Precambrian designation has been made primarily on apparent position in the rock column and on degree of metamorphism, both unreliable criteria.

The later Precambrian is exposed in sedimentary and metasedimentary sequences developed by static and low-grade metamorphism. These sequences compare with some of the well-bedded, thick, Basin Range late Precambrian rocks in which primitive fossils have been identified. (See Chapter 4.) Mojave formations designated late Precambrian on the basis of stratigraphic position occur in the Kingston Mountains (Pahrump series), the Silurian Hills (probable Pahrump equivalent), the Rand Mountains (Rand schist and Johannesburg gneiss), the El Paso Mountains (Mesquite schist), and the Sierra Pelona (Pelona schist). Although many geologists consider the Rand and Pelona schists identical, the Precambrian age assignment of the Pelona has been questioned. (See Chapter 8.)

Paleozoic

The Paleozoic system is not well represented in complete sequence in the Mojave block. Partial sections occur in some of the Mojave's eastern ranges, but the main Mojave contains few Paleozoic rocks. In the Marble Mountains, early Cambrian quartzite and fossiliferous shale lie unconformably on Precambrian granites. The province's lower Paleozoic is less than 5000 feet (1500 m) thick, even when all fragments from the various exposed units are pieced together. Nowhere is there a continuous section more than 2500 to 3000 feet (760–900 m) thick. Upper Paleozoic is recognized in the Ord Mountains and near Victorville (Oro Grande series, probably Mississippian) and in the Providence (Pennsylvanian) and Soda Lake (Permian) mountains. The rocks are primarily limestones, quartzites, and clastics, like the Paleozoic sections of the Basin and Transverse ranges. Maximum thickness appears to be about 10,000 feet (3000 m), in the Ord Mountains, compared with Basin Range thicknesses up to 36,000 feet (11,500 m).

Mesozoic

Mesozoic bedded rocks are practically unknown in the Mojave except in or east of the New York–Providence Mountains. Sedimentary units of the Colorado Plateau (Moenkopi, Triassic) are found west of the Colorado River (in Nevada), however, and correlations have been inferred with the Mojave Desert proper. Fragmentary Triassic fossils are reported from the Soda Lake Mountains, and volcanics near Barstow have been designated Triassic because of their stratigraphic position. Sediments carrying fossil wood no older than Cretaceous have been found in the McCoy and Palen mountains on the border of the Colorado Desert.

The Mojave's mountain blocks and hundreds of square miles of pediment are floored primarily by granitic intrusives of the Nevadan orogeny. Many plutons with composition similar to Sierra Nevada plutons have been described. Eastern Mojave plutons are younger and more silicic (quartz monzonites to granites) than western Mojave plutons, which include the silicic types plus gabbros and quartz diorites. Radiometric dates have established a middle Jurassic age for the silicic plutons. The gabbros and diorites are older, but presumably do not reflect a separate orogenic episode predating the Nevadan. Instead they probably equate with the forerunners of major intrusion identified in the Sierra Nevada.

Cenozoic

Cenozoic rocks appear throughout the Mojave Desert. Except for thin, restricted, lower Miocene marine sediments in western Antelope Valley, deposition is all nonmarine, with extensive thicknesses of Quaternary alluvium. Tuff, ash, and other volcaniclastics interbedded with lake-bed sediments and evaporites are widespread. Volcanic flows and flow breccias are common, with andesites and rhyolites in older extrusives and basaltic flows and cinder cones characterizing younger episodes. Accumulations were in localized basins. Correlation between these basins is difficult, because drainage integration between depositional areas was temporary and the evidence therefore unsubstantial. Some basins accrued 10,000 feet (3000 m) of deposits (practically all post-Eocene) from the pre-Cretaceous basement. Dozens of these isolated Cenozoic basins once existed, and some are evident today as playas. Most have been drilled for water, potash, nitrate, other evaporite deposits, and even petroleum. A 1967 publication lists well logs for over 200 exploratory holes drilled in 15 different basins in the western Mojave alone.

The oldest Cenozoic unit yet described is a thin (150 ft or 45 m maximum) mudstone, with limestone and red sandstone. It is found in limited exposure in only one locality on the pre-Cretaceous basement and is all unfossiliferous and nonmarine. Sometimes conformably overlying this unit, but usually lying unconformably directly on the older basement, is a sequence of pyroclastics up to 3700 feet (1130 m) thick. Included are flow breccias, tuff, and conglomerate, with some pumice, perlite, and opalite. This sequence is the Pickhandle formation and is present in all the basins studied in the western Mojave.

FAULTING

The Mojave block is approximately bounded by the San Andreas and Garlock faults. Both are at least Cenozoic, and some geologists think the San Andreas may have originated in the middle or early Mesozoic. The western Mojave Desert is broken by major faults that primarily parallel the San Andreas and seem to be truncated by the Garlock. Many faults undoubtedly occur in the eastern Mojave also, but since most of this area is underlain by rather uniform granitic rocks, the faults are notoriously difficult to map. Some faults are known positively, but many can only be inferred. Basin Range types of faults have been mapped in the eastern Mojave ranges. In the Clark Mountains, for example, thrust faults of considerable magnitude reflect several episodes of movement. These faults are believed to establish pre-Cenozoic faulting related to the Cordilleran orogeny. Our interest centers on post-Mesozoic faulting, however, since these structures are revealed most clearly in today's geomorphology. Principal faults of the Mojave are shown in Figure 5-3.

Garlock Fault

The Garlock fault extends from Frazier Mountain east for more than 150 miles (240 km). It is characterized by left-slip displacement that is variously estimated from 6 to 40 miles (9.6–64 km). Recent displacements up to a half-mile (0.8 km) are known. The Garlock shows the characteristic features of high-angle faults with major strike-slip component. The fault zone is narrowly confined through most of its length and is seldom more than a mile (1.6 km) wide.

East from its junction with the San Andreas, the Garlock forms two parallel segments that extend for nearly 20 miles (32 km) until they merge into a single structure. This single unit then extends another 20 miles (32 km) to Red Rock Canyon and the El Paso Mountains. As mentioned earlier, here, near the junction of the Garlock and the southwest-curving Sierra Nevada fault, the El Paso fault begins, parallel to the Garlock and within the Garlock zone of faulting.

The Garlock displaces the Sierra Nevada and Basin Range faults and is therefore younger. The Garlock may have been a boundary of the Mojave block as early as the Eocene. It was certainly a boundary of major movement by late Oligocene or early Miocene, because by then the Mojave had developed internal drainage, reversing the previous erosional pattern. Displacement on the Garlock had started by the opening of the Tertiary, with strong vertical components of uplift subsequently changing to dominantly strike-slip movement in middle Tertiary. It is suggested that the Garlock is a transform fault in the plate tectonic history of western North America.

San Andreas Fault

The San Andreas fault appears to form as sharp a boundary on the southwest of the Mojave Desert as the Garlock fault does on the north. In fact, however, the San Andreas lies about a mile (1.6 km) inside the foothill belt terminating the desert floor, and minor subparallel faults seem to be the usual boundary between the Mojave and Transverse Range provinces. The San Andreas fault is considered more fully in Chapter 11, but it is important to note here that the Garlock appears offset by the San Andreas where the two intersect at the western end of the Mojave block.

Internal Faults

The internal faults of the Mojave block show remarkable parallelism with the strike of the San Andreas fault and equally impressive divergence from the strike of the Garlock, against which the projected strikes of the internal faults seem to terminate. The inter-

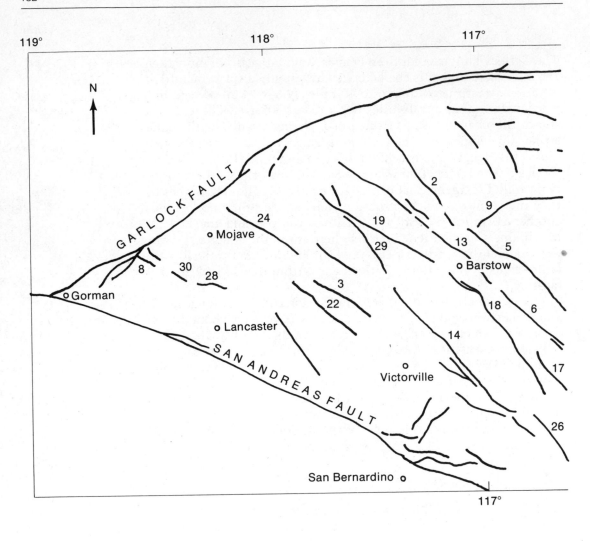

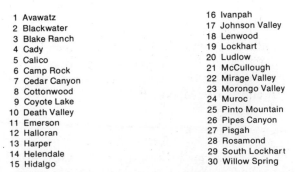

1	Avawatz	16	Ivanpah
2	Blackwater	17	Johnson Valley
3	Blake Ranch	18	Lenwood
4	Cady	19	Lockhart
5	Calico	20	Ludlow
6	Camp Rock	21	McCullough
7	Cedar Canyon	22	Mirage Valley
8	Cottonwood	23	Morongo Valley
9	Coyote Lake	24	Muroc
10	Death Valley	25	Pinto Mountain
11	Emerson	26	Pipes Canyon
12	Halloran	27	Pisgah
13	Harper	28	Rosamond
14	Helendale	29	South Lockhart
15	Hidalgo	30	Willow Spring

Figure 5-3. *Some known and inferred Cenozoic faults of the Mojave Desert. (Source: California Division of Mines and Geology and U.S. Geological Survey)*

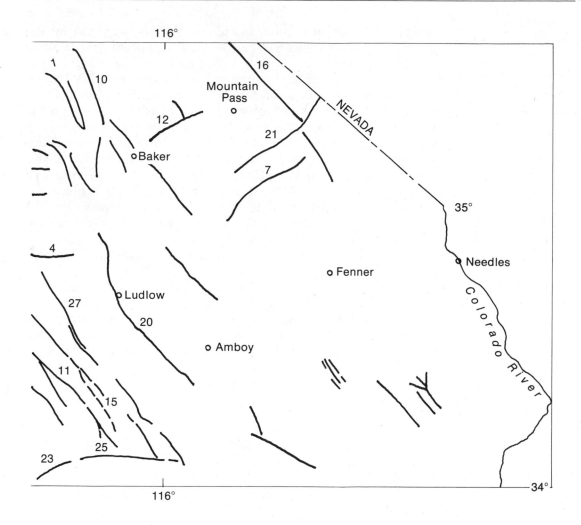

116°

16

Mountain
Pass

NEVADA

12

21

Baker

7

35°

4

Needles

Fenner

Colorado River

27

Ludlow

20

Amboy

11

15

25

23

116°

34°

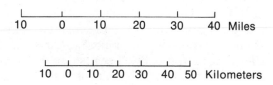

10   0   10   20   30   40   Miles

10   0   10   20   30   40   50 Kilometers

nal faults are probably more numerous than is suggested by the generalized map of Figure 5-3. Furthermore, it is probable that many individually named faults are actually parts of single fault trends that are continuous in the basement beneath the deep alluvium.

Important vertical displacement occurs on the Muroc, Blake Ranch, and Mirage Valley faults in the western Mojave and on the Johnson Valley, Pipes Canyon, and Spring faults in the central Mojave. Right-slip displacement of uncertain amount is inferred for the Cottonwood, Lockhart, and Blackwater faults in the west and for the Lenwood, Hidalgo, West Calico, and Pisgah faults in central and south-central Mojave. The extensive work of T. W. Dibblee has permitted the determination of strike-slip movement on the numerous internal faults. Strike-slip seems to follow vertical displacement in almost every case.

Vertical displacements are continuing, as shown by radiometrically dated Pleistocene basaltic flows that have been vertically displaced as much as several hundred feet. Thick alluvial deposits are cut by some of the internal faults to form present-day scarps, further demonstrating continuing vertical movement. Internal faulting also is reflected by uplifted bedrock surfaces of low relief. Such a surface is the Ivanpah upland of the eastern Mojave, which might well be extended to the less continuous but nonetheless impressive surface that bevels the pre-Cenozoic rocks of the western Mojave. A 1975 earthquake was accompanied by surface displacement between the Calico and Emerson faults.

GEOLOGIC HISTORY

The Mojave block shares the pre-Mesozoic geologic history of the Sierra Nevada and Basin Ranges. Degradation after the Nevadan orogeny reduced the Mojave and the Sierra to low relief by the close of the Mesozoic. Subsequently uplift was initiated, and by the opening of the Miocene the Mojave block had been lifted between 10,000 and 15,000 feet (3000–4550 m). This estimate is deduced from the amount of detritus shed west and south to form lower Tertiary units of the Transverse and southern Coast ranges. By early Miocene or late Oligocene, the Ivanpah erosional surface existed, rising gently eastward from the Garlock–San Andreas convergence. Depression then began, probably because of faulting, and transformed the Mojave block into an area of internal drainage. Thick Miocene, Pliocene, and Pleistocene nonmarine sections accumulated in local basins, a condition that persists today. Pleistocene lake sequences developed extensively during times of cool temperatures and low evaporation.

SUBORDINATE FEATURES

Cima Dome and Ivanpah Upland

The Ivanpah upland slopes from eastern elevations of 4500 to 5000 feet (1370–1500 m) west to the central Mojave at 1500 to 2000 feet (450–600 m). Its topography is irregular (probably the result of old mountain block roots and resistant Pliocene to Recent volcanic covers), but the surface is widely underlain by uniformly grained quartz monzonite of Nevadan age. The basement has low relief wherever younger rocks have preserved it, and apparently much of the Mojave block was planed before uplift to its present position.

On the eastern margin of the Ivanpah upland is Cima Dome, an upland segment of about 100 square miles (160 km²). Cima Dome has received considerable attention because it frequently has figured in attempts to understand the mechanisms of desert erosion, especially slope retreat of block-faulted rock units under desert regimens. In 1915, A. C. Lawson studied slope retreat and concluded that as erosion progressed, a fault scarp would retreat by being degraded essentially parallel to the original dip of the scarp slope, with gradual shift of the scarp declivity. Thus rock floors (pediments) would form at the base of the scarp, progressively widening as the erosional cycle advanced and panfans developed. Cima Dome was seemingly a prime example of the panfan concept. It was later found, however, that the dome is wholly bedrock and not covered with alluvial material.

In 1933, William Morris Davis tried to explain Cima Dome as a special case of mass wasting of a desert fault block, unique because the rock was of uniform texture and composition, producing a convex surface. This concept is not supported by field evidence.

R. P. Sharp has demonstrated that the fundamental premises of both Lawson's and Davis's ideas do not apply to Cima Dome. Instead, like adjacent domes, Cima appears to result from slight upwarping of the Ivanpah upland during or after a shallow cover of Pliocene and younger volcanics buried the upland. Subsequent stripping of the cover exposed the domal surface of the granite. The prominent regolith is presumably a two-cycle weathering product, developed partly during degradation of the pre-Pliocene erosional surface and partly during the present cycle.

Cima Volcanic Field

Marginal to Cima Dome and concentrated to the west and northwest are 26 volcanic cinder cones associated with basaltic flows up to 300 feet (90 m) thick. These late Tertiary and Pleistocene flows overlie much of the southwestern edge of Cima Dome and the surrounding alluvial fans and may once have covered as much as 250 square miles (400 km²). The younger volcanic cones to the south, on the higher levels of Cima Dome, are built on the basaltic flows. W. S. Wise has noted compositional variations in the flows and cones that suggest the origin of their magmas and associated magmatic contaminations. Commercial quarries produce such items as cinder block and road metal from the cinder cones.

Volcanic Cones: Barstow-Amboy Axis

From Barstow to Amboy, there is a conspicuous line of volcanic cones including Newberry Crater, Pisgah volcanic field and cone, Dish Hill (formerly Siberia Crater), and Amboy volcanic field and cone. Pisgah and Amboy are the best known, because of their prominence as desert landmarks, and their degree of dissection suggests that they are the youngest of the chain.

Newberry Crater is about 15 miles (24 km) southeast of Barstow. It is built on a base of older rocks and emerges from the flank of the Newberry Mountains. The crater is a large volcano with extensive lava flows extending toward the valley through which Interstate 40 passes.

About 5 miles (8 km) southeast of Newberry is Pisgah Crater, an extremely young cinder cone. Prior to the crater's eruption, basaltic lava emerged from fissures and flowed south and northwest in thin sheets up to 6 miles (9.6 km) from the vent sources. The Pisgah flows were highly fluid with typically vesicular tops. This suggests a high gas content that probably accounts for the viscosity. The vesicles have no significant filling, indicating lack of mineralizing solutions in the magma. The flows cover thick lake-bed clays that include bentonites. Both clays and volcanic products are mined. Superimposed on the Pisgah flows is the cone itself, which rises about 300 feet (90 m) above the thin volcanic platform. The cone is almost circular, with a distinct crater, and is composed of red and black cinders, some lava, and clinkery breccias.

Dish Hill (Siberia Crater) is the largest of a group of flows and cones between Ludlow and Amboy. Although not so spectacular as Pisgah or Amboy, the area has long been famous for volcanic bombs that have olivine or granitic cores. Mineral collections throughout the country contain olivine bombs from Siberia Crater, although specimens are becoming scarcer.

Southernmost of the Barstow-Amboy chain is Amboy Crater (Figure 5-4). This almost circular hill is probably less than 2000 years old and forms a prominent landmark on Bristol playa. The crater is breached on its western side and rises about 300 feet (90 m) above the playa on basaltic flows from 10 to 30 feet (3–10 m) thick. These flows cover about 5 square miles (8 km²) of the playa's north-

*Figure 5-4.
Amboy Crater, from
the east. Sand
dunes are forming
because the prevail-
ing northwesterly
winds are slowed
by the cone's eleva-
tion. A small playa
is in the center of
the crater. (Photo by
Spence Air Photos,
courtesy of De-
partment of Geog-
raphy, University of
California, Los
Angeles)*

ern margin. The variety of volcanic features—lava tubes and tunnels, pressure ridges, collapsed domes, and ropy lavas—in Amboy
and Pisgah volcanic fields is as great as any volcanic area south of
the Cascade Range.

Mojave River

The Mojave River, the only major stream crossing the Mojave block,
is intermittent through most of its course from its head in the San
Bernardino Mountains to its present terminus in Soda Lake (Figure
5-5). In earlier postglacial time, the river continued north and joined
the Amargosa River, flowing into Death Valley. The unusual variations in the river's channel pattern are at least partially due to the
complex local history of segments of the Mojave block.

The Mojave River has three widely separated areas of constric-
tion, where surface flow often occurs. The Victorville watergap, on
which the river is superimposed, is a mass of bedrock formerly
buried by alluvium (Figure 5-6). The bedrock now is being exhumed
because of local shift in baselevel. The volcanic barrier at Barstow
has deflected and impeded flow as the more durable volcanics along
the river channel are exhumed. At the Afton Canyon watergap,
poorly consolidated bedded sediments have been cut through to
crystalline rock. This segment of channel is thus superimposed.

Alternatively, the Afton Canyon watergap may have formed by downcutting of the outlet of glacial Lake Manix.

Since the Mojave shifted from external to internal drainage 15 to 20 million years ago, its channel variations are reflections of this complex geomorphic history. Local and regional erosion and deposition have alternated repeatedly.

Pleistocene Lake Manix

The modern Mojave River crosses the floor of the basin occupied by Lake Manix during glacial times. Covering about 200 square miles (320 km²), the lake was a major feature of the drainage connecting Manix, Cronise, Soda, and Silver lakes. These lakes may once have been connected with Bristol, Cadiz, and Danby playas, subsequently

Figure 5.5.   *Mojave River at Soda Lake playa during the flood of March 1938. Town of Baker is in lower right. (Photo by Spence Air Photos, courtesy of Department of Geography, University of California, Los Angeles)*

*Figure 5-6.
Mojave River
watergap at Victor-
ville (center), during
unusual flood con-
ditions. (Photo by
Spence Air Photos,
courtesy of De-
partment of Geog-
raphy, University of
California, Los
Angeles)*

draining into the Colorado River. The basin of Lake Manix has special relevance because there is some suggestion that prehistoric man
lived there. In addition, early Pleistocene vertebrate fossils (horses,
jackrabbits, camels, deer, pronghorns, and tapirs) have been recovered. These findings establish climatic characteristics of the
Mojave during the Pleistocene.

Kelso Dunes and Devil's Playground

The Kelso dunes are the highest and most prominent portion of the
Mojave's largest dune field, an area known by the forbidding name of
Devil's Playground. This dune field extends about 35 miles (56 km)
east from the lower end of Afton Canyon, where the Mojave River
enters the basin of Soda Lake. Though some sand may have been
added to the Devil's Playground from old beaches around Soda Lake,
most of it was derived from material carried by the Mojave River.
Apart from the Kelso dunes proper, the Devil's Playground does not
contain particularly distinctive dunes. The area primarily consists
of low dune groups and featureless sandy areas anchored by vegetation, although a few good examples of seif dunes and barchans also
are included.

The Kelso dunes are approximately in the center of a broad alluviated valley that imposes no obvious barriers to farther eastward movement of sand. Although east of the dunes the valley floor slopes upward, in traveling from the mouth of Afton Canyon to the Kelso area the sand already moved uphill at least 1500 feet (450 m), so a further climb would not seem to present a problem. Studies of sand behavior in the Kelso dunes show that the sand moves to and fro rather than in a consistent direction. Apparently, then, present conditions differ from those prevailing when the dunes accumulated. (This may have occurred as recently as a few thousand years ago or as much as 20,000 years ago.) Consequently, the Kelso dunes and much of the Devil's Playground are probably relics of past conditions now greatly altered. Furthermore, the dune sand has a light tan color rather than the white or pale gray that characterizes most river sand. This also suggests an older age for the dunes, because older sands often become brownish or even red as iron-bearing minerals stain surrounding grains with red and brown iron oxides. The four approximately parallel ridges of the dunes show little relation to existing wind patterns and may be another indication of origin under previous conditions. The southern ridge is the highest, rising nearly 550 feet (168 m) above the desert floor, and is nearly 4 miles (6.4 km) long.

Two stream channels cross the eastern portion of the Kelso dunes. The larger is Cottonwood Wash, which lies about 100 feet (30 m) below the sand bordering the channelway. Neither Cottonwood nor Winston Wash to the east reveals any bedrock under the higher parts of the dunes. The channels are cut primarily in alluvial deposits, so it is unlikely that the height of the dunes results from a buried bedrock core.

The Kelso dunes have several unusual features. Among California desert dunes, they are the only known barking dunes. When certain dunes are disturbed by the activity of people sliding down slip faces or rapidly shuffling feet in the sand, low-pitched barking noises can be heard. There are also reports of sudden, apparently spontaneous initiations of low booming noises from the dunes. As yet, explanations accounting for the phenomenon of dune noise are somewhat inconclusive.

Most Kelso dune sand is composed of quartz and feldspars, as is dune sand in many places. The Kelso dunes also contain an appreciable quantity of dark, heavy minerals often concentrated by the wind into streaks and patches. The dark minerals are dominated by magnetite, probably derived from the iron ores of Afton Canyon. Minor amounts of such minerals as zircon, ilmenite, monazite, rutile, and cassiterite are also present. In recent years, the presence of these heavy minerals has prompted efforts to mine the sand, but no commercial production has yet occurred.

SPECIAL INTEREST FEATURES

Providence Mountains (Formerly Mitchell's Caverns) State Park

Addition of Mitchell's Caverns to the California park system has assured preservation of a long-known but obscure cavern in the Providence Mountains. Mitchell's Caverns are limestone solution features in the thick Bird Spring formation of Permo-Carboniferous age. The caverns are dry and small, but varied. Many former lead-silver mines are located in the Providence Mountains, and several are near Mitchell's Caverns.

Trilobites of Marble Mountains

The Marble Mountains east of Amboy include a famous collecting locality, the lower Cambrian Latham shale of red, green, and gray fissile mudstones. This unit is not confined to the Marble Mountains, but other exposures are less accessible. In the southern Marble Mountains, the Latham shale is 50 to 75 feet (15–24 m) thick and is underlain by the (Cambrian) Prospect Mountain quartzite. The shale is highly fossiliferous, with brachiopods and trilobites the principal forms. A prominent trilobite at the Marble Mountain locality is *Fremontia fremonti.* Specimens with body and carapace up to 8 inches (25 cm) long have been collected.

An unusual feature of the Marble Mountain locality is that the underlying Prospect Mountain quartzite is exceedingly thin, from 30 to 50 feet (10–15 m) thick. Elsewhere, this quartzite is up to 1100 feet (350 m) thick. Consequently, the unconformity between Precambrian granites and the Cambrian is quite distinct. The quartzite layers contain small whitish quartz pebbles, particularly in the lower 10 feet (3 m). This lowermost basal conglomerate grades upward into sandstone (quartzite) and then into the Latham trilobite-bearing shale.

Vulcan Iron Mine

Iron ore reserves of commercial grade have been known in California since at least 1914. The largest deposits are in the Eagle Mountains of the Colorado Desert, but significant deposits occur in the middle Cambrian Bonanza King limestone of the southern Providence Mountains. Magnetite and hematite were processed from these ores at the Vulcan mine from 1942 to 1947, when the property was abandoned. Besides piles of waste and low-grade ore, the mine has an enormous glory hole, an open pit several hundred feet deep where visitors can see the relations of the ore body to the limestone country rock. Many minerals are present in the pit and on the dumps, including magnetite, hematite, pyrite, pyrrhotite, chalcopyrite, calcite, dolomite, serpentine, limonite, and epidote.

Bristol Playa

Near the town of Amboy is Bristol Lake, a large playa partially covered by flows from the Amboy volcanic center. Just below its surface, the playa has a saline water body from which evaporite minerals are mined. One of the mining operations consists of trenching the playa as much as 20 feet (6 m), stacking the playa clays in rows along the trenches, and allowing the trenches to fill with brines from the waters in the playa. Evaporation precipitates salts like table salt and calcium chloride, which can then be harvested. Operations similar to those at Amboy are carried on sporadically in other Mojave playas, notably Koehn and Danby.

Mountain Pass Rare Earth Deposit

Mountain Pass is the highest point on Interstate 15 between Barstow and Las Vegas, where the road crosses the southern Clark Mountains. The area possesses one of the most unusual mines of North America. Discovered in 1949, in Precambrian rocks, is a deposit of rare earth metals initially determined as cerium, lanthanum, and neodymium. All are present in the mineral bastnaesite. Exploration also has developed large reserves of vein material that include thorium and europium. The property is currently a major producer. Interestingly, the previously overlooked rare earth ores were discovered adjacent to and in the workings of gold mines.

Randsburg Mining District

The Rand Mountains adjoin the Garlock fault immediately south and east of the El Paso Mountains. Their rocks include several Precambrian formations, with some Paleozoic, Mesozoic, and Tertiary intrusive rocks and surficial volcanics. This complex has yielded gold, silver, and tungsten, each producing a mining "boom" in the district.

Gold—The Yellow Aster Several famous gold mines produced the Rand's first boom. Towns that sprang up were Randsburg and Johannesburg, obviously reflecting hope of rivaling the great South African mining areas. About the only thing in common, however, is the Precambrian age of some rock units, which in South Africa contain gold and in the Randsburg district generally do not. The Rand ore was mineralized granitic rock that invaded Precambrian(?) schists of the Rand schist formation. Gold was disseminated in the granite and found on contacts with the schist.

Gold production from the Yellow Aster is estimated at $10 million dollars. The Yellow Aster's glory hole, several hundred feet deep, dominates the landscape of hillside Randsburg and was mined by open pit methods. Last major production in the Yellow Aster was in 1917–1918. Some placer gold mining resumed in 1930, when squatters eked out a bare livelihood by mining the gravels of the alluvial fans derived from the outcrops of the gold areas. No significant production occurred.

Silver—The Kelly Rand A chance discovery in 1919 of high-grade silver ore in the eastern Rand district led to development of the Kelly Rand silver mine. Production was from rich ores that occurred in irregular veins and blocks, permitting stoping of large bodies of rock from 250 to 300 feet (80–90 m) below ground. Large underground "caverns" (stopes) were created, sometimes resulting in surface cave-ins. The primary mineral was miargyrite, an unusual sulfantimonide of silver related to the more common pyrargyrite. Associated minerals were a few sulfides such as stibnite and marcasite. The ore was apparently formed from magmatic waters accompanying shallow rhyolitic intrusives. Though figures vary, when production ceased in 1928, ore valued at about $13,580,000 had been processed. Gold ores were discovered at the 1500-foot (450 m) level of the Kelly mine, but production was only a by-product of silver mining.

Tungsten—The Atolia Mines Scheelite, a principal ore of tungsten, was recognized in 1903 in a Nevadan granitic pluton on the southeastern slopes of the Rand Mountains. The World War I demand for tungsten for steel manufacture promoted both lode and placer production. Production subsequently languished, but revived during World War II. Small production by leaseholders has been almost continuous, but except for the peak periods total production has been minimal. The placer ground is characterized by rows of conically shaped mounds that resemble the tips or wastepiles of the tin mines of Cornwall, England. The Atolia tips are the residues from placer mining in the gravels derived from the bedrock with scheelite-bearing veins. Scheelite has a specific gravity of 6 and is therefore readily separated from associated minerals by either wet or dry placer mining. Scarcity of water in the Rand district necessitated only dry placer methods.

Kramer Borate Deposits

In 1925, a large deposit of borate minerals was discovered in the Kramer district, about 35 miles (56 km) east of Mojave and 1 mile (1.6 km) north of Boron (Figure 5-7). Many unusual and several new borate minerals have been described from this deposit. The major ore body is composed of tincal, a hydrous borate of soda, and kernite,

Figure 5-7. *Open pit mine at Kramer, near Boron. (Photo by Spence Air Photos, courtesy of Department of Geography, University of California, Los Angeles)*

a hydrous borate of soda with 7 water molecules instead of the 10 of tincal. Until 1925, California's commercial production of borax had been from the Death Valley colemanite deposits. Colemanite, a hydrous borate of lime, requires processing in which calcium is replaced by sodium to make borax; this is a relatively expensive process compared with that required to convert tincal or kernite to borax. By 1928, borax mining operations were concentrated in the Kramer district, where the operation continues today. The present mine is open pit, with a huge annual production of raw ore that is bedded in an ancient playa. The source of the boron is still debated. Nevertheless, it has been suggested that volcanic emanations from basaltic lavas underlying the borate layers penetrated the clays of the lake beds and formed the borate minerals. There is little evidence to support this view, however.

Calico Mountains

The restored ghost town of Calico in the Calico Mountains was one of southern California's early silver camps. Silver was discovered in

1881, and production of high-grade ore continued until 1896, with negligible activity subsequently. The productive ores (argentite and cerargyrite) came from mineralized fault zones, veins, and irregular bodies developed by shallow intrusives. The primary mining areas were in a mineralized belt extending about 5 miles (8 km) along the south base of the range.

Prior to 1914, colemanite was mined on the northeastern side of the Calico Mountains from folded lake beds, and today visitors are impressed with the enormous waste piles that hug the slopes. The colemanite was found in narrow seams, veins, and geodes in the clay shale and constituted only a small percent of the volume of material excavated. Hand sorting of much of the ore was required.

REFERENCES

General

Bassett, A. M., and D. H. Kupfer, 1964. A Geological Reconnaissance of the Southeastern Mojave Desert, California. Calif. Div. Mines and Geology Spec. Rept. 83.

Dibblee, Thomas W., Jr., 1967. Areal Geology of the Western Mojave Desert, California. U.S. Geological Survey Prof. Paper 522.

——— and D. F. Hewett, 1970. Geology of the Mojave Desert. Mineral Information Service (now California Geology), v. 23, pp. 180–185.

Geologic Map Sheets of California. Kingman (1961), Trona (1963), Needles (1964), Bakersfield (1965), Los Angeles (1969), and San Bernardino (1969). Calif. Div. Mines and Geology.

Hewett, D. F., 1954a. General Geology of the Mojave Desert Region, California. In Geology of Southern California. Calif. Div. Mines and Geology Bull. 170, pp. 5–20.

———, 1954b. A Fault Map of the Mojave Desert Region. In Geology of Southern California. Calif. Div. Mines and Geology Bull. 170, pp. 15–18.

Miller, William J., 1946. Crystalline Rocks of Southern California. Geol. Soc. Amer. Bull., v. 49, pp. 417–446.

Special

Beeby, David J., and Robert L. Hill, 1975. Galway Lake Fault. California Geology, v. 28, pp. 219–221.

Evans, James R., 1966. California's Mountain Pass Mine Now Producing Europium Oxide. Mineral Information Service (now California Geology), v. 19, pp. 23–32.

———, 1974. Relationship of Mineralization to Major Structural Features in the Mountain Pass Area, San Bernardino County, California. California Geology, v. 27, pp. 147–157.

Mojave Desert

Gale, Hoyt S., 1951. Geology of the Saline Deposits, Bristol Dry Lake, San Bernardino County, California. Calif. Div. Mines and Geology Spec. Rept. 13.

Gardner, Dion L., 1954. Gold and Silver Mining Districts in the Mojave Desert Region of Southern California. *In* Geology of Southern California. Calif. Div. Mines and Geology Bull. 170, pp. 51–58.

Hulin, C. D., 1925. Geology and Ore Deposits of the Randsburg Quadrangle, California. Calif. Div. Mines and Geology Bull. 95.

Lamey, C. A., 1949. Vulcan Iron-Ore Deposit, San Bernardino County, California. Calif. Div. Mines and Geology Bull. 129, pp. 87–95.

MacDonald, Angus A., 1970. The Northern Mojave Desert's Little Sahara. Mineral Information Service (now California Geology), v. 23, pp. 3–6.

Merriam, Charles W., 1954. Rocks of Paleozoic Age in Southern California. *In* Geology of Southern California. Calif. Div. Mines and Geology Bull. 170, pp. 9–14.

Morgan, Vincent, and Richard C. Erd, 1969. Minerals of the Kramer Borate District, California. California Geology: Part 1, v. 22, pp. 143–153; Part 2, v. 22, pp. 165–172.

Olson, J. C., and L. C. Pray, 1954. The Mountain Pass Rare-Earth Deposits. *In* Geology of Southern California. Calif. Div. Mines and Geology Bull. 170, pp. 23–29.

Parker, Ronald B., 1963. Recent Volcanism at Amboy Crater, California. Calif. Div. Mines and Geology Spec. Rept. 76.

Schaller, Waldemar T., 1929. Borate Minerals of the Kramer District, Mojave Desert, California. *In* Shorter Contributions to General Geology. U.S. Geological Survey Prof. Paper 158-I, pp. 137–170.

Sharp, Robert P., 1954. The Nature of Cima Dome. *In* Geology of Southern California. Calif. Div. Mines and Geology Bull. 170, pp. 49–52.

———, 1966. Kelso Dunes, Mojave Desert, California. Geol. Soc. Amer. Bull., v. 77, pp. 1045–1074.

Weber, F. Harold, Jr., 1966 and 1967. Silver Mining in Old Calico. Mineral Information Service (now California Geology): v. 19, pp. 71–80; v. 20, pp. 3–8, 11–15.

Chapter Six

Colorado Desert

*We must accept that we can never see the
past in fine detail.*

Konrad Krauskopf

As indicated in Chapter 5, there is a certain arbitrariness in distinguishing between the Colorado and Mojave deserts. Nevertheless, the Colorado corresponds approximately to the "low desert" and the Mojave to the "high desert" of southern California weather reports. This distinction is based chiefly on general altitude, but also reflects climate and indigenous vegetation. Figure 6-1 gives the locations of some of the important features of the Colorado Desert.

For purposes of this book, the Colorado Desert province is bounded on the east by the Colorado River, on the south by the Mexican border, and on the west by the Peninsular Ranges. The northern border lies along the southern edge of the eastern Transverse Ranges, approximating the San Bernardino–Riverside county line. In the northwestern corner of the province is the Coachella Valley, tucked in between the highest parts of the Peninsular and Transverse ranges at San Gorgonio Pass. The Orocopia Mountains are assigned by the authors to the Colorado Desert province, though some consider them the eastern unit of the Transverse Ranges. The ranges of eastern Riverside County are included in the Colorado Desert, though they could be placed in the Mojave province with equal logic. In short, the boundaries of the Colorado Desert, like those of other provinces, are somewhat artificial.

GEOGRAPHY

Much of the Colorado Desert lies at low elevation. The Colorado River valley at the Riverside–San Bernardino county line is 350 feet (107 m) above sea level, and southeast at Winterhaven elevation is only 130 feet (40 m). The province's largest low area is the Salton Basin, divided into the Imperial Valley in the south and the Coachella Valley in the north. The Salton Sea, an inadvertently man-made lake with many natural but long-vanished predecessors,

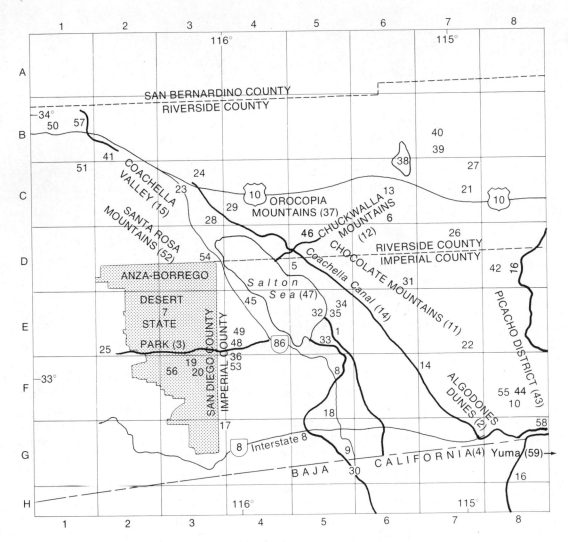

Figure 6-1. Place names: Colorado Desert.

occupies the central part of the basin. It has a surface elevation of about 235 feet (72 m) below sea level. The total portion of the basin lying below sea level extends from near Indio south for about 90 miles (144 km) and is about 25 miles (40 km) wide. This nearly flat depression includes practically all the towns in Imperial County and has developed into one of California's most productive agricultural areas (Figure 6-2).

Apart from a narrow band along the Colorado River and the northeastern quarter of Imperial County, drainage in the Colorado Desert is internal. In eastern Riverside County, much of the drainage ends in the broad Chuckwalla Valley, which contains playas separated from one another by sand dunes. The largest playa is Palen

 1 Alamo River—E–G, 5–6
 2 Algodones Dunes (Imperial Sand Hills)—F–G, 7–8
 3 Anza-Borrego Desert State Park—D–G, 2–3
 4 Baja California—G–H, 1–8
 5 Bat Cave Butte—D5
 6 Black Butte—C6
 7 Borrego Valley—D–E, 2–3
 8 Brawley—F5
 9 Calexico—G5
10 Cargo Muchacho Mountains—F8
11 Chocolate Mountains—D–E, 5–7
12 Chuckwalla Mountains—C–D, 5–6
13 Chuckwalla Valley—C6
14 Coachella Canal—C–G, 3–8
15 Coachella Valley—C2–3
16 Colorado River—D–G, 8
17 Coyote Mountains—G3
18 El Centro—F5
19 Fish Creek (Split Mt. Canyon)—F3
20 Fish Creek Mountains—F3
21 Ford Dry Lake (see Figure 5–1)
22 Gavilan Wash—E7
     Imperial Sand Hills (see Algodones Dunes)—F–G, 7–8
23 Indio—C3
24 Indio Hills—C3
25 Julian—E2
26 Little Chuckwalla Mountains—D7
27 McCoy Mountains—C7
28 Mecca—C3
29 Mecca Hills—C4
30 Mexicali—G5–6
31 Mt. Barrow—D6
32 Mullet Island—E5
33 New River—E–H, 5–6
34 Niland—E5
35 Obsidian Butte—E5
36 Ocotillo Wells—E–F4
37 Orocopia Mountains—C4–5
38 Palen Dry Lake—B–C 6
39 Palen Mountains—B7
40 Palen Pass—B7
41 Palm Springs—B2
42 Palo Verde Mountains—D8
43 Picacho District—E–F8
44 Pilot Knob—F8
45 Salton City—E4
46 Salt Creek—D5
47 Salton Sea—D–E, 4–5
     Sand Hills (Imperial) (see Algodones Dunes)—F–G, 7–8
48 San Felipe Creek—E2–4
49 San Felipe Hills—E4
50 San Gorgonio Pass—B1
51 San Jacinto Peak—C1
52 Santa Rosa Mountains—C–D, 2–3
     Split Mt. Canyon (see Fish Creek)—F3
53 Superstition Hills—F4
54 Travertine Rock—D3
55 Tumco—F8
56 Vallecito Mountains—F3
57 Whitewater River—B1–2
58 Winterhaven G8
59 Yuma (Arizona)—G8

N

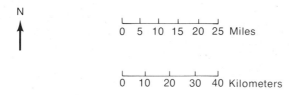

Dry Lake southwest of the Palen Mountains, but Ford Dry Lake, southwest of the McCoy Mountains, is almost as large.

The Colorado Desert's major drainage is into the Salton Sea. For example, the northeastern side of the Peninsular Ranges and the southeastern face of the Transverse Ranges are drained by the Whitewater River and its tributaries, which reach the northern end of the Salton Sea not far from Mecca. Salt Creek drains the southern slope of the Orocopia Mountains and the northern end of the Chocolate Mountains. The most important western drainage is San Felipe Creek, with headwaters in the Peninsular Ranges near Julian about 50 miles (80 km) west of the Salton Sea. The Imperial Valley is drained by the New and Alamo rivers. Together these rivers account

Figure 6-2.
Salton Sea and Im-
perial Valley.
(Photo courtesy of
National Aeronau-
tics and Space Ad-
ministration)

for most of the flow into the Salton Sea, for both cross extensively irrigated farmland and carry water from seepage after irrigation. Their channels were greatly widened and deepened between 1904 and 1907, when most of the Colorado River was entering the valley through man-made openings in the levee south of the international border.

Salton Trough

The dominant feature of the Colorado Desert province is the Salton Trough, a large structural depression that extends from near Palm Springs about 180 miles (290 km) south to the head of the Gulf of California. The term Salton Trough refers to the entire basin, from San Gorgonio Pass to the Gulf of California, whereas the term Salton Basin applies to only the region draining directly into the Salton Sea. The lower portion of the trough, entirely in Mexico, is occupied by the delta of the Colorado River. In addition, much of the sediment fill in the Imperial Valley is deltaic, because on occasion the river has flowed into the California portion of the basin.

When William Phipps Blake, the first geologist known to enter the Salton Basin, crossed the area in 1853, he correctly surmised that the basin's floor lay below sea level. He further suggested that construction of the deltaic barrier by the Colorado River had isolated the northern extremity of the Gulf of California, allowing it to evaporate and form the extensive salt flats he observed. (There was no Salton Sea at that time.) Later investigations confirmed the existence of elevations below sea level, but they also showed that the basin's history was considerably more complicated than Blake had realized.

STRATIGRAPHY

The oldest rocks exposed in the Colorado Desert are Precambrian crystalline gneisses, anorthosites, and schists. These are known from the Chocolate, Cargo Muchacho, Palo Verde, Orocopia, Chuckwalla, and Little Chuckwalla mountains and from Pilot Knob, an isolated hill near the international border. These ancient rocks have been intruded by several younger plutonic bodies, ranging from late Paleozoic to middle Cenozoic. In fact, this district contains the youngest plutonic rocks so far known in California. Plutonic rocks in the Chocolate Mountains have yielded radiometric ages of 23 to 31 million years (early Miocene to late Oligocene). The Orocopia schist was thought to be Precambrian when it originally was named by W. J. Miller in 1944. Based on correlations with the Pelona schist of the Transverse Ranges, however, it is probably considerably younger, perhaps even Cretaceous.

Most sedimentary and volcanic rocks in the Colorado Desert are Cenozoic and generally rest on the eroded surfaces of the older rocks. The oldest Cenozoic deposits are marine Eocene beds in the Orocopia Mountains. This range also contains the only Oligocene rocks found thus far in the province. These are continental deposits resembling the Vasquez formation in the Soledad Basin of the central Transverse Ranges. Although early Cenozoic sedimentary rocks appear only sporadically, post-Oligocene sequences are thick, widespread, and well displayed. This often accounts for notable scenery, as in Anza-Borrego Desert State Park. In southwestern Imperial Valley, 16,500 feet (5030 m) of late Cenozoic sedimentary rocks are exposed, in northwestern Imperial Valley 18,700 feet (5700 m), and in Coachella Valley 8600 feet (2650 m). Beneath the valley floor, just south of the international border, geophysical evidence has indicated that these same sediments reach a thickness of more than

21,000 feet (6400 m). Most of these beds are nonmarine, but the Mio-Pliocene Imperial formation is a widespread marine unit sometimes nearly 4000 feet (1200 m) thick.

Pliocene rocks west of the Salton Sea include the Canebrake conglomerate, which was deposited near the mountains at the edge of the basin. Toward the basin's center, the Canebrake grades into its finer-grained counterpart, the Palm Spring formation. Much of the desert floor west of the Salton Sea is underlain by this soft, nonmarine formation, which has been tightly folded and beveled by erosion (Figure 6-3). The Palm Spring formation also contains abundant petrified wood, plus numerous zones of sandstone concretions. In some zones, the concretions are only a few inches long and usually dumbbell-shaped; in other zones, they are a foot or more across and usually spherical.

Sediments deposited in Lake Cahuilla (forerunner of the present Salton Sea) during the late Pleistocene and possibly the early Holocene form a nearly horizontal cover on many exposures of the older rocks. This cover is normally only a few feet thick, but in places it is as much as 300 feet (90 m) thick. Like the Palm Spring

Figure 6-3. *The folded Palm Spring formation makes intricate patterns in the barren rocky desert west of the Salton Sea. Tule Wash barchan dune is in the right foreground. (Photo by John S. Shelton)*

Component	Salton Sea		Ocean
	(in parts per thousand)		
Chloride (Cl)	9.033		18.971
Bromide (Br)	—		0.065
Sulfate (SO₄)	4.139		2.649
Carbonate (CO₃)	—		0.071
Bicarbonate (HCO₃)	0.232		0.140
Fluoride (F)	0.002		0.001
Calcium (Ca)	0.505		0.400
Strontium (Sr)	—		0.013
Potassium (K)	0.112		0.380
Boron (B)	0.005	(H₂BO₃)	0.026
Sodium (Na)	6.249		10.556
Magnesium (Mg)	0.581		1.272
Lithium (Li)	0.002		—
Silica (SiO₂)	0.021		—
TOTAL	20.900		34.482

Table 6-1
Comparative Composition of the Salton Sea and the Ocean

formation, the Lake Cahuilla beds are soft, weakly consolidated siltstones and clays readily cut by streams whenever there is any runoff. Sufficient runoff occurs only two or three times every five years or so. When it does, stream channels are cleared of wind-deposited sand and steep banks are swiftly undercut, releasing blocks of the Palm Spring formation or the overlying lake beds. These blocks are rolled along by the water, becoming rounded in the process, and commonly acquire a coating of gravel and small rocks. When the water subsides, the channels often retain hundreds of such armored mudballs.

Although the Mio-Pliocene Imperial formation represents the final, large-scale marine incursion into the basin, Pleistocene marine mollusks occur at several localities. These may reflect a brief, relatively recent marine invasion. The shells are all found on the surface, not in the lake beds, at 150 to 250 feet (45–80 m) below sea level. Their significance is strongly debated. The wide distribution, occurrence of articulated pelecypod shells, and lack of attached matrix indicate that the animals lived in one of the later stages of Lake Cahuilla. Moreover, although they are marine species and may have been introduced into water with salinity similar to the ocean's, such a chance introduction would not require invasion by the ocean. Furthermore, these marine species are mixed with freshwater forms. During World War II, marine barnacles were introduced into the Salton Sea on the floats of seaplanes. The barnacles could live in the lake because its salinity is about two-thirds that of the ocean, although the salt mix is somewhat different (Table 6-1).

The various lakes that occupied the basin have left recognizable deposits, some quite thick. For example, the Pliocene Borrego formation in the Borrego badlands west of the Salton Sea is 6000 feet (1800 m) thick.

STRUCTURE

The Colorado Desert shows the northwesterly structural trends characteristic of most geologic provinces of California (Figure 6-4). The faults of the eastern Peninsular Ranges facing the Salton Basin have a more northerly trend than the main faults entering the basin, however. Some, like the San Jacinto fault, follow well-defined courses in the Peninsular Ranges, cutting obliquely across the mountains to either dissipate completely or vanish beneath the basin deposits. The San Jacinto and its branches enter Borrego Valley and cut across the Ocotillo badlands near Ocotillo Wells before disappearing beneath lake beds. During the 1968 earthquake, low scarps developed along this segment of the San Jacinto. Similarly, the more complicated Elsinore fault system disappears as it enters the valley near the Coyote Mountains.

The San Andreas fault zone extends along the northeastern side of the basin from the upper Coachella Valley to probably the southeastern corner of Imperial County. There is some uncertainty regarding the fault's location southeast of Bat Cave Butte, a low range of hills just east of the Salton Sea near the Imperial-Riverside county line. Beyond this point, the San Andreas has no obvious surface expressions. Some geologists believe the fault continues southeast beneath the Algdones dune belt, but this has not yet been confirmed. There are also differing opinions regarding which (if any) faults in the Colorado Desert expression of the San Andreas system are the San Andreas proper. Nevertheless, since Pliocene time the Salton Basin portion of the presumed San Andreas seems to have been offset several miles right laterally, and post-Eocene offset may total 180 miles (290 km). Topographic evidence of faulting (offset streams, truncated alluvial fans) is strikingly well displayed in the Salton Basin (Figure 6-5).

In 1940, the Imperial Valley and surrounding regions were shaken by a strong earthquake. A fault previously concealed beneath the valley's alluvium was outlined by the resulting ground breakage, which extended from near Brawley south across the international boundary. Low scarps were produced and as much as 15 feet (5 m) of right-lateral separation occurred, offsetting rows of trees, roads, and presumably even the international boundary (Figure 6-6). This fault is now known as the Imperial fault, and its trace is still evident despite 35 years of road repair and cultivation that have obliterated its more obvious features. The Imperial fault parallels other major faults in the basin.

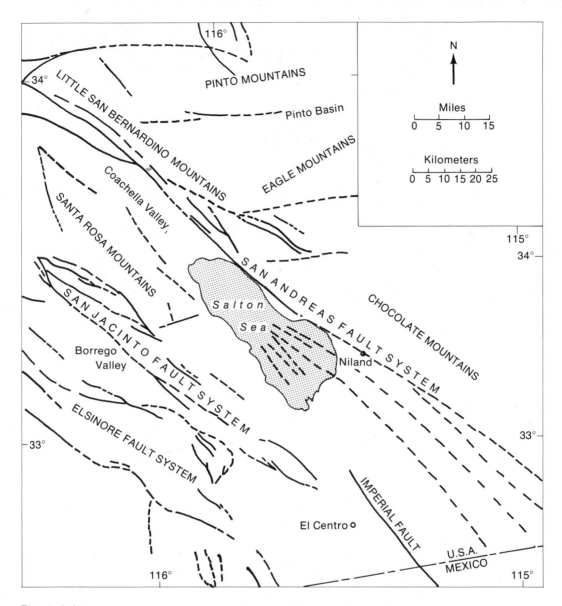

Figure 6-4.
Some known and inferred Cenozoic faults of the Colorado Desert. (Sources: California Division of Mines and Geology, U.S. Geological Survey, and Department of Geology, University of California, Riverside)

Folding is prominent in the Colorado Desert, particularly within younger rocks close to major faults. The Indio and Mecca hills, for example, are young anticlinal structures that contain numerous tightly buckled small folds especially near faults. Low hills in western Imperial Valley, such as the San Felipe and Superstition chains, are likewise anticlinal and show intense deformation adjacent to faults.

Colorado
Desert

Figure 6-5.
View to the north-
west along the San
Andreas fault, near
Indio. Intensely
folded Palm Spring
formation is on the
right. (Photo by
Spence Air Photos,
courtesy of De-
partment of Geog-
raphy, University of
California, Los
Angeles)

Pliocene sedimentary rocks exposed in the flat, barren country west and southwest of Salton City display a wide variety of small-scale folds that are outlined by thin sandstone beds or beds containing abundant sandstone concretions. The smaller folds were produced chiefly during movement along faults. Most of the faults have affected young sediments and provide good evidence of the area's continuing tectonism.

The Salton Basin itself is underlain by thick mainly or entirely Cenozoic sedimentary materials, primarily land-laid nonmarine deposits. The great thickness demonstrates that considerable sinking of the basin floor occurred as sediments accumulated. Continuing tectonism is also reflected, as the beds were crumpled, eroded, and exposed around the basin's margin. The general form of the basin's sedimentary fill is synclinal, with the upturned edges of many units exposed in the surrounding hills.

As plate tectonic theory has been applied to the eastern Pacific and western North America, evidence has accumulated to suggest that sea-floor spreading is widening the Gulf of California, particularly its northern end. This conveniently accounts for the thick sediment fill and persistent subsidence in the Salton Basin. As Baja

Figure 6-6.
Offset trees near Calexico. (Photo by Spence Air Photos, courtesy of Department of Geography, University of California, Los Angeles)

California and the Peninsular Ranges drift northwest, away from mainland North America, the opening rift (floored by thin, stretched continental crust) will continue trapping sediments delivered by the Colorado River and other sources. In addition, thick prisms of sediment are inclined to subside as water is squeezed from them, as they compact under their own weight, and as the crust below sags under the accumulating load.

SUBORDINATE FEATURES

Most mountains in the Colorado Desert province lie between the Salton Basin and the Colorado River, though the northern tip of Sierra Cocopah extends from Mexico into California west of Calexico and several low hills occur in the western Salton Basin. As far as bedrock geology is concerned, the Colorado River and the international border are somewhat irrelevant boundaries; the ranges of western Arizona and northern Baja California are similar in many ways to the ranges considered here. Topographically all the Colorado Desert ranges resemble those of the Mojave Desert, although the Colorado's are generally slightly lower and drier and usually are separated by broader valleys.

Orocopia Mountains

East of the town of Mecca and just north of the Salton Sea are the Orocopia Mountains. This range is comparable to the neighboring Chuckwalla and Chocolate mountains, but in addition it has characteristics whose interpretations have contributed substantially to our understanding of California's geologic history. The range exhibits a variety of probably Precambrian rocks, including gneisses along its eastern edge and intrusives like gabbro, diorite, anorthosite, and syenite. Many small bodies of titaniferous magnetite and ilmenite are also evident. These Precambrian rocks form a belt lying between two major faults, the Clemens Well on the northeast and the Orocopia on the southwest. As this belt crosses the range from northwest to southeast, its intrusive rocks cut the older gneisses. Interestingly, this entire suite of Precambrian crystalline rocks is almost perfectly duplicated in the San Gabriel Mountains, but on the opposite side of the San Andreas fault and more than 100 miles (160 km) to the northwest. Lithology, degree of metamorphism, and field relationships are so alike that these two terrains are regarded as parts of a single original mountain mass and have been cited as evidence for about 130 miles (210 km) of right slip on the San Andreas. J. C. Crowell has been primarily responsible for establishing these conclusions.

This Precambrian complex is cut by younger intrusive rocks, mostly of Mesozoic age but with some of possible Paleozoic age. Again, the relationships are strikingly similar to those in the San Gabriel Mountains.

West of the Orocopia thrust fault is the overridden plate of the Orocopia schist; the upper, overriding plate is the complex of Precambrian and Mesozoic rocks just described. So far the Orocopia schist has not yielded any direct evidence of its age. As indicated previously, however, it lithologically resembles the Pelona schist of the San Gabriel mountains. Most geologists believe the two units were originally part of the same depositional and metamorphic terrain—probably eugeosynclinal rocks consisting mainly of Cretaceous graywackes. Perhaps even more notable than the lithologic similarity is the matching structural pattern seen in the San Gabriel and Orocopia mountains. In both cases, a complex of Precambrian gneisses and intrusives forms the upper plate of a thrust. Likewise, in the nearby Chocolate Mountains, Precambrian crystalline rocks have been thrust over Orocopia schist. In the presumed position of all three ranges prior to offset, the Orocopia and Pelona schist terrains would be in close proximity to one another. It seems likely that present rock distribution reflects a single depositional basin disrupted and separated by long-term right slip on the San Andreas fault system.

The Orocopia Mountains are the only Colorado Desert range east of the San Andreas fault known to contain Eocene and

Oliogocene sedimentary beds. The Eocene rocks are marine and cover about 26 square miles (67 km²) in the range's northeastern section. The section is about 4800 feet (1460 m) thick and consists of a coarse conglomerate with clasts up to 30 feet (10 m) in the areas where the formation rests on underlying granitic rocks. The upper part is generally finer-grained, but includes many boulder beds.

Above the Eocene section is a thick unit of nonmarine beds and associated basaltic and andesitic flows. Not only do these beds and the Vasquez formation of the Soledad Basin look alike, but also radiometric dating of their upper, volcanic members has yielded quite similar ages: 24 and 25 million years for the Vasquez formation and 22.5 million years for the Diligencia formation of the Orocopia Mountains. The lower units may be Oligocene, but age determination is not available. Both formations apparently were deposited in relatively arid interior basins, probably close to one another. So far, however, evidence does not indicate that deposition was in the *same* basin.

Youngest deposits in the Orocopia Mountains are Pleistocene sandstones and gravels found on the range's eastern slopes and around its western flank. As is typical of desert regions, recent alluvial deposits surround the range and blanket the adjacent valley floor.

Chocolate Mountains

The Chocolate Mountains extend from Salton Creek, which separates them from the Orocopia Mountains, 60 miles (96 km) southeast at least as far as Gavilan Wash. Some authorities extend them another 20 miles (32 km) to the Colorado River and include the Picacho Mountains, which constitute the southeastern end of what is really a continuous range. Highest elevation is at Mount Barrow (2475 ft or 755 m) in the southeastern end. The mountains are seldom more than 10 miles (16 km) wide.

As mentioned before, the Chocolate Mountains, like the Orocopia Mountains, have a Precambrian crystalline basement thrust over Orocopia schist. These Precambrian gneisses and associated rocks are intruded by at least five distinct granitic plutons. The oldest pluton is apparently Permian, but most are Mesozoic. The Precambrian gneisses and associated rocks above and the Orocopia schist below the thrust fault, plus the thrust itself, are all cut by quartz monzonite stocks dated radiometrically at 23 million years—the youngest granitic intrusives hitherto dated in California. Volcanic rocks of similar age and possibly related to the quartz monzonite intrusives are widely distributed in the Chocolate Mountains. Miocene fanglomerates and basaltic flows overlie the young intrusives and associated volcanics and are themselves overlain by Pliocene and younger nonmarine sedimentary rocks probably deposited in an arid environment.

Palen and McCoy Mountains

These two ranges in eastern Riverside County have been studied in some detail and have proved to be less complex than the ranges along the eastern Salton Basin. The Palen Mountains extend almost due north from the Chuckwalla Valley for about 12 miles (19 km) and have a maximum elevation of 3000 feet (900 m). The McCoy Mountains, about the same length and only slightly lower, lie southeast of the Palen Mountains and form the northeastern wall of the Chuckwalla Valley.

Oldest rocks present in either range are the intensely faulted and folded marble, quartzite, and gypsum of the late Paleozoic Maria formation. The only exposure is in the northern Palen Mountains at Palen Pass. Maria formation rocks are intruded by a quartz porphyry that seems to underlie most of both ranges, though it is not often exposed. The porphyritic character of this rock suggests it was a shallow-depth intrusive exposed and eroded during the late Mesozoic.

Resting unconformably on the quartz porphyry are about 23,000 feet (7000 m) of sandstone, mudstone, and conglomerate. All are assigned to the McCoy Mountains formation, which forms the bulk of both ranges. A few rather poorly preserved plant fossils have been recovered from the formation's upper section, but they indicate only that the rocks are Cretaceous or younger. At the southern ends of both ranges, the intrusive quartz porphyry has been thrust over the younger McCoy Mountains formation, demonstrating tectonism of probable Cenozoic age.

Fish Creek Mountains

The Fish Creek Mountains are in western Imperial Valley, south of San Felipe Creek. They extend about 10 miles (16 km), with a maximum elevation of 2334 feet (712 m) near their southeastern end. They are separated from the larger Vallecito Mountains to the west by the gorge of Fish Creek, which is locally known as Split Mountain Canyon.

The basement of the Fish Creek Mountains is a mixture of gneisses, marbles, and granitic rocks. These rocks are of uncertain age, but probably are partially equivalent to the Cretaceous crystalline rocks forming much of the Peninsular Ranges to the west. It is not illogical, in fact, to interpret both the Fish Creek and Vallecito mountains as merely extensions or outliers of the Peninsular Ranges, although the former are true desert ranges and are surrounded on three sides by the Colorado Desert.

During at least the first half of the Cenozoic, the crystalline basement was unroofed and deeply eroded. By Miocene time, conditions had changed and deposition of nonmarine sediments began. This presumably occurred under arid or semiarid conditions, because coarse-grained fanglomerate accumulated on the exposed basement at the base of mountains that rose steeply to the west. These fanglomerates are now assigned to the Split Mountain formation. A fine cross section about 2700 feet (820 m) thick is displayed in the 250-foot (80 m) high cliffs of lower Split Mountain Canyon in Anza-Borrego Desert State Park. The rocks have a reddish clayey matrix and include beds of huge, angular boulders, some more than 10 feet (3 m) across. Some layers are composed almost exclusively of fresh granitic rubble and careful examination is needed to establish their sedimentary origin. Other layers show clear evidence of large-scale mudflows.

Resting on the Split Mountain fanglomerate is the Alvorson andesite, probably of late Miocene age. In the eastern Fish Creek Mountains and above the Alvorson andesite there is a gypsum deposit up to 100 feet (30 m) thick, locally capped by a thin bed of celestite. (Gypsum is hydrous calcium sulfate, and celestite is a strontium sulfate.) Although there has been only sporadic interest in the celestite, the gypsum has been quarried for many years and sometimes has yielded large plates of its nearly transparent variety, selenite.

Much of the ancestral Salton Basin, including the Fish Creek area, was invaded by a shallow sea near the end of the Miocene or early in the Pliocene. During this time, the clays, silts, sands, and oyster beds of the Imperial formation were widely deposited. Probably the thickest section of the formation is on the southern side of the Fish Creek Mountains, where about 3700 feet (1130 m) have been eroded into low, barren hills often capped with dark brown oyster beds.

Before the Pliocene closed, the sea was excluded from the Salton Basin, possibly by construction of the Colorado River delta across the head of the Gulf of California. Nonmarine conditions again prevailed. The Palm Spring formation was deposited in the center of the basin, grading west into its coarse-grained equivalent, the Canebrake conglomerate. As mentioned earlier, the Palm Spring formation contains abundant petrified wood, thought to be ironwood, a desert tree still extant throughout the Colorado Desert. Where the Palm Spring formation is exposed, the resistant gray wood often weathers out of the soft enclosing siltstones and claystones.

Sand Dunes

Although comparatively little of the state's arid regions contain sand dunes, sand deposits are more extensive and varied in the Colorado than in other California deserts. The largest Colorado dune tracts are the Algodones dunes or Imperial sand hills, which are about 45 miles (72 km) long and extend about 5 miles (8 km) into Mexico, to the edge of the Colorado River flood plain. The dunes are sometimes more than 5 miles (8 km) wide, but average width is nearer 3 miles (4.8 km). Their highest peaks rise nearly 250 feet (80 m) above the desert floor. The dune chain is made up of overlapping, slightly arcuate ridges lying at right angles to the length of the dune field and alternating with sandy-floored depressions (Figure 6-7). Toward the southern end of the chain, the depressions become larger, up to a mile (1.6 km) by a half-mile (0.8 km) in area. Some are free of sand, and faint traces of old drainage channels may be visible on the gravel floors. Presumably these channels predated the dunes and existed when drainages developed from the uplands in the eastern Salton Basin. In a few of these flat-floored and sand-starved hollows, swarms of small barchan dunes have formed. The entire dune chain is migrating southeast in response to the strong northwesterly winds that rake the area, especially in winter and spring.

Some geologists believe these dunes represent a long plume of sand blown inshore from the beaches of former Lake Cahuilla. Others think that groundwater moving up along the buried extension of the San Andreas fault (which may pass beneath the dunes) either trapped blowing sand directly or promoted a greater growth of vegetation that then trapped blowing sand. However they formed, these dunes represent perhaps the largest tract of desert dunes in North America and for many years imposed a serious barrier to travel between the Imperial Valley and Arizona. In 1914, a narrow plank road was built across the dunes to link El Centro and Yuma, but sand was continually being blown from beneath or on top of it. In the late 1920s, the first paved road was built across the dunes. It proved successful, although sand had to be scraped from the road surface frequently and dunes sloping down from the road had to be oiled to prevent sand being removed from under the road's surface. The same problems persist today, although a four-lane divided highway now crosses the area.

Until the middle 1930s, the canal system servicing the Imperial Valley collected water from the Mexican part of the Colorado River. The system flowed west around the southern end of the Algodones dunes some 40 miles (64 km) before turning north into the United States. Early canal builders were unable to construct the canal entirely on the American side of the border across the dunes. The modern canal system does cross the dunes, but requires constant dredging to remove dune sand and silt contributed by the Colorado River itself.

Other dunes occur in the Imperial and Coachella valleys, though none compares with the large Algodones chain. Well-formed barchan dunes are found west of the Salton Sea, north of San Felipe Creek. Some are moving 100 feet (30 m) per year toward the Salton Sea, though others move more slowly. Small dunes tend to move more rapidly than large ones, causing the configuration of dune fields to change as small dunes overtake and merge with larger ones.

*Figure 6-7.
Algodones dunes,
from the south.
Note the areas of
sand starvation in-
side the dune clus-
ters. (Photo by
Spence Air Photos,
courtesy of De-
partment of Geog-
raphy, University of
California, Los
Angeles)*

Extensive tracts of dune sand exist in the Coachella Valley east of Palm Springs. Perhaps because of the great volume of new sand brought into the valley by the Whitewater River and its tributaries, these dunes generally lack the well-defined geometry of those seen in the Imperial Valley. Nevertheless, the windiness of the upper Coachella Valley, coupled with the availability of large volumes of sand, provides many good examples of wind abrasion, deposition, and erosion.

Low rocky ridges extend from the base of San Jacinto Peak into the flat floor of San Gorgonio Pass at right angles to the strong winds that regularly sweep through this gap into the valley below. On the windward side of each of these spurs, windblown sand and gravel have etched the hard granitic rocks. On the lee sides, plants often grow, but their tops are perfectly even with the ridge. New shoots are trimmed off by the sharp, windblown sand as soon as they venture above the protecting rock barrier. William Phipps Blake observed these features during his visit in 1853 and correctly described the general cause of the persistent winds. His words are instructive.

> They both [the San Francisco Golden Gate and San Gorgonio Pass] appear to be great draught-channels from the ocean to the interior, through which the air flows with peculiar uniformity and persistence, thus supplying the partial vacuum caused by the ascent of heated air from the surface of the parched plains and deserts.

In connection with the effects of sandblasting in San Gorgonio Pass, he says:

> I had before me remarkable and interesting proofs of the persistence and direction of this air-current, not only in the fact that the deep sand-drift was on the east side of the spur, but in the record which the grains of sand engrave on the rocks in their transit from one side to the other.[1]

The highway traveler coming into Indio from the west often encounters windblown sand and gravel despite the many shelter belts of trees and sand fences. Fence posts, highway signs, telephone poles, bottles, and just about anything that is exposed is subjected to frequent sandblasting. Many automobiles have undergone severe damage during a single sandstorm. Windbreaks alleviate the problem, but do not eliminate it. The blowing wind still manages to seek out loose sand, and the processes of erosion, abrasion, and deposition continue.

SPECIAL INTEREST FEATURES

Lake Cahuilla and Its Fossil Shorelines

On the west side of the Salton Sea, a feature resembling a high-water mark appears along the base of the Santa Rosa Mountains. From Travertine Rock on the Riverside-Imperial county line, tiny gastropod and pelecypod shells can be seen littering the desert below the water line. Travertine Rock itself is a granitic knob, with its lower portion covered with a pale brown, spongelike crust of travertine. This lime deposit was secreted by algae that once inhabited a freshwater lake—ancient Lake Cahuilla. Above the highest lake level, algae were unable to deposit and so the rocks are darker colored, owing to a coat of desert varnish. The old, highest lake level is thus plainly marked for miles along the base of the Santa Rosa Mountains.

Most of the high shoreline does not reach the bases of the other mountains that rim the Salton Basin, for the shoreline's average elevation is only about 30 to 40 feet (9–12 m) above sea level (Figure

[1]William Phipps Blake *in* R. S. Williamson, Report of Explorations in California for Railroad Routes (1853–54). Thirty-third Congress, Second Session; Senate Executive Document n. 78, v. 5, part 2, ch. 8, pp. 91–92.

6-8). Its existence is manifested in other ways, however. Old beaches, sand spits, and bay mouth bars can be seen along the base of the Santa Rosa Mountains, but the finest examples occur along the Coachella canal between Niland and Mecca. Smooth, curved beaches bridge former headlands that once entered the lake. Where flat pebbles of Orocopia schist were carried to the old lake by streams draining the Orocopia and northern Chocolate mountains, shingle beaches of neatly stacked, smooth, flat schist pebbles now stretch for several miles. Elsewhere, waves in Lake Cahuilla cut low cliffs into soft sedimentary rocks such as the Palm Spring formation; good examples can be seen near Niland and in the San Felipe badlands.

The complete history of Lake Cahuilla and its predecessors is yet to be resolved. Nevertheless, it is likely that the lake filled on several occasions, when distributaries of the Colorado River changed their courses across the delta. Sometimes these distributaries entered the closed Salton Basin to the north, at other times they flowed into the Gulf of California, and sometimes they did both simultaneously. Between fillings, evaporation quickly reduced the lake level, often leaving a salty crust on the basin floor.

It seems certain that Lake Cahuilla did not follow the pattern of the ancient lakes of the Mojave Desert and Basin Ranges to the

Figure 6-8.
Ancient shorelines of Lake Cahuilla, near Niland. Salton Sea is on the left (west). The zigzag pattern along the railroad is made by flood revetments. (Photo by Spence Air Photos, courtesy of Department of Geography, University of California, Los Angeles)

north; these seem to correlate with wetter periods associated with glaciation. Lake Cahuilla may well have responded to wetter conditions in the high ranges nearby, but the lake was almost certainly as much a consequence of the Colorado River floods, which are independent of glacial cycles. The last lake filling prior to formation of the present Salton Sea has been dated at about 300 years ago.

Salton Sea

The present Salton Sea is a product of the twentieth century and the result of real estate promotions and farming development in the last decade of the nineteenth century. By 1900, more than a thousand people had settled in the Imperial Valley, and the available water supply was insufficient for farming and development. Beginning that year, a canal system was constructed from a point on the Colorado River in Mexico. By 1902, 400 miles (640 km) of canals and lateral branches were in service, and nearly 100,000 acres (40,500 ha) had been prepared for cultivation. By the end of 1904, the canals had silted up badly, the river was low, and settlers were clamoring for more water. To increase flow into the canal system, the control gates on the river were bypassed by making cuts in the bank and building wing dams from shore to deflect water into the canals. Several of these openings promptly silted up, but did provide some badly needed water. Plans for new control gates were completed in November 1904, but were not approved before the next major flood.

The first important flood to occur during this perilous situation was in January 1905, followed by three large flash floods in February. By then the river was out of control, and considerable water was entering the canal system. The engineers in charge hoped to close the gaps in the west bank during the normal period of low water before the late spring floods. This period failed to materialize, and during the late summer attempts to plug the gaps were curtailed by continued high water.

Early in the summer of 1905, about 16 percent of the river was entering the canal system, and by October virtually the entire Colorado River was flowing into the valley. The Southern Pacific Railroad tracks had been inundated, and the company built a new barrier. Unfortunately, a violent flash flood in November again allowed much of the river to enter the valley. In early 1906, new control gates were completed, and the river was finally contained on 4 November 1906. On 10 December, still another violent flash flood swept down the river, and the entire flow once again poured into the canal system and from there into the channels of the Alamo and New rivers and into the valley. On 11 February 1907, the breach was sealed for the last time. The level of the Salton Sea then stood at 198 feet (60 m) below sea level, its highest point.

Ironically, many canal branches were left totally dry because the flooding had widened and deepened the channels of the New and

Alamo rivers, which previously had been tiny gulleys across a nearly featureless desert floor. These two rivers now became prominent channels cut into the valley's soft alluvium, sometimes measuring a quarter of a mile (0.4 km) wide and 50 feet (15 m) deep. They, rather than the canals, then carried much of the water across the Imperial Valley and into the Salton Sea. Subsequently these rivers also became valuable drainage channels for saline waste irrigation water.

Spectacular examples of headward erosion developed as the river channels were widened, deepened, and lengthened. During 1905 and 1906, the process began at the shore of the Salton Sea and rapidly extended about 50 miles (80 km) up the New River to Mexicali. The erosion was so rapid near Mexicali that banks were undercut in a few hours, destroying many buildings. These processes were less dramatic on the Alamo River, but even here headward and lateral erosion eventually extended upstream about 30 miles (48 km).

By 1925, evaporation had reduced the lake level to almost 250 feet (80 m) below sea level. Beginning by the late 1930s, as agricultural development was augmented and more water was brought in from the Colorado River, net inflow into the Salton Sea reversed the lake's shrinkage pattern. The present level is about 235 feet (72 m) below sea level. This is probably near the stable level because as the lake rises and its area increases, annual evaporative loss approaches a balance with annual inflow, which is itself limited by the amount of water that California may legally withdraw from the Colorado River.

Recent Volcanic Features

Near Niland, at the southeastern end of the Salton Sea, are some volcanic features that include several low pumice and obsidian domes. Five low volcanic hills have been formed, rising 100 to 150 feet (30–45 m) above the basin floor. These hills extend along a 4-mile (6.4 km) axis from southwest of Mullet Island to Obsidian Butte and are young (presumably Quaternary) rhyolitic domes of pumice and obsidian. Several still produce warm gases, and at Mullet Island the gases are acidic enough to decompose the rocks. Mud volcanos, mud pots, and boiling springs lie along a northwest-southeast line almost at right angles to the trend of the volcanic domes. Principal gases produced are steam, carbon dioxide, and hydrogen sulfide.

Beginning in 1927, attempts were made to recover the carbon dioxide from wells. In the 1930s, the operation produced gas from depths of 200 to 700 feet (60–710 m), but the field was abandoned in 1954 because of the rise of the Salton Sea. In 1957, geothermal exploration was initiated, and about a dozen wells have been drilled. The hot brines obtained from these wells are sometimes extremely high in table salt, calcium chloride, and several base metals. If economically feasible, in future the steam may be used for electric power generation and for partially evaporating the brines.

Mining in the Cargo Muchacho Mountains

The Cargo Muchacho Mountains are small, somewhat isolated hills that rise from the flat desert floor in the southeastern corner of the Colorado province. Most of the surrounding desert is less than 600 feet (180 m) above sea level, so the range is prominent although it is only 2221 feet (677 m) high and about 8 miles (13 km) long and 6 miles (9.6 km) wide.

The rocks of the Cargo Muchacho Mountains are generally similar to those of the Chocolate Mountains and include Precambrian gneisses and Mesozoic granitic intrusives. For some reason, however, mineralization was more intense in the Cargo Muchacho district, and gold has been mined here intermittently since 1780. One particularly well known gold-producing area was at Tumco, now a ghost town. The town's name is an acronym for The United Mining Company, which operated the local mines about the turn of the century. In the early 1900s, Tumco had a population of 2000. There were four saloons on the main street—known as Stingaree Gulch—but only a few buildings remain today.

In the southern part of the range, the mineral kyanite has been mined. Like other aluminum silicates, kyanite is used in making high-temperature ceramic materials. Annual production was about 100 tons (80 metric tons) in the early 1940s, but the mines have been inactive in recent years.

REFERENCES

Biehler, S., and R. W. Rex, 1971. Structural Geology and Tectonics of the Salton Trough, Southern California. *In* Geological Excursions in Southern California. Univ. Calif. Riverside Campus Museum Cont. no. 1, pp. 30–42.

Crowell, John C., ed., 1975. San Andreas Fault in Southern California. Calif. Div. Mines and Geology Spec. Rept. 118.

Dibblee, T. W., Jr., 1954. Geology of the Imperial Valley Region, California. *In* Geology of Southern California. Calif. Div. Mines and Geology Bull. 170, pp. 21–28.

Hamilton, Warren B., 1969. Geology of the Colorado Desert. California Geology, v. 22, pp. 96–98.

Koenig, James B., 1967. The Salton-Mexicali Geothermal Province. Mineral Information Service (now California Geology), v. 20, pp. 75–81.

Miller, William J., 1944. Geology of the Palm Springs–Blythe Strip, Riverside County, California. Calif. Jour. Mines and Geology, v. 40, pp. 11–72.

Chapter Seven

Peninsular Ranges

*In science the important thing is to modify
and change one's ideas as science advances.*
                                   Claude Bernard

The Peninsular Ranges are one of the largest geologic units in western North America. They extend 125 miles (200 km) from the Transverse Ranges and the Los Angeles Basin south to the Mexican border and beyond another 775 miles (1250 km) to the tip of Baja California. The total province varies in width from 30 to 100 miles (48–160 km), includes—according to some interpretations—the offshore area, and is bounded on the east by the Colorado Desert and the Gulf of California. The locations of important features are shown in Figure 7-1.

The Peninsular Ranges contain extensive pre-Cretaceous igneous rocks associated with Nevadan plutonism and a few remnants of roof rock that establish continuity with the better-known pre-Nevadan history of the Sierra Nevada. The Cretaceous marine is well represented. Post-Cretaceous rocks form a restricted veneer of volcanic, marine, and nonmarine sediments, although adjacent to the Peninsular Ranges (for instance, in the Los Angeles Basin) post-Cretaceous marine sections up to 40,000 feet (12,200 m) thick are found.

GEOGRAPHY

The Peninsular Ranges are a northwest-southeast oriented complex of blocks separated by similarly trending faults. Highest elevations are in the San Jacinto–Santa Rosa Mountains of the easternmost block: San Jacinto Peak is 10,805 feet (3296 m), and summits in the Santa Rosa Mountains average 6000 feet (1800 m). Toward the Pacific are the Agua Tibia, Laguna (Cuyamaca), and Santa Ana mountains, all with lower elevations. The escarpment between the San Jacinto Mountains and the adjacent Coachella Valley is one of the boldest in North America, especially near Palm Springs where the San Jacinto summit rises abruptly 10,000 feet (3000 m) above the

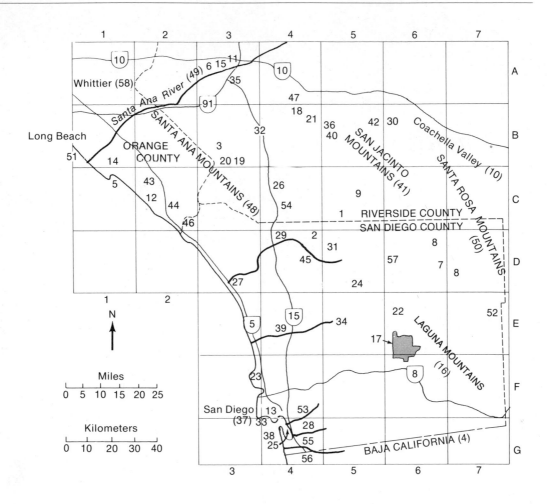

Figure 7-1. Place names: Peninsular Ranges.

1 Aguanga—C5
2 Agua Tibia Mountains—D4
3 Alberhill—B3
4 Baja California—G4–7
5 Balboa—C1
6 Bloomington—A3
7 Borrego (Hot) Springs—D6
8 Borrego Valley—D7
9 Cahuila—C5
10 Coachella Valley—B6–7
11 Colton—A3
12 Corona del Mar—C2
13 Coronado (Strand)—F4
14 Costa Mesa—B1
15 Crestmore (Quarries)—A3
16 Cuyamaca (Laguna) Mountains—E–F, 6–7
17 Cuyamaca Rancho State Park—E6
18 Eden Hot Springs—B4
19 Elsinore—B3
20 Elsinore Lake—B3
21 Gilman Hot Springs—B4

22 Julian (Mining Area)—E6
   Laguna (Cuyamaca) Mountains (see 16)—E–F, 6–7
23 La Jolla—F3
   Lake Elsinore (see 19)—B3
24 Mesa Grande—D5
25 Mission Bay—G4
26 Murrieta Hot Springs—C4
   Newport (Bay)–Balboa Harbor (see 5)—C1
   Newport Beach (see 5)—C1
27 Oceanside—D3
28 Otay River—G4
29 Pala—D4
30 Palm Springs—B6
31 Palomar Mountain—D5
32 Perris—B3–4
33 Point Loma—G3–4
34 Ramona—E5
35 Riverside—A3
36 Saboba Hot Springs—B5
37 San Diego—F4
38 San Diego Bay—G4

39 San Dieguito River—F4
40 San Jacinto Hot Springs—B5
41 San Jacinto Mountains—B–C, 5–6
42 San Jacinto Peak—B5
43 San Joaquin Hills—C2
44 San Juan Capistrano—C2
45 San Luis Rey River—D3–5
46 San Onofre—D2
47 San Timoteo Canyon—A4
48 Santa Ana Mountains—B–C, 2–3
49 Santa Ana River—A–B, 1–4
50 Santa Rosa Mountains—C–D, 6–7
51 Seal Beach—B1
52 Split Mountain—E7
53 Sweetwater River—F4
54 Temecula—C4
55 Tia Juana River—G4
56 Tijuana (Mexico)—G4
57 Warner Hot Springs—D6
58 Whittier—A1

Bold face of San Jacinto Peak, with Palm Springs in the center and San   *Figure 7-2.*
Gorgonio Pass in the far right. (Photo courtesy of U.S. Geological Survey)

valley (Figure 7-2). The boundary between the Santa Rosa Mountains and the Imperial Valley is less defined. If the boundary is a fault, it apparently has been inactive for some time.

The general cross section of the Peninsular Ranges somewhat resembles that of the Sierra Nevada, since each range has a gentle westerly slope and, normally, a steep eastern face. The western side of the Peninsular Ranges is composed of discrete blocks that slope progressively lower to the west and are produced by the breaks of major fault zones.

Drainage of California's Peninsular Ranges is primarily by the San Diego, San Dieguito, San Luis Rey, and Santa Margarita rivers. Short streams drain the rest of the province. Those flowing east enter the Imperial Valley or local interior drainage basins in depressed fault blocks; those flowing west enter the Pacific. Rainfall is 10 to 30 inches (250–760 mm) annually, and most streams are intermittent.

In the Baja California section of the Peninsular Ranges, essential geographic units are: (1) the (northern) Sierra San Pedro Martir, the Sierra Juarez, and some minor ranges, all of which are direct southeasterly extensions of the California mountain blocks; (2) the Vizcaino Desert; (3) the Sierra de la Giganta; and (4) the southern tip, the Sierra Victoria. Elevations are generally low, but reach 10,126 feet (3088 m) at La Providencia Peak in the northern San Pedro Martir.

ROCKS

Pre-Cretaceous

Metasedimentary and metavolcanic rocks occupy restricted areas in the Peninsular Ranges. Near Riverside there are restricted outcrops of late Paleozoic limestone (from which cement is processed). Some older rocks are exposed in the San Jacinto and Santa Rosa mountains, where altered schist and gneiss may reach total thicknesses of 22,000 feet (6700 m). Adjacent to the Los Angeles Basin, rocks up to 20,000 feet (6100 m) thick have been mapped as the Bedford Canyon series in the Santa Ana Mountains and southward. Fragmentary fossils from this series have been identified as Jurassic. Thick volcanics (Santiago), up to 2300 feet (700 m) in the Santa Ana Mountains, overlie the Bedford Canyon rocks. Similar volcanics inland from San Diego may be equivalent. These volcanics are considered Jurassic also.

Igneous rocks are age equivalents of Nevadan plutonism. In declining age, typical Peninsular Range intrusives are gabbro, quartz diorite or tonalite (the commonest), and granodiorite. The entire plutonic complex of the Peninsular Ranges is known as the southern California batholith and includes plutons like the San Marcos gabbro, the Bonsall tonalite, and the Woodson Mountain granodiorite. Typically all units have widespread dark inclusions of older rocks. Radiometric dating suggests Nevadan plutonic emplacements, from 115 to 120 million years old. Younger dates, to 70 million years, have been determined in some instances, but the intrusives are definitely part of the batholithic development grouped in the Nevadan orogeny.

Peninsular Range rocks are distinctly less silicic and more calcic than typical Sierran intrusives. Nevertheless, the abundance of Nevadan intrusives may imply a relationship between Peninsular and Sierran histories—albeit a tenuous one because the thick Paleozoic sections of the Sierra are not repeated in the Peninsular Ranges. Possibly an extensive Paleozoic sequence has been eroded away, but no evidence substantiates this. Precambrian history is not recorded in Peninsular Range rocks or structure, whereas the Transverse Ranges immediately to the north include undoubtedly Precambrian igneous intrusives and metamorphic gneisses. Likewise, latest Paleozoic intrusives of the Transverse Ranges are not identified in the Peninsular Ranges. Overall, then, it appears that the homogeneity of their Mesozoic plutonic rocks and their restricted Paleozoic preclude grouping the Peninsular Ranges with any other province.

Post-Cretaceous

All post-Cretaceous rocks lie unconformably on crystalline basement. Lengthy, widespread erosion of the crystalline units contributed substantially to Cenozoic deposits in the Los Angeles, Imperial, and offshore basins and to thick sediments in localized interior basins like the Elsinore Trough.

Early Tertiary rocks are confined to coastal margins, reaching a maximum thickness of 4500 feet (1370 m) in the Santa Ana Mountains. Only 600 feet (180 m) of marine sediment occur in the San Diego area, overlain by about 1000 feet (300 m) of mostly nonmarine sediments (Poway conglomerate). These are small thicknesses when compared with those of adjacent basins like the Los Angeles, Ventura, and some offshore areas. Nonmarine Oligocene up to 3000 feet (900 m) thick is found in the Santa Ana Mountains, but the Peninsular Range province was rejuvenated by significant uplift at the close of the Oligocene, further restricting sedimentation.

Late Tertiary beds are also limited in extent. Little Tertiary volcanism occurred in the Peninsular Range block (in contrast to the widespread Tertiary volcanism of the Sierra Nevada). A few late Tertiary marine and continental deposits are found in intramontane valleys. A well-known formation from this period is the San Onofre breccia, which is composed of irregular blocks of a Franciscan type of rock. The San Onofre is distributed from the Santa Monica Mountains and offshore islands of the Transverse Ranges to Oceanside and the Los Coronados Islands (Baja California) in the Peninsular Ranges. Pliocene nonmarine rocks, dated by vertebrate fossils, are thick and widespread in the northern Peninsular Ranges; for example, the Mount Eden and San Timoteo Canyon beds of siltstone, sandstone, and conglomerate reach a thickness of 7000 feet (2150 m).

Quaternary deposits include fluviatile and lacustrine sediments in the interior and restricted volcanic and marine terrace deposits along the coast. Apparently none of the province was high enough to develop Pleistocene glaciers, as no evidence to corroborate their existence has yet been discovered.

STRUCTURE

Faults dominate the structure of the Peninsular Ranges (Figure 7-3). Moreover, recent movement has facilitated mapping of the chief through-going faults. Topographic expressions are clear, and major escarpments often can be traced for tens of miles.

Major faults are the San Jacinto and related branches within the San Jacinto zone and the Elsinore and associated faults within the Elsinore zone. The San Jacinto originates in the Transverse Ranges, bifurcating from the San Andreas and crossing province lines. The fault continues southeast through the Peninsular Range province

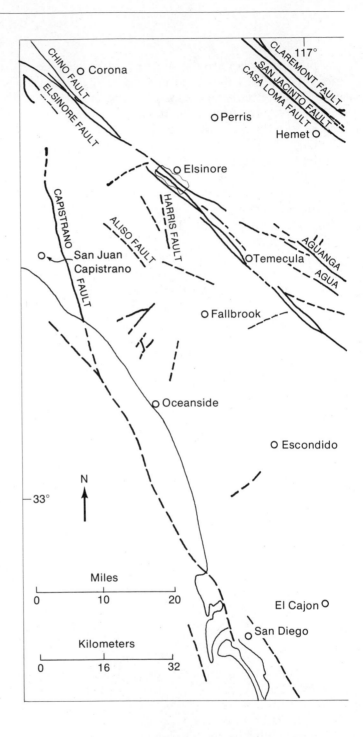

Figure 7-3.
Some known and
inferred Cenozoic
faults of the
Peninsular Ranges.
(Sources: California
Division of Mines
and Geology and
U.S. Geological
Survey)

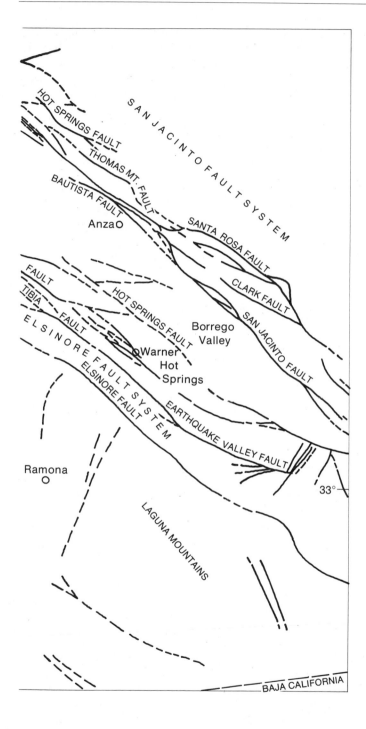

HOT SPRINGS FAULT

S A N   J A C I N T O   F A U L T   S Y S T E M

THOMAS MT. FAULT

BAUTISTA FAULT

AnzaO

SANTA ROSA FAULT

CLARK FAULT

FAULT

TIBIA

FAULT

HOT SPRINGS FAULT

Borrego
Valley

SAN JACINTO FAULT

E L S I N O R E   F A U L T   S Y S T E M

OWarner
Hot
Springs

ELSINORE FAULT

EARTHQUAKE VALLEY FAULT

Ramona
O

33°

LAGUNA MOUNTAINS

BAJA CALIFORNIA

and into the Colorado Desert. It appears that as they leave the
Peninsular Range province southward, the San Jacinto and Elsinore
zones converge and eventually join. The Elsinore may become the
eastern fault of the Sierra Juarez scarp in Baja California. The San
Jacinto fault follows the Gulf of California, probably on the Gulf's
east side.

San Jacinto Fault Zone

Within the Peninsular Ranges, the primary named faults of the San
Jacinto zone are, from north to south, the San Jacinto, Claremont,
Hot Springs, Thomas Mountain, Bautista, Santa Rosa, and Clark.
Many smaller, unnamed faults anastamose between the major zones
of faulting.

Movement is post-Nevadan, since the southern California
batholithic rocks of Nevadan age are all displaced. Evidence of older
movement includes crushed zones in dominantly plutonic rocks,
fault-line and eroded fault scarps, fault contacts transgressing
metasedimentary inclusions, and linear topographic trends paralleling known faults. Evidence of more recent movement includes epicenter swarms, earthquakes (San Jacinto 1918 and Borrego Valley
1968), and alignment of hot springs. Sags with small playas and sag
ponds with springs of permanent water exist along some faults, and
several fault-line valleys are being lengthened and deepened by
modern streams.

As a rule, amount and type of movement have not been definitely established. Nevertheless, all offsets appear to be right slip,
and accumulated movement since middle Cenozoic is probably several miles. Evidence for really substantial right slip is inconclusive,
however.

On the other hand, evidence for vertical movement is compelling. Recent movements have had strong dip-slip patterns that produced scarps and scarplets. Known faults of the San Jacinto zone
move relatively up on the east and down on the west. The Perris
Plain, with general elevation of 1500 to 2000 feet (450–600 m), is a
major topographic feature between the San Jacinto (northeast) and
Elsinore (southwest) fault zones. The plain is an undulating surface
eroded on primarily plutonic igneous rocks and lies 7000 feet
(2150 m) below the summits of the San Jacinto Mountains (Figure
7-4). Likewise the Borrego lowland, near sea level, lies 7000 feet
(2150 m) below the Santa Rosa Mountains. These features clearly
reflect dominantly vertical displacement in the San Jacinto fault
zone. Vertical displacement of similar magnitude presumably is
expressed along the Banning (or South Banning) fault on the northeast side of the San Jacinto Mountains.

More detailed study is needed before movement patterns for the
San Jacinto zone can be appraised realistically. Furthermore, even
partial reliance on uncorroborated but tempting interpretations can

only hinder our eventual understanding. For instance, geometric relationships with the San Andreas that seem obvious from patterns on a structural map of California may sometimes exert strong psychological pressure on geologists to infer major strike-slip movement, even though supporting evidence is actually weak or nonexistent.

Elsinore Fault Zone

The major named faults of the Elsinore zone are the Elsinore, Aguanga, Agua Tibia, Earthquake Valley, and Hot Springs. (Unfortunately, the latter is not the same as the San Jacinto zone fault of the same name; "hot springs" faults are almost as common in California as "rattlesnake" canyons.) Many interconnecting faults

Figure 7-4
Looking northeast at San Jacinto Peak from the Hemet–San Jacinto Valley. The linear ridge pattern reflects internal faulting. (Photo by Spence Air Photos, courtesy of Department of Geography, University of California, Los Angeles)

also are present. Evidence for faulting within the Elsinore zone is similar to that for the San Jacinto zone except that more demarcations between distinct rock types are mappable and topographic expression is less marked. In addition, more horst-graben features are apparent. (These structures are small topographically, but impressive geologically.)

Parallel to the San Jacinto and subparallel to the San Andreas, the Elsinore fault zone predictably has been assigned right-slip movement. On a small scale, such movement may be demonstrable, but apparently dip slip also has persisted throughout the zone's history. In the Elsinore graben proper, vertical movements of several thousand feet are known (Figure 7-5). Buried parallel faults are separated by at least 2000 feet (600 m) of nonmarine Quaternary sediment (shown by well records). On the Agua Tibia fault, landslides and rubble from adjacent uplands bury the fault's trace for several miles.

The faults of the Elsinore zone disappear northwest beneath sediments of the Los Angeles Basin, where the Elsinore apparently bifurcates into the Chino fault (northeast) and the Whittier fault (southwest). These two faults apparently do not reach the Malibu Coast–Hollywood–Raymond–Cucamonga fault trend and more significantly do not reveal strike-slip movement.

Presently available evidence suggests that the vertical movements so prominently recorded in the Plio-Pleistocene along the Elsinore zone reflect the entire history of the fault's movement, with strike-slip movement being minimal or nonexistent. Confirmation of this will establish that the Elsinore fault zone is not directly related to the San Andreas system. Moreover, some geologists think that high-altitude satellite photography demonstrates that the San Andreas fault terminates at the southern entrance to Cajon Pass, possibly against the Cucamonga fault. It also has been suggested that the Perris Plain is bounded on the north by the Cucamonga fault. Along this fault, Precambrian and associated Mesozoic units are thrust south across Pleistocene alluvium, indicating recent tectonism. In this area, the Cucamonga fault dips steeply to the north.

GEOMORPHOLOGY

The Peninsular Ranges differ from other areas of California in that, for the most part, they have been under an erosional regimen since the Nevadan orogeny. This has resulted in a major surface—presumably a peneplain that has been uplifted primarily by faulting, with continued erosion developing present topography.

Because of its faulting, the province possesses strong linear elements trending northwest-southeast. A stepped topography, with

Figure 7-5. Elsinore graben: view to the northwest from Temecula. San Gabriel Mountains with snow-capped Mount San Antonio (Baldy) are in the right distance, and the Santa Ana Mountains are on the left. (Photo by Spence Air Photos, courtesy of Department of Geography, University of California, Los Angeles)

each tread successively higher and narrower to the east, culminates in the Laguna Mountains (Cuyamaca Peak) at nearly 6000 feet (1800 m). Streams flow on the treads in broad, aggraded valleys, followed by narrow canyon segments where the risers are dissected. Occasionally, risers like horsts separate two adjacent treads at approximately similar elevations. In such cases, streams have cut defile canyons through the risers, suggesting that streams are antecedent or superimposed. Gross topographic expression suggests structural control by faults, systems of master joints, and patterned fabrics inherent in the plutonic rocks. The drainage patterns are distinctively trellised between treads and dendritic on each tread. The reason for the imperfect rectilinear drainage pattern has not yet been established.

SUBORDINATE FEATURES

San Jacinto–Santa Rosa Mountains

The highest block of the Peninsular Ranges is the easternmost San Jacinto Mountains, which include the second highest peak in southern California, San Jacinto Peak (10,805 ft or 3296 m). The San

Jacinto Mountains and their southern continuation, the Santa Rosa
Mountains, contain a thick section of banded gneiss and widespread
quartz diorite plutons along the saddle between them. Both ranges
are a giant squeeze-up block (horst) caught between the San Jacinto
and Banning faults.

The San Jacinto Mountains are well watered, and snows on San
Jacinto Peak normally last into the summer. The Santa Rosa block is
almost waterless, however, with few springs and no permanent
streams.

Elsinore Trough

The Elsinore Trough is the linear, low-lying block northeast of the
Santa Ana Mountains and southwest of the Perris Plain. It extends
from Corona on the northwest about 30 miles (48 km) southeast and
has a maximum width of 3 miles (4.8 km). The Elsinore graben was
formed by vertical movements on the Elsinore and subordinate
faults. Its lowest part contains Lake Elsinore (maintained by water
from the Colorado River aqueduct) and is at the base of a particularly
steep eroded scarp. The trough includes one of the most representa-
tive sections of post-Nevadan rocks within the Peninsular Ranges.
Its geology is shown by the cross section of Figure 7-6.

The trough contains lower Tertiary strata that originated in
brackish water from which lignite, ceramic-quality clays, and
glass-sand have been mined. The clay and glass-sand deposits are at
Alberhill, where there are extensive pits and some processing plants.
Lignite is no longer mined, but other products are still extracted,
primarily from the district's lower-grade clays.

*Figure 7-6. Cross section of the Elsinore fault zone. Depth shown in feet. (Source:
California Division of Mines and Geology)*

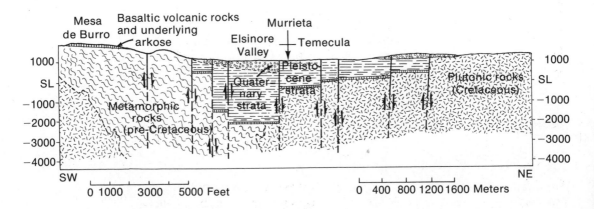

SPECIAL INTEREST FEATURES

Cuyamaca Rancho State Park

In eastern San Diego County, on the summits of the Laguna Mountains (elevations to 6000 ft or 1800 m), recreational facilities incorporating the area's flat-topped topography have been developed as Cuyamaca Rancho State Park. The Laguna Mountains are bounded on the east by the Elsinore fault, and a bold eroded scarp of granitic crystalline rocks extends for 15 to 20 miles (24–32 km). The rocks are Cretaceous igneous plutonics, except for a few remnants of older metamorphic rocks into which the plutonics intruded. Occasionally, pegmatitic dikes rather like those of the Pala and Mesa Grande gem areas occur. The Laguna Mountain pegmatites have yielded few minerals of special interest, however.

Julian Mining Area

Julian is the center of a mining district noted during the nineteenth century for its gold and nickel deposits. (The Banner Queen was one of the famous gold mines.) Along the Elsinore fault where igneous intrusives cut the metamorphic sequences of the Jurassic(?) Julian schist, nickel mineralization occurred in veins and pockets. Since it is usually in short supply and with the U.S. demand depending on foreign sources, nickel always has excited interest. Principal ore was the nickel-bearing iron sulfide pyrrhotite, which occurred abundantly. Percentage of nickel was low and inconsistent, however, and though there have been several mining revivals, Julian is maintained today only to serve tourists and local ranchers. Old mines abound, but few may actually be seen by visitors.

Gem Areas

Pegmatitic dikes occur in two adjacent areas in the upper drainage of the San Luis Rey River and constitute the world-famous Pala and Mesa Grande gem localities. The network of related dikes was formed from mineralizers associated with granitic intrusions of the Nevadan orogeny, either by direct crystallization of the granitic magma or by replacement of the invaded rock by mineralizing solutions. The dikes invaded joints and zones of weakness in wall rocks, crosscutting older formations and in some cases the granitic parent rocks. Radiometric dating suggests 100 to 110 million years ago (Cretaceous) as the time of emplacement.

The dikes average about 10 feet (3 m) thick and are often more than a mile (1.6 km) long. Podlike thicker zones in the dikes contain cavities often the size of a small room, where giant crystals of pink, green, and yellow transparent tourmaline grew. Individual crystals from 5 to 20 inches (2–10 cm) in diameter and up to 3 feet (1 m) long have been collected. Other minerals extracted include beryl, kunzite (a gem variety of spodumene first discovered at Pala), gem garnet, and occasional topaz. Such concentrations of rare minerals seldom occur, and the reason for this unique and prolific development in the Peninsula Ranges is unknown.

Mining heyday was from 1900 to 1910, when gem tourmaline was highly prized. Decline of interest in tourmaline forced closure of most mines, but production of specimen materials continues. Many museums feature San Diego County gem minerals; some of the best are at Harvard University in Massachusetts and Balboa Park in San Diego. Pegmatites are widespread in the Peninsular Ranges, and in addition to the two most famous areas, the localities of Ramona, Aguanga, Cahuilla, and the Santa Rosa Mountains have yielded gem material. Useful nongem minerals from the region are commercial quantities of feldspar, lithium products, and quartz.

Crestmore Quarries

The Crestmore limestone quarries in western Riverside County have provided one of the world's largest suites of contact metamorphic minerals. By 1966, 138 minerals had been described, including several new species, and new compounds still are being discovered. The minerals are generally unspectacular in appearance, but they are scientifically important and knowledge of their chemistry has enhanced understanding of many ore- and mineral-forming processes. Occurrence is in a limestone-marble sequence of presumably late Paleozoic age that was intruded by silica-rich rocks of the Nevadan plutonic episode. Silicification of the limestone produced the exotic chemical trace elements that have combined to yield the unusual minerals. Suites of Crestmore minerals are displayed in mineral collections worldwide.

Hot Springs of the San Jacinto–Elsinore Fault Zones

Hot and mineral springs have been known in the Peninsular Ranges for some time. Both Indians and Spanish used the hot springs, and Americans developed them as mineral spas. This aspect has waned, however, because the medicinal value of most waters is now discounted.

Along the San Jacinto fault zone at the west base of the San Jacinto Mountains are Gilman, Eden, Saboba, and San Jacinto hot springs. Other less well known but nonetheless sizable springs rise along the zone's southern extensions into Borrego Valley. Water volume and temperature vary considerably in such springs, suggesting that juvenile water is diluted by meteoric sources.

The Elsinore zone contains springs also. The town of Elsinore was known originally for its sulfur water and its spa. Warner Springs is one of the province's largest springs, and Murrieta Hot Springs near Temecula is one of its oldest spas.

Both fault zones have many smaller springs along traces of their branch faults.

Newport Bay

Newport Bay is located at the southeastern end of the Los Angeles coastal plain, with the San Joaquin Hills a short distance to the east. The bay itself consists of two main parts. The lower portion parallels the coast and is sheltered from the open ocean by a sand spit on which the towns of Newport Beach and Balboa are situated. The upper portion is a winding estuary extending about 5 miles (8 km) inland from the lower bay, widening toward its upper end (Figure 7-7).

It is probable that the upper portion is the drowned channel of the Santa Ana River, which previously emptied into the bay. In 1915, because of severe silting that resulted from flooding in the Santa Ana, the river was diverted into a man-made outlet about 1.5 miles (2.4 km) north of Newport harbor. Like most estuaries and salt marshes along the California coast, this bay is a temporary feature that, barring man's interference or sea level change, would be rapidly converted to dry land (Figure 7-8).

Newport Bay differs from some coastal estuaries because it is rimmed with steep cliffs 90 to 100 feet (27–30 m) high, particularly around the lower bay and on the east side of the upper bay. Cliffs along the bay's west side, at the edge of flat-topped Costa Mesa, gradually lose height inland and merge with the Santa Ana River plain near the head of the bay. As a result, the bay is really a steep-sided, water-filled valley set down into slightly uplifted mesa lands that extend west from the San Joaquin Hills. The river-cut estuary extends across a block that has been elevated as much as 100 feet (30 m) above the general level of the main river plain to the north and west. If the present topographic pattern had existed when the Santa Ana River first established its course, the river would have avoided elevated Costa Mesa and entered the sea north of Newport Bay. It is therefore likely that Costa Mesa was elevated after the river had developed a well-established channel to the sea. During uplift, the river maintained its channel even though it had to cut a gorge across a slowly rising upland. If this interpretation is correct, the river-cut channel today forming Newport Bay is antecedent.

Newport Bay is dotted with islands of various sizes and shapes.

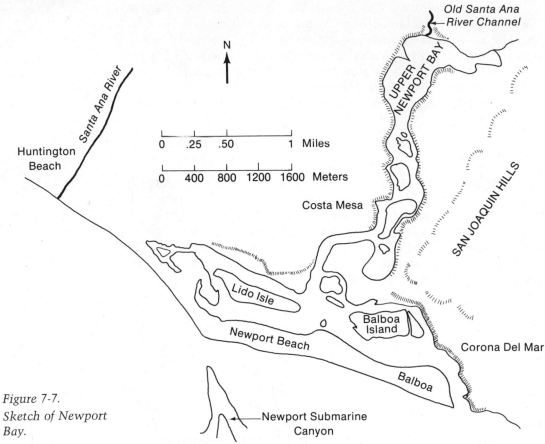

Figure 7-7. Sketch of Newport Bay.

Many are now so extensively urbanized that examining their composition and structure is difficult. With few exceptions, all the islands result from dredging. Bay floor materials dredged to provide boating channels have been dumped on shallow mud banks to make islands.

The perceptible bulge in the bay's sand spit is probably the only remaining natural irregularity on the coast between Seal Beach and Corona del Mar. This bulge is a direct result of the presence of Newport submarine canyon, which approaches shore at the point of the bulge. In the late nineteenth century, a railroad pier was built into the head of the canyon. The deeper water was an advantage to shipping, and it had been observed that waves were usually lower over the canyon than elsewhere. The presence of submarine canyons close to shore always reduces the height of waves over the canyons and in the adjacent surf. This reduction in wave energy also favors deposition of sand transported along the beach by longshore currents, contributing to formation of seaward bulges in the beach. Similar bulges on low sandy coasts occur elsewhere in California and may be caused by reefs, islands, or breakwaters a short distance offshore. Like submarine canyons, if they are close enough to the beach these features also reduce wave height and encourage sand deposition.

San Onofre–Oceanside Coast

From the southern Orange County line to Oceanside, the coast exhibits prominent marine terraces that stretch inland several miles to elevations of at least 800 feet (240 m). The broadest and most conspicuous terrace, which extends intermittently from near San Onofre to south of Oceanside, is about 60 to 100 feet (18–30 m) above sea level.

Few, if any, marine terraces in California predate the Pleistocene. The San Onofre–Oceanside terraces are cut by stream channels younger than the terraces themselves, but even the channels reflect important geologic changes since formation. Furthermore, in a geologically active area like California, so far as terrace origin is concerned it is not yet possible to distinguish convincingly between the effects of a fluctuating Pleistocene sea level and the effects of

Figure 7-8.

Newport Bay in 1948. (Photo by Spence Air Photos, courtesy of Department of Geography, University of California, Los Angeles)

vertical tectonic movement. In constructing even the simplest Quaternary history of the San Onofre–Oceanside coast, this limitation must be kept in mind.

During relatively stable intervals in the Quaternary, marine erosion quickly cut terraces into the coastal margin of the Peninsular Ranges. The width of these terraces was a function of the lengths of the stable intervals, but it was also affected by such considerations as degree of exposure to waves, slope of the sea floor, and rock resistance. Generally, the longer the stable interval the wider the resulting terrace—although there are many exceptions. During rapid *relative* uplift of the land, terrace cutting gave way to emergence of the sea bottom. During the stable period that followed, the exposed sea bottom was leveled into a new terrace with a cliff at its landward side. Once the cliffs were elevated beyond the reach of the sea, land erosion and mass movement processes began to attack them.

During the final Pleistocene glacial stage, sea level was lowered about 300 feet (90 m). This forced streams crossing the terraces to cut deep gullies and to extend themselves across the newly exposed sea floor. This modification doubtless kept pace with sea-level changes, for streams adjust quickly to new circumstances.

As the postglacial sea level rose, stream courses were shortened and the lower ends of the channels were flooded by the rising ocean, forming estuaries. Streams seeking the sea and tidal action in lagoons oppose the final closing and filling of estuaries, however. Quiet estuaries still are being filled by river sediment while coastal currents straighten the shoreline by building bay mouth bars across estuary entrances. The effects of these competing processes are evident along the San Diego County coast. Some former lagoons are now filled completely; others are nearly filled but persist as salt marshes; and still others retain open-water lagoons of varying sizes. Most have bay mouth bars separating them from the open ocean. During the past 50 years or so, practically all these features have been subjected to some degree of human management. In few cases has nature been allowed to proceed unmolested.

Coronado Island and Silver (Coronado) Strand

San Diego Bay is the only sizable natural harbor on the California coast south of San Francisco. It owes its existence to Point Loma, a mountainous ridge of marine Cretaceous sedimentary rock rising almost 400 feet (120 m) above sea level and forming a prominent barrier west of the city of San Diego and the northern part of the bay (Figure 7-9). Because Point Loma shelters much of the bay from westerly and northwesterly winds and waves, a long, low, curving sand spit has gradually built northward from the mouth of the Tia Juana River. At the north end of the spit are two low islands that are

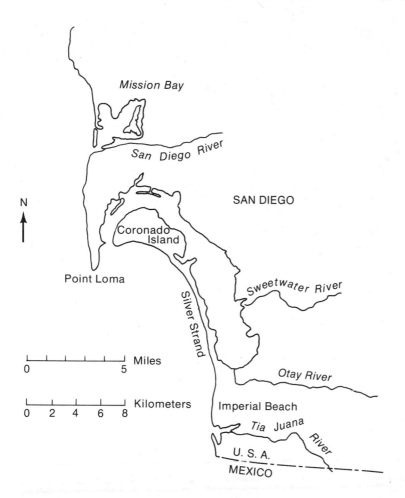

*Figure 7-9.
Sketch of Coronado
Island–Silver
Strand.*

probably remnants of the Nestor terrace, which is exposed around the city of San Diego at an elevation of about 25 feet (7.5 m). These two islands, once known as North Island and South Island, have been joined by the sand spit, which also connects them to the mainland on the south. North Island is the site of a naval air station, and South Island is the site of the city of Coronado. The inlet that once separated the two islands was called Spanish Bight, but it has since been filled.

Without human interference in the area, San Diego Bay would have filled with sediment delivered by the Otay, Sweetwater, and San Diego rivers. (The latter has been diverted into a channel along the south side of Mission Bay.) In addition, it is likely that the northward drift of beach sand that connected Coronado (South) Island with the mainland and Coronado and North islands together eventually would have blocked or nearly blocked the harbor entrance. Breakwaters, channel maintenance, and tidal action prevent this happening.

REFERENCES

General

Beal, C. H., 1948. Reconnaissance of the Geology and Oil Possibilities of Baja California, Mexico. Geol. Soc. Amer. Mem. 31, p. 138.

Hertlein, L. G., and U.S. Grant, 1954. Geology of the Oceanside–San Diego Coastal Area, Southern California. *In* Geology of Southern California. Calif. Div. Mines and Geology Bull. 170, pp. 53–64.

Jahns, R. H., 1954. Geology of the Peninsular Range Province, Southern California and Baja California. *In* Geology of Southern California. Calif. Div. Mines and Geology Bull. 170, pp. 29–52.

Larsen, E. W., and others, 1951. Crystalline Rocks of Southwestern California. Calif. Div. Mines and Geology Bull. 159.

Peterson, G. L., and others, 1970. Geology of the Peninsular Ranges. Mineral Information Service (now California Geology), v. 23, pp. 124–127.

Sharp, R. V., 1967. San Jacinto Fault Zone in the Peninsular Ranges of Southern California. Geol. Soc. Amer. Bull., v. 78, pp. 707–730.

Woodford, A. O., and others, 1954. Geology of the Los Angeles Basin. *In* Geology of Southern California. Calif. Div. Mines and Geology Bull. 170, pp. 65–82.

Special

Burnham, C. W., 1954. Contact Metamorphism at Crestmore, California. *In* Geology of Southern California. Calif. Div. Mines and Geology Bull. 170, pp. 61–70.

Clark, W. B., 1970. Gold Districts of California. Calif. Div. Mines and Geology Bull. 193.

Creasey, S. C., 1946. Geology and Nickel Mineralization of the Julian-Cuyamaca Area, San Diego County, California. Calif. Jour. Mines and Geology, v. 42, pp. 15–29.

Devito, F., and others, 1971. Contact Metamorphic Minerals at Crestmore Quarry, Riverside, California. *In* Geological Excursions in Southern California. Univ. Calif. Riverside Campus Museum Cont. no. 1, pp. 94–125.

Donnelly, M. G., 1935. Geology and Mineral Deposits of the Julian District, San Diego County, California. Calif. Jour. Mines and Geology, v. 30, pp. 331–370.

Engel, Rene, 1959. Geology of the Lake Elsinore Quadrangle, Calif. Div. Mines and Geology Bull. 146.

Jahns, R. H., 1954. Pegmatites of Southern California. *In* Geology of Southern California. Calif. Div. Mines and Geology Bull. 170, pp. 37–50.

———— and L. A. Wright, 1951. Gem- and Lithium-Bearing Pegmatites of the Pala District, San Diego County, California. Calif. Div. Mines and Geology Spec. Rept. 7A.

Kennedy, Michael P., 1973. Sea-Cliff Erosion at Sunset Cliffs, San Diego. California Geology, v. 26, pp. 27–31.

Murdoch, Joseph, 1961. Crestmore—Past and Present. Amer. Miner., v. 46, pp. 245–257.

Sutherland, J. C., 1935. Geological Investigation of the Clays of Riverside and Orange Counties, Southern California. Calif. Jour. Mines and Geology, v. 31, pp. 51–87.

Chapter Eight

Transverse Ranges

*Let us leave a few problems for our children
to solve; otherwise they might be so bored.*
                                    Tom F. W. Barth

Unlike California's other geologic provinces, the Transverse Ranges form a conspicuously east-west trending unit. A few short ranges with east-west trend do occur in other provinces, but they are all exceptions to the state's general structural alignment. It is worth noting that in North America as a whole, apart from Alaska, there are few prominent east-west mountain belts. Most of the continent is dominated by structures that trend north-south.

The Transverse Ranges include California's highest peaks south of the central Sierra Nevada, the only Precambrian rocks in the coastal mountains of the United States and probably North America, and four of the eight islands off the southern California coast. The province extends about 325 miles (520 km) from Point Arguello and San Miguel Island on the west into Joshua Tree National Monument on the east, where it merges with the Mojave and Colorado deserts. Along the Ventura–Los Angeles county line, the province reaches a maximum width of nearly 60 miles (96 km); it narrows to about 40 miles (64 km) at its western end. Figure 8-1 gives the locations of important Transverse Range features.

From northwestern Ventura County east to Cajon Pass, the San Andreas fault system forms the northern boundary of the province. The fault deviates markedly from its usual north-south trend, suggesting that it too was influenced by the forces that produced this oddly aligned province.

The province subdivides into individual ranges with intervening valleys. Several of these units are topographically distinctive and easily distinguished from one another. On the other hand, there are some complex groups of ridges and valleys *not* readily distinguishable. The ranges of the province are separated by alluviated, broadly synclinal valleys, narrow stream canyons, prominent faults, and sometimes by downwarps of phenomenal magnitude. Although the geologic column for this province is incomplete, it does include rocks of all ages except possibly some parts of the early Paleozoic.

Despite their comparatively small area, the Transverse Ranges seem to incorporate a greater spectrum of rock types and structure than any other province in the state.

SANTA YNEZ MOUNTAINS

The Santa Ynez Mountains form a continuous, south-facing rampart along the Santa Barbara coast from Point Arguello east for nearly 70 miles (110 km) to near Ojai, where they merge with the mountains of northern Ventura County. For much of this distance the range crowds the shoreline, leaving scant room for the coastal plain. Figure 8-2 is a view of the western Transverse Ranges.

From Point Arguello to Gaviota Pass, the range is generally less than 2000 feet (600 m) high, although Tranquillon Mountain and several peaks near Point Arguello are slightly higher. The mountains rapidly gain height east of Gaviota Pass, reaching 4298 feet (1311 m) at Santa Ynez Peak. San Marcos Pass, near Santa Barbara, occupies a low saddle formed by a synclinal fold that obliquely crosses the axis of the range. East of San Marcos Pass, the mountains rise again, reaching maximum elevation at Divide Peak (4690 ft or 1430 m) close to the Santa Barbara—Ventura county line.

The Santa Ynez Mountains are an anticline, with a major fault along their axis. The south limb is so much more prominent than the north, however, that many geologists prefer to describe the range as a south-dipping homocline with beds strongly overturned to the north.

Rocks and Geologic History

Oldest rocks exposed in the Santa Ynez Mountains are parts of the Franciscan formation. Here, as in the Coast Ranges to the north, the Franciscan consists of a highly mixed assemblage of deep-water marine sedimentary rocks such as radiolarian chert and graywacke, plus altered basalt, serpentinite, and ultrabasic crystalline rock thought to be derived from oceanic crust. These Franciscan rocks probably make up the basement in this part of the Transverse Ranges and are assigned a late Jurassic to late Cretaceous age.

The processes that formed the Franciscan assemblage evidently terminated during the late Cretaceous, although the region remained beneath the sea and sandstones and shales were deposited. Prevailing conditions of the early Tertiary are unknown, because no Paleocene rocks are exposed in the Santa Ynez Mountains and none has turned up in well records.

Middle and upper Eocene rocks are all marine and are well represented in the range, attaining thickness of more than 10,000 feet (3000 m). Most of the modern range has been carved from these

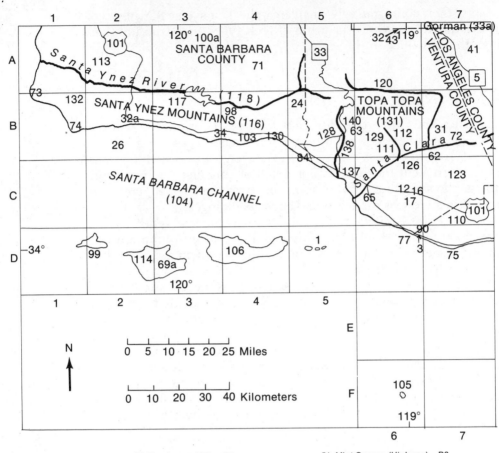

1 Anacapa Island (The Anacapas)—D5
2 Arcadia—C10
3 Arroyo Sequit—D6
4 Baldwin Hills—D8
5 Ballona Creek—D8
6 Beverly Hills—D8
7 Big Bear Lake—C13
8 Big Tujunga Canyon—C9
9 Blackhawk Canyon—B13
10 Cabazon—D14
11 Cajon Pass—B–C, 12
12 Camarillo—C6
13 Castaic—B8
14 Catalina Island—F8
   Channel Islands: see
   Anacapa (1)—D5
   Santa Cruz (106)—D4
   Santa Rosa (114)—D2–3
   San Miguel (99)—D2
15 Coachella Valley—E15
16 Conejo Grade—C6
17 Conejo Oil Field—C6
18 Corona—D11
19 Coyote Hills—D10
20 Cucamonga Peak—C11
21 Cushenbury Grade—C14
22 Desert Hot Springs—D15
23 Devils Punchbowl—B10
24 Divide Peak—B5

25 Dominguez Hills—E9
26 Dos Cuadras Oil field—B2
27 Downey—D9
28 Duarte—C10
29 Eagle Rock—C9
30 Elysian Hills (Park)—D9
31 Fillmore—B7
31a Fontana—C–12
32 Frazier Mountain—A6
32a Gaviota Pass—B2
33 Glendale—C9
33a Gorman—A7
34 Goleta—B3
35 Grayback (Mt. San Gorgonio)—C14
36 Huntington Beach (Mesa)—E10
37 Joshua Tree National Monument—D15
38 La Brea (Tar Pits)—C8
39 La Habra Valley—D10
40 Lake Arrowhead—C12
41 Liebre Mountain—A7
42 Little San Bernardino Mountains—D15
43 Lockwood Valley—A6
44 Long Beach—E9
45 Lopez Canyon—C8
46 Los Angeles Harbor—E9
47 Los Angeles River—C–E, 9
48 Lucerne Valley—B13
49 Lytle Creek—C11
50 Mill Creek—C–D, 12–13

51 Mint Canyon (Highway)—B8
52 Mojave River—B12
53 Monrovia—C10
54 Montebello—D9
55 Montrose—C9
56 Morongo Valley—D14–15
57 Mt. San Antonio (Old Baldy)—C11
   Mt. San Gorgonio (Grayback) (see 35)—C14
58 Mt. Pacifico—B9–10
59 Mt. Wilson—C9
60 Newhall—B8
61 Newport Bay (Mesa)—E10
62 Oak Ridge—B7
63 Ojai (Valley)—B5
63a Old Woman Springs—B13
64 Orange County—E–F, 10–11
65 Oxnard—C6
66 Pacific Palisades—D8
67 Palmdale—B9
68 Palos Verdes (Peninsula) (Hills)—E8–9
69 Pasadena—C9
69a Picacho Diablo—D3
70 Pico Canyon Oil Field—B8
71 Pine Mountain (Big Pine Mountain)—A4
72 Piru—B7
72a Playa del Rey—D8
73 Point Arguello—A–B, 1

Figure 8-1. Place names: Transverse Ranges.

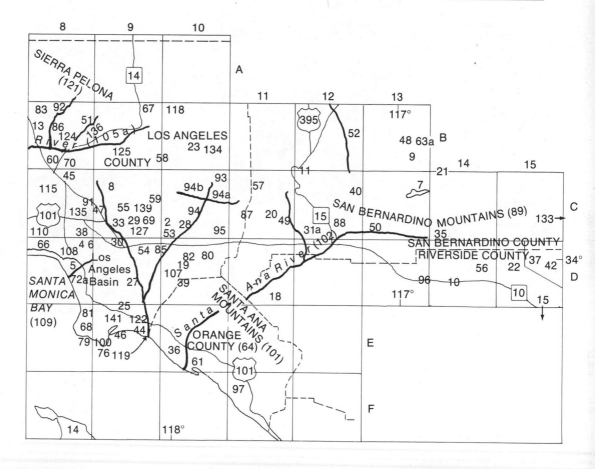

Figure 8-2. *View to the southeast from the western Santa Ynez Mountains. The steeply dipping beds of the south flank of the range form strike ridges (durable beds) and intervening strike valleys (weaker beds) that parallel the coastline. The narrow coastal plain widens to the east and borders the Santa Barbara Channel. In spite of protection afforded by the offshore islands (background), the channel waters are usually rough because the channel is open to the west and receives strong prevailing and storm winds. (Photo by Mark Hurd Aerial Surveys)*

rocks, and the more resistant sandstones form nearly all the higher peaks. Toward the top of the Eocene section, where the Coldwater sandstone occurs, nearshore conditions are revealed by oyster beds and occasional streaks of red shale, presumably washed into the shallow marine waters from nearby land.

By Oligocene time, the sea had withdrawn from the eastern region, leaving a broad coastal plain on which were deposited the sands, gravels, and silts of the Sespe formation. Although green, tan, and buff beds are also evident, its prominent red color makes the Sespe especially distinctive. It can be traced from west of San Marcos Pass into central Ventura County. A few vertebrate remains such as horse teeth have been found, providing dates for the formation.

In the west, marine conditions continued during the Oligocene, and initially sands resembling the Coldwater formation were deposited in the shallow seas. Even though the shoreline changed position, as shown by interfingering of the land-deposited Sespe on the east and the marine beds to the west, the western Santa Ynez Mountains record continuous marine conditions through the Oligocene.

Before the Oligocene closed, the sea again spread over the Santa Ynez area. The first marine unit from this invasion is the Vaqueros formation. This sequence contains much broken shell material and some well-preserved, heavy-shelled scallops (pectens), which endured harsh and abrasive conditions on a gravelly or pebbly shore.

As Miocene time began, the sea advanced north or northeast over the Santa Barbara–Ventura area, bringing deeper-water conditions to the site of the Santa Ynez Mountains. The sands and gravels of the Vaqueros were replaced by the fine silts and soft clays of the Rincon formation. Exposures of the Rincon frequently weathered to expansive and unstable clay soils that subsequently have plagued many structures built on the formation.

By middle Miocene, somewhat unusual conditions developed over much of what is now coastal California between San Francisco Bay and Orange County. These conditions involved development of probably elongate marine basins, some perhaps over a mile (1.6 km) deep, that were similar to the deep basins of the Gulf of California. Evidently the shoreline was sufficiently distant from the Santa Ynez and Santa Monica mountains of today that much of the material accumulating on the sea bottom was of organic rather than detrital origin. Occasionally there were influxes of terrestrial matter, including thin layers of volcanic ash, but most of the material was ooze derived from the tiny diatoms living at the sea's surface. Their remains were mixed with fine clays transported from the shore, eventually making the upper part of the Monterey formation. In addition, the upper beds of the Monterey are sometimes nearly saturated with bituminous material that permeates the rock and seeps from every fracture, crack, and bedding plane. Fresh samples have a strong and unmistakable petroliferous odor.

During earlier Monterey deposition, widespread volcanic activity broke out in California. In the extreme western Santa Ynez Mountains, rhyolitic and basaltic eruptions produced the volcanic rocks near Tranquillon Mountain and in the Santa Rosa Hills to the north. Apart from these exposures, volcanism in the Santa Ynez Mountains is only evident from the thin ash beds preserved in the Monterey and overlying Sisquoc formations. Elsewhere in the Transverse Ranges, particularly in the western Santa Monica Mountains, Miocene volcanic activity was extremely important, and lava flows are thick and prominent.

The depositional environment evidently changed little as Sisquoc formation deposits succeeded those of the Monterey. The Pliocene may have begun before the last Sisquoc beds were laid down, but placement of the Mio-Pliocene boundary is highly disputed. The Pliocene is preeminently a time of change in the whole paleogeography of southern California.

By Pleistocene time, the Coast Range or Pasadenan orogeny had elevated many previously marine areas, in particular most, if not all, of the Santa Ynez and Santa Monica mountains and the offshore islands. During this orogeny, the Santa Ynez Mountains started to rise, perhaps as an anticlinal uplift or as a tilted block along the Santa Ynez fault to the north. In addition to the vertical uplift, there is considerable evidence for left slip on the fault. Uplift of the Santa Ynez Mountains produced little coastal plain initially. Marine conditions prevailed to the base of the mountains through the Pliocene and into the early Pleistocene, when the richly fossiliferous Santa Barbara formation was laid down in shallow nearshore waters. Not until middle Pleistocene time, near the climax of the Pasadenan orogeny, did today's coastal plain fully emerge.

Evidence is compelling that the orogeny, or at least its associated faulting, continues today, because earthquakes are common in this part of the state and near Santa Barbara several Oligocene Sespe beds are in fault contact with late Pleistocene gravels. Moreover, recent surveys across some faults in the city of Santa Barbara imply that continuing creep is occurring.

MOUNTAINS OF CENTRAL VENTURA COUNTY

In Santa Barbara County, the Transverse Ranges can be distinguished fairly easily from the southern Coast Ranges along the Santa Ynez fault zone. South of the fault, the Santa Ynez Mountains, the Santa Barbara Channel, and the Channel Islands all trend clearly east-west, whereas to the north the fold axes, faults, and topographic grain of the San Rafael Mountains show the distinctive northwest trend characteristic of the Coast Ranges as a whole. In Ventura County, these distinctions are less clear. Here the rocks and structures of the southern Coast Ranges have changed their trend to merge with the east-west alignment of the Transverse Ranges.

The mountainous part of Ventura County, north of the Ojai and Santa Clara River valleys, is dominated by four prominent east-west faults (Figure 8-3). Two of these, the Big Pine and the Santa Ynez, are considered primarily faults of left-lateral separation along which there has been appreciable vertical slip. Some geologists think the Pine Mountain fault is a thrust or reverse fault, but others consider it the same type as the Big Pine and Santa Ynez. Nearly everyone, however, agrees that the southernmost of these faults, the San Cayetano, is a north-dipping thrust, at least where it faces the Santa Clara River valley.

Rocks and Geologic History

The Topa Topa, Pine, and Santa Ynez Mountains are carved chiefly from Eocene rocks, although several wedges of late Cretaceous rocks are exposed along the Santa Ynez fault in the Topa Topa Mountains. Along the prominent east-west valley occupied by upper Sespe Creek, an elongate block of rocks has been folded and down-faulted along the Pine Mountain fault so that Oligocene and Miocene rocks are preserved in a topographic depression. Other younger rocks, including extensive exposures of the Sespe formation, occur along the southern Topa Topa Mountains and around Ojai Valley. North of Ojai Valley, rocks and structural features resemble those in the Santa Ynez Mountains. For example, the overturned or vertical beds first encountered near Santa Barbara continue as prominent features beyond Ojai Valley and are well exhibited in the Matilija overturn (Figure 8-4). This steeply dipping overturn is clearly displayed in cross section along Wheeler Gorge.

Pliocene marine rocks appear infrequently in the Santa Ynez Mountains, but they constitute an important part of the coastal hills from Rincon Point east along the south side of Sulphur Mountain. The same rocks occur sporadically on either side of the Santa Clara River valley as far east as the San Fernando Valley. Below the valley floor between Santa Paula and Fillmore, the Pliocene beds are 12,000 to 14,000 feet (3650–4250 m) thick (Figure 8-5). These Pliocene beds are famous geologically for their small-scale structural features, which are thought to indicate deep-water turbidity current deposition. Included are load casts, graded beds, slump structures, rip-ups, current features, and convoluted bedding. All are well displayed in a classic locality along Santa Paula Creek between Santa Paula and Ojai.

Marine deposition continued into Pleistocene time, with some beds nearly 4000 feet (1200 m) thick. These young rocks are exposed in the lower foothills of the Topa Topa and Sulphur mountains and in the hills behind Ventura. Some are no older than middle Pleistocene, yet they have been tilted as much as 70° from the horizontal, showing that the Pasadenan orogeny did not begin here until middle or late Pleistocene.

Evidence of recent orogenic activity is also provided by the appreciably warped and uplifted late Pleistocene marine terraces. The highest of these is more than 1000 feet (300 m) above sea level. Studies by W. C. Putnam showed that some are arched so that they slope down to the north and away from the ocean, a direction opposite to the one existing when they were cut. In addition, stream terraces along the Ventura River prove that the present canyon of the river is antecedent, one of the few undisputed instances in the United States. Both the marine and stream terraces are interrupted by recent fault scarps.

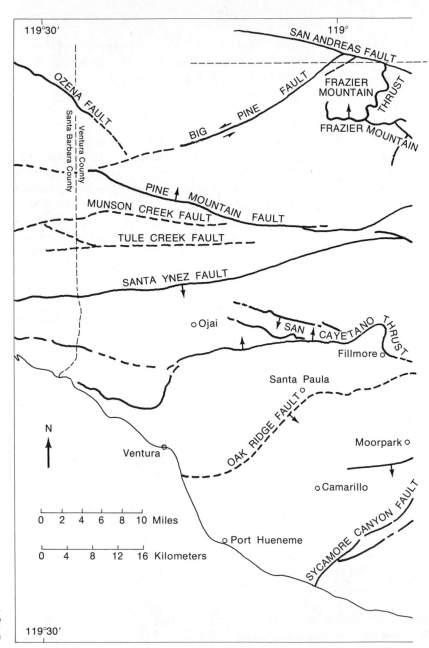

Figure 8-3.
Some known and inferred faults of the central Transverse Ranges. (Source: California Division of Mines and Geology)

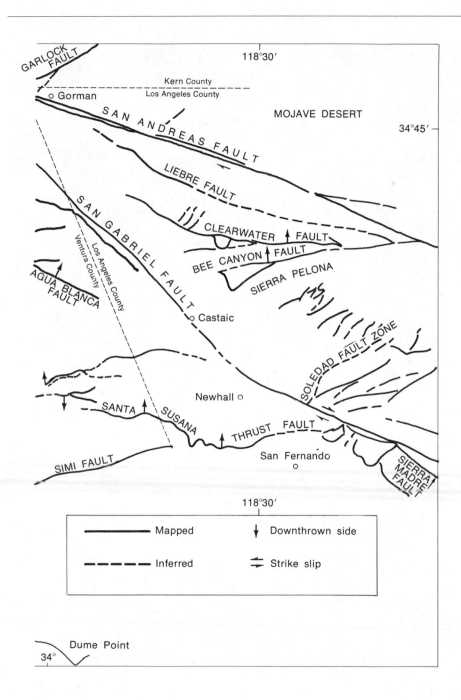

Mapped

Inferred

Downthrown side

Strike slip

Figure 8-4. Looking east across the Matilija overturn north of Ojai. (Photo by Spence Air Photos, courtesy of Department of Geography, University of California, Los Angeles)

SANTA MONICA MOUNTAINS AND THE CHANNEL ISLANDS

The Santa Monica Mountains and their offshore extensions are the most geologically varied part of the entire Transverse Range province. The area involved extends about 120 miles (192 km) west from Elysian Park in Los Angeles to San Miguel Island. The mountains are not so high as the Santa Ynez and the ranges of central Ventura County. They reach only 3111 feet (949 m) at Sandstone Peak on the mainland and 2450 feet (747 m) at Picacho Diablo on Santa Cruz Island.

Rocks and Geologic History

The Santa Monica Mountains have a granitic and metamorphic basement more akin to that of the Sierra Nevada than to the Franciscan basement of the Santa Ynez Mountains and the Coast

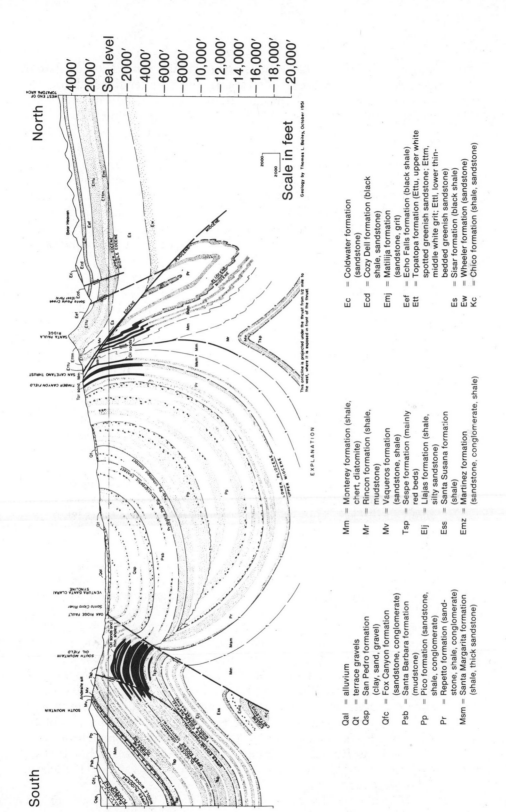

Figure 8-5.

Structure section across part of Ventura Basin, showing extraordinary thicknesses of the strata, particularly the Pliocene. Rock types and structures are typical of much of the western Transverse Range sedimentary pattern. (Source: California Division of Mines and Geology)

EXPLANATION

Qal = alluvium
Qt = terrace gravels
Qsp = San Pedro formation (clay, sand, gravel)
Qfc = Fox Canyon formation (sandstone, conglomerate)
Psb = Santa Barbara formation (mudstone)
Pp = Pico formation (sandstone, shale, conglomerate)
Pr = Repetto formation (sandstone, shale, conglomerate)
Msm = Santa Margarita formation (shale, thick sandstone)

Mm = Monterey formation (shale, chert, diatomite)
Mr = Rincon formation (shale, mudstone)
Mv = Vaqueros formation (sandstone, shale)
Tsp = Sespe formation (mainly red beds)
Elj = Llajas formation (shale, silty sandstone)
Ess = Santa Susana formation (shale)
Emz = Martinez formation (sandstone, conglomerate, shale)

Ec = Coldwater formation (sandstone)
Ecd = Cozy Dell formation (black shale, sandstone)
Emj = Matilija formation (sandstone, grit)
Eef = Echo Falls formation (black shale)
Ett = Topatopa formation (Ettu, upper white spotted greenish sandstone; Ettm, middle white grit; Ettl, lower thin-bedded greenish sandstone)
Es = Sisar formation (black shale)
Ew = Wheeler formation (sandstone)
Kc = Chico formation (shale, sandstone)

Geology by Thomas L. Bailey, October 1951

Scale in feet

2000
2000

Ranges. One of the oldest rocks present is the Santa Monica slate of the central Santa Monica Mountains. With its regular bedding, the slate is thought to be of marine origin. It has yielded Jurassic fossils and resembles rocks in the Santa Ana Mountains that also contain Jurassic marine fossils. Perhaps correlative with the Santa Monica slate is the unfossiliferous Santa Cruz Island schist, which forms an elongate ridge south of the Santa Cruz Island fault. Unlike the Santa Monica slate, this rock appears to have been derived chiefly from volcanic rocks and to a lesser extent from sandstones and siltstones. Both formations have been intruded by granitic rocks of similar age and origin. These granitic rocks are probably from the same Mesozoic event that produced the widespread granitic rocks of the Peninsular Ranges and the Sierra Nevada.

Upper Cretaceous rocks are exposed in the eastern Santa Monica Mountains and on San Miguel Island. About 6500 feet (1980 m) of marine sandstone and shale are found on San Miguel and about 3000 feet (900 m) of primarily marine conglomerate in the eastern Santa Monica Mountains. The lower unit of this conglomerate sequence is a reddish land-laid deposit, which shows that the late Cretaceous shoreline lay in this area. Cretaceous rocks have been encountered in wells on Santa Cruz Island and presumably are concealed beneath younger rocks on Santa Rosa Island.

Paleocene rocks are entirely marine. Their exposures in the Santa Monica chain are limited to shales and limestones in the eastern Santa Monica Mountains and sandstones on Santa Cruz and San Miguel islands.

Dominant in the Santa Ynez Mountains, Eocene deposits are widely scattered in the Santa Monica Mountains, but not extensively or thickly. They occur on southwestern Santa Cruz Island and sporadically on Santa Rosa Island and reach a thickness of nearly 2000 feet (600 m) on San Miguel Island.

A distinctive unit of the mainland Transverse Ranges is the previously discussed nonmarine Sespe formation, a late Eocene through early Miocene flood-plain deposit distinctive for its prevailing red color and lenticularity. The Sespe is widespread, and its occurrence indicates either that much of the western Transverse Range province was elevated or that a shallow marine area was simply blanketed by sediments until it became a low coastal plain. In either case, a short-lived land mass was formed, interrupting the marine conditions that had prevailed since the Cretaceous. Before the Oligocene ended, the sea again invaded land, covering most of the site of the western Transverse Ranges by the opening of the Miocene. Sespe rocks occur in the central Santa Monica Mountains and on Santa Rosa Island. They are reported also from a well at the eastern end of Santa Cruz Island, where they include Franciscan schist fragments.

Miocene rocks are the most extensively exposed of any Cenozoic rocks in the Santa Monica chain. They are abundant on

Anacapa, Santa Cruz, Santa Rosa, and San Miguel islands. During the Miocene, the Santa Monica–Channel Islands area was the site of a deep marine trough into which 15,000 feet (4550 m) of sedimentary rocks were deposited. These vary from the coarse San Onofre breccia with its large angular blocks of Franciscan schist to the deep-water diatomaceous, dolomitic, and cherty shales of the Monterey formation. The basin seems to have deepened during the Miocene: earlier units are often coarse, shallow-water marine conglomerates and breccias, and later deposits are fine-grained, deeper-water shales.

These Miocene beds are particularly interesting because many were organically rich muds when deposited and consequently are the most probable source for much of the petroleum and natural gas found in the Transverse Range province. As in the Santa Ynez Mountains, the Monterey formation of the Santa Monica Mountains is rich in marine diatom remains and often contains pockets of bituminous material. Where it is buried and fractured, the Monterey sometimes acts as reservoir rock for gas and oil.

During Miocene time, extensive volcanic activity broke out in this part of the Transverse Range province. Andesitic, diabasic, and basaltic flows, sills, and dikes plus a few silicic rocks were extruded. Many are of submarine origin, as demonstrated by the common occurrence of pillow structure in lava flows. The volcanic rocks total nearly 10,000 feet (3000 m) in thickness in the western Santa Monica Mountains, 8000 feet (2440 m) on Santa Cruz Island, and 2400 feet (730 m) on San Miguel Island. Lesser amounts occur on Anacapa and Santa Rosa islands.

A previously mentioned middle Miocene rock that occurs in the Santa Monica chain is the San Onofre breccia. This unusually interesting rock consists of a coarse-grained breccia and conglomerate with prominent clasts of blue glaucophane schist, green schist, gabbro, and limestone. It is extremely poorly sorted and, although marine, apparently was poured into its depositional site from nearby highlands with little opportunity for rounding or sorting of the blocks. The rock is so unusual and distinctive that its occurrences have provided a key for establishing Miocene paleogeography of southern California. It appears that present exposures on Santa Rosa, Santa Cruz, and Anacapa islands at Points Mugu and Dume were removed from their original source areas in the Los Angeles Basin by post-Miocene, left-lateral movement along the Malibu Coast and presumably related Santa Cruz Island fault systems.

Post-Miocene rocks in the Santa Monica chain are mainly Pleistocene and younger marine and nonmarine terrace materials. Locally thick marine terrace deposits appear on San Miguel and Santa Cruz islands, and thick-bedded, coarse, nonmarine gravels probably deposited as broad alluvial fans occur along the Malibu coast. These make up the high, prominent cliffs that have generated destructive landslides at Pacific Palisades and Santa Monica.

VENTURA AND SOLEDAD BASINS

A large syncline forms the structure that is known as the Ventura Basin in the west and the Soledad Basin in the east. This structure is about 120 miles (192 km) long and includes the Santa Barbara Channel between the Channel Islands and the Santa Ynez Mountains (Figure 8-2) and the inland area between the Santa Monica and Topa Topa mountains. The primarily marine Ventura Basin joins the nonmarine Soledad Basin near the San Gabriel fault. The Soledad Basin extends almost 30 miles (48 km) farther east, north of the San Gabriel Mountains toward the San Andreas fault.

Within the Ventura Basin are some prominent anticlinal hills, some higher than the Santa Monica Mountains to the south. The highest, the Santa Susana Mountains, enclose the west and northwest San Fernando Valley. Other ridges are Sulphur Mountain, between the Ojai and Santa Clara River valleys, and the South Mountain–Oak Ridge complex, which joins the Santa Susana Mountains to the east.

The Ventura Basin is famous for its remarkably thick section of mostly marine sedimentary rocks, which total more than 58,000 feet (17,700 m). The lower 8000 feet (2440 m) are Cretaceous, but the remaining 50,000 feet (15,000 m) of Cenozoic include perhaps the thickest accumulation of Pliocene deposits in the world. The section is as follows.

Pleistocene	6000–7000 ft (1800–2150 m)
Pliocene	12,000–14,000 ft (3650–4250 m)
Miocene	8000–10,000 ft (2440–3000 m)
Oligocene (nonmarine Sespe)	2500 ft (760 m)
Eocene and Paleocene	17,000 ft (5200 m)
Cretaceous	8000 ft (2400 m)

The Ventura Basin began to subside in the Cretaceous, and by the end of the Eocene was nearly filled. During Oligocene time, some 2500 feet (760 m) of land-laid deposits accumulated, but early in the Miocene rapid subsidence resumed and produced a marine embayment in which water may have been 5000 feet (1500 m) deep. Some of the hills south of the Santa Clara River valley were folded and uplifted, but later were submerged beneath a Pliocene sea. Apart from the Santa Barbara Channel, the entire area was subjected to strong uplift, folding, and faulting during the middle Pleistocene (Pasadenan) orogeny. This produced today's topography and created the structures in which the region's prolific oil fields have developed. The Santa Barbara Channel was also folded and faulted, but apparently was not uplifted above sea level. The seaward end of the Rincon anticline is planed off, however, presumably by wave erosion.

Most of the valleys in the Ventura Basin are synclines. For instance, the Santa Clara River valley is structurally a large, thick, synclinal fold. The San Fernando Valley is topographically more striking, but is a broader, flatter syncline incorporating fewer Cenozoic sediments. The other valleys are smaller counterparts of these larger features.

The portion of the Ventura Basin east of the San Gabriel fault is usually called the Soledad Basin. It is dominated by rough, hilly country drained mainly by the Santa Clara River and represents the landward margin of the marine Ventura Basin. The Soledad Basin contains mostly middle and late Cenozoic nonmarine sedimentary rocks that rest on the crystalline basement of the San Gabriel Mountains to the south and the Sierra Pelona to the north. This nonmarine extension of the Ventura Basin has been studied extensively by Richard H. Jahns.

During Oligocene time, the Soledad Basin was sometimes cut off from direct access to the sea and developed closed basins in which saline lakes briefly formed. Coarse sands and gravels eventually filled the basin, and drainage to the sea presumably was reestablished since later deposits show no evidence of deposition in closed interior basins. The Soledad Basin deposits include the coarse clastics that make up the prominent hogback ridges at Vasquez Rocks and the lake-sediment borax deposits near Tick Canyon.

When the borax lakes were developing, volcanics were being extruded into the Soledad Basin, forming dikes, sills, flows, and breccias. By Miocene time, the basin had filled sufficiently so that lakes were no longer significant in the landscape. During Miocene subsidence, the sea reached into the western Soledad Basin and persisted there into the early Pliocene. By the end of the Pliocene, the sea had withdrawn, and the nonmarine Saugus beds mark the onset of land conditions that have prevailed since.

Although movement on some of the area's major faults may have started early in the Cenozoic or even before, most faulting and folding is Miocene or later. In fact, many folds and faults are related to the onset of the most recent orogenic episode, near the end of the Pliocene. Strong evidence of continuing tectonic activity is the 3-foot (1 m) vertical movement on the faults in northeast San Fernando Valley during the 1971 earthquake (Figure 8-6).

SIERRA PELONA

The Sierra Pelona forms a block terminated on the west by San Francisquito Canyon and on the north by the San Francisquito and San Andreas faults. The southern boundary lies more or less along the old Mint Canyon (Sierra) highway. Adjacent Liebre Mountain to the northwest is similar in rocks and structure.

The Sierra Pelona is composed almost entirely of Pelona schist. This formation is still undated, but is regarded by many as Precambrian. As indicated previously, however, a strong case can be made

Figure 8-6.   *Fault scarp in Lopez Canyon, showing 3-foot displacement from the 1971 San Fernando earthquake. (Photo by V. A. Frizzell, courtesy of U.S. Geological Survey)*

for Cretaceous age. This is based on studies of the Pelona and its close correlative, the Orocopia schist of the Colorado Desert. In a pattern resembling that in the Orocopia Mountains, in the San Gabriel Mountains the Vincent thrust fault has moved Precambrian and Cretaceous crystalline rocks over Pelona schist, causing some metamorphism of the schist. In turn, the thrust and the schist are cut by late Cretaceous plutonic rocks.

RIDGE BASIN

This narrow basin lies between the San Gabriel and San Andreas faults near their juncture at Frazier Mountain. It is a down-dropped wedge of sedimentary rocks about 33,000 feet (10,000 m) thick, of which Miocene and Pliocene beds account for 29,000 feet (8850 m). Only the lower 2000 feet (600 m) of Miocene are marine; the other beds are all nonmarine. Remarkably, the vertical thickness of this column appears to be much greater than the width of the basin in which it accumulated. A particularly striking unit is the Violin breccia, about 27,000 feet (8250 m) thick but less than a mile (1.6 km) across. This and other units in the Ridge Basin accumulated at the base of an active fault scarp along the San Gabriel fault zone. Deposition continued as displacement occurred on the fault, providing a complete record of the area's late Cenozoic structural history. This unusual sequence of rocks can be seen along Interstate 5 between Castaic and Gorman.

SAN GABRIEL MOUNTAINS

The San Gabriel Mountains are a high, rugged block located between the Los Angeles Basin and the Mojave Desert. They form a continuous feature some 60 miles (96 km) long and up to 24 miles (39 km) wide, roughly along a north-south line that passes through Mounts Wilson and Pacifico. The Sierra Madre fault zone forms the range's southern boundary. The eastern boundary is the San Andreas fault zone, which crosses through Cajon Pass and separates the San Gabriel Mountains from the similar but higher San Bernardino Mountains (Figure 8-7). The San Gabriel Mountains face the Soledad Basin on the northwest and the San Fernando Valley on the west. Separated by a depressed fault block from the San Gabriel Mountains proper are the Verdugo Hills, which form the eastern boundary of the San Fernando Valley. The geology of the Verdugo Hills is essentially similar to that of the San Gabriel Mountains.

Figure 8-7. Cajon Pass (center foreground), between the San Gabriel and San Bernardino Mountains. Note the linear valley (center) carved along the San Andreas fault. Beheaded alluvial fans now being exhumed are shown in the upper right corner. (Photo by Fairchild Aerial Surveys, courtesy of Department of Geography, University of California, Los Angeles)

Because the San Gabriel Mountains have experienced considerable uplift in recent geologic time, the range has become a deeply dissected, rugged horst. Stream canyons are steep-sided and up to 3000 feet (900 m) deep. Many peaks exceed 7000 feet (2100 m) in elevation, the highest being Mount San Antonio (Old Baldy) at 10,080 feet (3074 m). The southern and western flanks are steep and bold where they face the lowlands of the Los Angeles Basin and the San Fernando Valley. The north face is less dramatic, although equally steep. The difference results from the greater elevation of the Mojave Desert, with its floor nearly 4000 feet (1200 m) high at the base of the range.

Faults

The strong influence of faulting is shown in the streams of the San Gabriel Mountains, which occupy deep and narrow gorges. To the north, the Santa Clara River has cut Soledad Canyon roughly along the trace of the Soledad fault. In the east, Lytle Creek maintains a straight course along the San Jacinto fault, a major branch of the San Andreas. Especially striking is the strong control on the East and West forks of the San Gabriel River, which follow the San Gabriel fault for nearly 25 miles (40 km).

Right slip on the San Andreas has produced prominent topographic features along the north and northeastern sides of the San Gabriel Mountains. Large alluvial fans built by streams flowing off the north slope have been displaced, giving a series of in-facing bluffs from Cajon Pass to west of Valyermo. Beheaded stream channels are frequently exposed on the tops of these displaced fans; sometimes the streams that adjusted to separation on the fault show prominent right-angle bends clearly displaying the direction of fault movement.

Prominent alluvial fans also occur on the range's south flank, forming a nearly continuous coalescing alluvial apron from Pasadena to Cajon Pass. Movement on the Sierra Madre fault zone has been primarily vertical, so the fans and their source streams are still closely associated. Topography has been somewhat obscured by intensive urban development, however.

Although most evidence suggests that late Cenozoic slip along the San Andreas and San Jacinto fault zones has been right lateral, only substantial vertical slip can account for the current elevation of the San Gabriel Mountains 2000 to 4000 feet (600–1200 m) above the Mojave Desert. As recently as the 1971 San Fernando earthquake, more than 3 feet (1 m) of elevation occurred along several high-angle reverse faults in Lopez Canyon, presumably increasing the difference in elevation between the San Gabriel Mountains and the San Fernando Valley. Similar abrupt slopes along the Soledad and Sierra Madre faults also show recent vertical uplift.

The San Andreas and San Jacinto fault systems merge between Palmdale and Valyermo, but toward the southeast they diverge until they are about 6 miles (9.6 km) apart. The right-slip San Gabriel fault disappears near Cucamonga Peak, where it probably merges with the San Jacinto fault zone. The Vincent thrust is well exposed north of Mount San Antonio, but dips south under the highest part of the range. The Vincent is much older than the faults already mentioned because it is cut by late Mesozoic granitic rocks.

Rocks

A feature of the San Gabriel Mountains that is critical to understanding the geologic history of North America is the presence of ancient crystalline rocks, particularly in the northwest of the range. Included are extensive exposures of anorthosite, which is known almost exclusively from Precambrian terrains (and from lunar sites). The only other California exposure is in the Orocopia Mountains. The San Gabriel anorthosites have been dated as 1022 million years old. Other ancient rocks are the Mendenhall gneiss (1045 million years old) and the augen gneisses (1700 million years old). These rocks are not so old as the ancient rocks of the Lake Superior region, but they are still considered early Precambrian.

Large exposures of older metamorphic rocks also exist in the northeast San Gabriel Mountains. These are generally assigned to the Pelona schist formation. As mentioned earlier, however, the Pelona's Precambrian age assignment has been questioned.

Mesozoic granitic rocks dominate the San Gabriel Mountains and constitute perhaps 70 percent of the exposed rocks. The few radiometric dates that exist for these rocks tend to be about 70 to 84 million years (late Cretaceous), but at least one is only 61 million years (Paleocene). Dates from the Mount Lowe and Parker Mountain granitic rocks yielded ages of 220 million years (Triassic) and 245 million years (Permian). Apparently the granitic rocks were intruded in at least two, and possibly four, episodes. They range from granites, through granodiorites and quartz monzonites, to syenites and diorites.

Cenozoic beds are located only along the range's margins. The oldest exposure is the anomalous marine Paleocene assigned to the Martinez formation, which occurs near the Devil's Punchbowl and near Cajon Pass. At the Punchbowl, the beds lie between the San Jacinto and San Andreas faults, here about 2 miles (3.2 km) apart. The Cajon Pass exposures form little patches northeast of the San Andreas fault. Paleocene marine beds do not reoccur northeast of the San Andreas fault until the San Joaquin Valley. Apart from a Miocene exposure at the western tip of Antelope Valley, the Martinez beds and some Oligocene Vaqueros near Cajon Pass are the youngest marine beds known north of the San Gabriel Mountains or in the Mojave Desert proper. Above the Martinez and Vaqueros beds are the massive conglomeratic sandstones exposed at Cajon Pass as the middle to late Miocene Cajon beds and at the Punchbowl as the early to middle Pliocene Punchbowl formation. The Punchbowl formation contains fossils of land vertebrates.

LOS ANGELES BASIN

The term Los Angeles Basin means different things to different people. To those concerned with air pollution, the term refers to the atmosphere of the lowland areas surrounding the Transverse and Peninsular ranges in the Los Angeles region, where temperature inversions and photochemical smog are recurrent phenomena. Some geologists limit the Los Angeles Basin to the coastal plain between the Santa Monica Mountains on the north, the Puente Hills and Whittier fault on the east, the Santa Ana Mountains and San Joaquin Hills on the south, and the Palos Verdes Peninsula and the shoreline on the west. The U.S. Geological Survey, on the other hand, takes a broader view, dividing the basin into four blocks that contain both uplifted portions and synclinal depressions (Figure 8-8).

Southwestern Block

The southwestern block is the seaward part of the Los Angeles Basin. It is bounded on the east by the Newport-Inglewood zone of deformation, which can be traced from Beverly Hills to Newport Bay where it strikes offshore. This structural trend, a combination of faults and folds, is expressed as a chain of low, en echelon anticlinal hills.

The distinguishing feature of the southwestern block is its basement. Although actually exposed only in the Palos Verdes Hills, it has been encountered in numerous oil wells at depths of 5000 to 14,000 feet (1500–4250 m) below sea level. These basement rocks belong to the Catalina schist facies of the Franciscan formation and are chiefly green chlorite and blue glaucophane schists. They have no known base, are always in fault contact with other basement rocks, and are of undetermined age. The oldest rocks in depositional contact with them are Miocene. Based on lithologic affinities with dated Franciscan in the Coast Ranges, a late Jurassic to late Cretaceous age is probable.

The main structural features of the southwestern block are the anticlinal Palos Verdes Hills that have been raised along a steep reverse fault, several anticlinal ridges in the basement rocks over which younger sediments have been draped, and intervening broad synclines. The anticlinal structures of the younger rocks have formed important traps for petroleum and natural gas. For example,

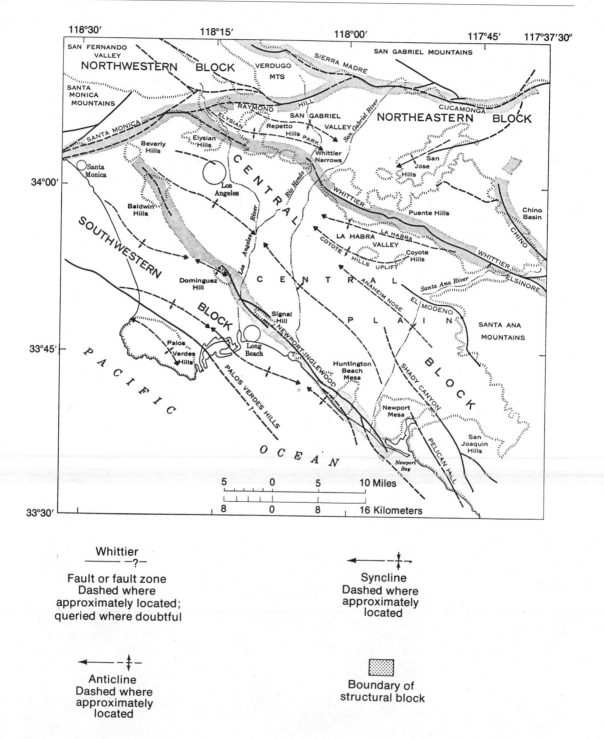

Figure 8-8.

Map of the Los Angeles Basin. (Source: U.S. Geological Survey)

the Wilmington field is the most productive field in California and the second most productive in the United States. The sedimentary blanket in this block is quite thick, up to 20,500 feet (6250 m). It is all post-Oligocene and almost entirely marine.

Displacements on the Newport-Inglewood faults have both vertical (to 4000 ft or 1200 m) and horizontal (at least 5000 ft or 1500 m) components. The upper surfaces of the basement rocks typically have about 4000 feet (1200 m) of separation across the fault zone, but overlying sediments show less vertical separation the younger they are. Movement seems to be Miocene at least and is still progressing, as indicated by arching of late Pleistocene and younger strata and by recent seismicity—the Newport-Inglewood fault zone caused the 1933 Long Beach earthquake. The zone displays mainly right slip like that observed on the San Andreas, although of smaller magnitude. This movement probably produced the en echelon wrinkles that are reflected on the surface as Baldwin, Dominguez, and Signal hills (Figure 8-9). The surface expression of folding is more prominent than the expression of faulting. Several of these anticlinal hills overlie up-faulted blocks of basement rocks, however, so their origin may be more complex than first appears.

Figure 8-9. Signal Hill in 1941. (Photo by Spence Air Photos, courtesy of Department of Geography, University of California, Los Angeles)

Central Block

The central block of the Los Angeles Basin includes the low portions of the Los Angeles coastal plain from Beverly Hills southeast to central Orange County (the Downey Plain), the Coyote Hills uplift, the La Habra Valley, the San Joaquin Hills, and Newport and Huntington Beach mesas. The Santa Ana Mountains may be included also, but they are more conventionally placed in the Peninsular Range province. The block's main portion is occupied by the Downey Plain, a broad synclinal sag about 10 to 14 miles (16–22 km) wide.

There are several folds within the coastal plain. One that lacks surface expression is the anticlinal Anaheim Nose. Another is represented by the Coyote Hills uplift, extending from the low hills near Santa Fe Springs southeast to the Coyote Hills proper, which stand nearly 500 feet (160 m) above the adjacent lowland. The synclinal trough of La Habra Valley lies northeast of the Coyote Hills uplift and is bounded on the northeast by the Whittier fault zone. The Whittier zone also forms the eastern side of the central block from Montebello to near Corona, where it merges with the Elsinore fault. Northwest from Montebello, the presence of the Whittier fault is uncertain. Here the block's eastern boundary is marked by the Elysian and Repetto hills.

The distinctive basement rocks underlying the central block are exposed only in the Santa Ana Mountains, although they have been encountered in some oil wells around the block's margins and are probably equivalent to those seen in the eastern Santa Monica Mountains. They consist of slightly metamorphosed sedimentary Jurassic rocks that have been intruded by late Cretaceous granitic plutonic rocks of the southern California batholith. No transition is discernible between the basement rocks of the central and southwestern blocks. This suggests that basement rocks of quite different origin have been brought into contact with one another, probably by appreciable right slip on the fault zones.

Younger rocks resting on the basement are most completely exposed on the western slopes of the Santa Ana Mountains. Here at least 32,000 feet (9700 m) of marine and nonmarine late Cretaceous to Pleistocene sedimentary rocks occur, plus some Miocene volcanics. The older rocks in this sequence are missing from the central part of the block, although Miocene, Pliocene, and Pleistocene beds alone total more than 22,000 feet (6700 m) thick. Where older sediments occur, near the junction of the Rio Hondo and the Los Angeles River, total thickness is 32,000 to 35,000 feet (9700–10,700 m). The basement surface is bowed downward from the sides and ends of the central block, making the basement 13,000 feet (4000 m) below sea level at the ends and 14,000 feet (4250 m) along the Newport-Inglewood fault zone. In the deepest part of the Los Angeles Basin, the basement surface lies more than 30,000 feet (9150 m) below sea level.

Northeastern Block

The northeastern block is situated between the Whittier fault zone and the base of the San Gabriel Mountains and is separated from the northwestern block by the Raymond Hill fault. This block is a deep synclinal basin that contains mostly marine Cenozoic sedimentary rocks, but includes some thick Miocene volcanics in the east. Basement lies as much as 12,000 feet (3650 m) below the surface in the central part of the San Gabriel Valley, and in the eastern Puente Hills more than 22,000 feet (6700 m) of Cenozoic sediment covers basement rock.

Northwestern Block

The northwestern block embraces the eastern Santa Monica Mountains and the San Fernando Valley. It is bounded on the south by the Santa Monica and Raymond Hill faults, on the east and northeast by the San Gabriel Mountains, and on the west and north by the ranges usually included in the Ventura Basin portion of the Transverse Ranges. The San Fernando Valley is a broad syncline with the eastern Santa Monica Mountains an adjacent anticline. No faulting of consequence separates the Santa Monica Mountains from the San Fernando Valley, but the Santa Monica block has been appreciably uplifted with respect to the other blocks of the Los Angeles Basin.

Geologic History

Unraveling the geologic history of the Los Angeles Basin is a complicated process. It involves not only vertical movements of great magnitude (such as more than 20,000 ft or 6100 m of subsidence in the central basin since the middle Miocene), but also substantial strike-slip movement on the Newport-Inglewood zone and boundary faults like the Malibu Coast–Santa Monica. This latter fault zone may have undergone as much as 50 miles (80 km) of left slip since the Eocene.

At the beginning of the late Cretaceous, an extensive erosional surface developed across the older rocks and was subsequently covered by later Cretaceous marine sediments from an advancing sea. Rocks of this age are known in the Santa Ana and eastern Santa Monica mountains. A similar pattern probably applies to the Paleocene and Eocene rocks, although they are buried so deeply that they have not even been encountered in wells drilled in the central basin. Late Eocene, Oligocene, and early Miocene nonmarine sedimentary beds occur in both the Santa Ana and Santa Monica mountains, and the presumption is that they also extend across the basin.

The great relief of the present basin floor began to evolve in late Miocene time, when sizable vertical movements began to exert control on the pattern of later deposition. During the late Miocene, the sea advanced over the Los Angeles Basin from south to north, eventually covering the basement highs at Palos Verdes and other parts of the southwestern block. By the close of the Miocene, the sea had reached the base of the San Gabriel Mountains and flooded most of the Los Angeles Basin, although a shoal existed at the site of today's Anaheim Nose. There is evidence that the Los Angeles and Ventura basins were connected at this time and that marine conditions prevailed over the intervening area. Modern ranges and valleys did not exist. Furthermore, although the great thickening of marine Miocene in the central Los Angeles Basin is reflected by well records, drills have not yet penetrated to the basement (often more than 20,000 ft or 6100 m below the surface).

During the Pliocene, the rate of sinking accelerated in the central basin and some sediments were deposited in as much as 6000 feet (1800 m) of water. Concurrently, the basin's margins were undergoing marked uplift, and rocks from both the surface and oil wells show many unconformities that record continuing tectonism. The central basin continued to receive large volumes of sediment from the northeast, and the basin's sea became steadily shallower. By the close of the epoch, more than 10,000 feet (3000 m) of deposits had been laid down. An unconformity near the top of the Pliocene section shows that deposition was interrupted by deformation about this time. As the Pliocene closed, land areas included an island formed by the Palos Verdes Hills, much of the Santa Ana and Santa Monica mountains, the Puente Hills, and probably some small islands along the Inglewood fault zone.

In early Pleistocene time, the Palos Verdes Hills sank below sea level, the central basin continued to receive marine deposits as did the San Joaquin Hills and the San Gabriel Valley, and the San Gabriel Mountains and the Puente-Repetto Hills were rising. The Santa Monica Mountains seem to have persisted as a lowland.

Deposition began to outpace subsidence, and by middle Pleistocene the shoreline probably lay along the southern margin of the Santa Monica chain and along the Whittier fault zone. Exceptions were two embayments: one at Whittier and one at the base of the Santa Ana Mountains and north of the San Joaquin Hills. Near the center of the basin are more than 3000 feet (900 m) of Pleistocene land-laid beds, some deposited in nearshore lagoonal environments along a low-lying coast. To accommodate such a thickness with today's low elevations, there must have been substantial subsidence until effectively the present. The region experienced its last deformational episode, the Pasadenan orogeny, by the middle Pleistocene. This was expressed in the central basin by more subsidence, but in the surrounding areas by considerable uplift.

By late Pleistocene, the Palos Verdes Hills began to rise along the Palos Verdes fault, producing the marine terraces described more fully later. Lowerings of sea level, partially caused by continental glaciation, made the entire coastal plain emerge and allowed streams to cut channels as much as 250 feet (76 m) deep at the present shore. (The land area at that time was thus more extensive than today's.) The hills along the Newport-Inglewood uplift were developed, and some of the hills in the eastern basin were uplifted. For example, the San Joaquin Hills rose as much as 1000 feet (300 m) during the late Pleistocene and Recent. The Whittier fault zone is thought to have about 3 miles (4.8 km) of right separation, 1 mile (1.6 km) of which appears to be late Pleistocene and Recent. This last orogenic episode is apparently continuing, as indicated by historical earthquakes and folding of young deposits in such areas as Signal Hill, the central basin, and the Coyote Hills. There seems to be good evidence too that the Gaffey anticline, north of San Pedro, is still rising.

The thick marine sediments of the Los Angeles Basin are richly fossiliferous, especially within Pleistocene sections. The San Pedro and Timms Point beds from the Palos Verdes Hills area are particularly rich. In one study, a single exposure of a 1-foot (0.3 m) layer of Palos Verdes sand near Playa del Rey yielded more than a million shells. Nonmarine deposits are less fossiliferous, but do contain numerous vertebrate remains. Particularly remarkable is the array of animal remains recovered from tar pits at Rancho La Brea.

The Los Angeles Basin is one of the world's most prolific petroleum-producing areas. Fields like Signal Hill, Huntington Beach, Santa Fe Springs, Wilmington, those on the Newport-Beverly uplift, and several dozen smaller ones have produced more than 5 billion barrels (700 million metric tons) of oil. They continue to be important producers, with nearly a billion barrels (140 million metric tons) of reserves still in the ground.

Problems of Los Angeles Basin Geology

Several problems must be solved before an adequate history can be established for the area incorporating the Los Angeles Basin, the northern Peninsular Ranges, and the western Transverse Ranges. Among these problems are the following.

1. Distribution of the early middle Miocene. In particular, the widely scattered occurrences and source area of the San Onofre breccia must be explained. It seems probable that present occurrences were once much closer both to one another and to a source area containing glaucophane schist.

2. Juxtaposition of Catalina schist and granitic basement along the Newport-Inglewood fault zone.

3. Development of deep marine basins in the Los Angeles area during Miocene time.

4. Establishment of the present geographic outlines of the area in Pliocene time.

5. Late Pleistocene uplift of the entire Los Angeles Basin including the Palos Verdes Hills and at least some of the offshore islands.

6. Relationships in time and origin of the right-slip and left-slip faults that today cross the basin in unexplained patterns.

SAN BERNARDINO MOUNTAINS

The San Bernardino Mountains are the loftiest range south of the Sierra Nevada in either California or Baja California. The highest peak is Mount San Gorgonio at 11,502 feet (3508 m) (Figure 8-10). This high range has an extensive upland plateau of modest relief that is situated at 6500 to 7500 feet (1980–2300 m) elevation. On this surface are the dams constructed to impound Lakes Arrowhead and Big Bear, plus several smaller lakes.

Figure 8-10.

Looking east across the northern San Jacinto Mountains at Mount San Gorgonio. San Gorgonio Pass separates the San Jacinto Mountains from the San Bernardino Mountains in the background. The flat-floored valley is at the base of the San Jacinto eroded fault scarp. Note the facets. The dark spots along the base of the range are springs emerging from the San Jacinto fault. (Photo by Spence Air Photos, courtesy of Department of Geography, University of California, Los Angeles)

The flanks of the range are cut by deep, narrow canyons similar to those in the San Gabriel Mountains. The San Gabriel Mountains, however, lack the distinctive rolling upland of the San Bernardino unit. If such a plateau ever did exist in the San Gabriel area, it has been destroyed by headward stream erosion. The San Bernardino Mountains are a larger and higher mountain mass and so perhaps were able to retain more of their rolling upland area. Whatever the case, headward erosion will eventually consign today's elevated plateau to oblivion.

Figure 8-11.   Some known and inferred faults of the eastern Transverse Ranges. (Sources: California Division of Mines and Geology, U.S. Geological Survey, and Department of Geology, University of California, Riverside)

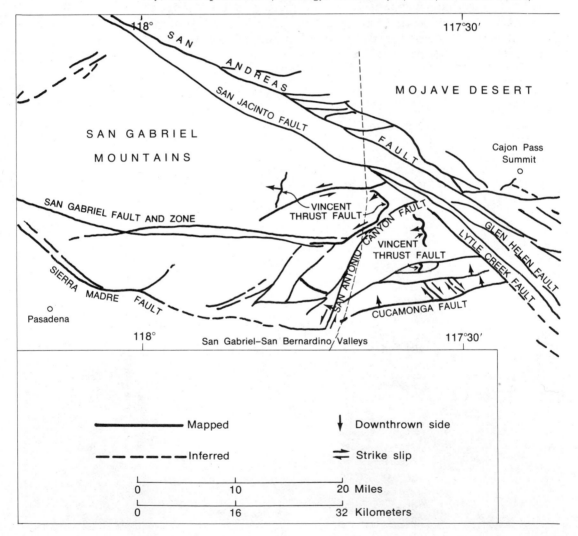

The San Bernardino Mountains are about 65 miles (105 km) long, with maximum width of 30 miles (48 km). They are bounded on the west and southwest by Cajon Pass and the San Andreas fault, on the north by the Mojave Desert, on the east by Twentynine Palms Valley, and on the southeast by Morongo Valley. Morongo Valley lies along the Morongo Valley fault and separates the San Bernardino Mountains proper from the Little San Bernardino Mountains. (Figure 8-11 shows the faults of the eastern Transverse Ranges.)

On the south, the San Andreas fault branches near the city of San Bernardino. The north branch, which is closely related to if not identical with the Mission Creek and Mill Creek faults, crosses the

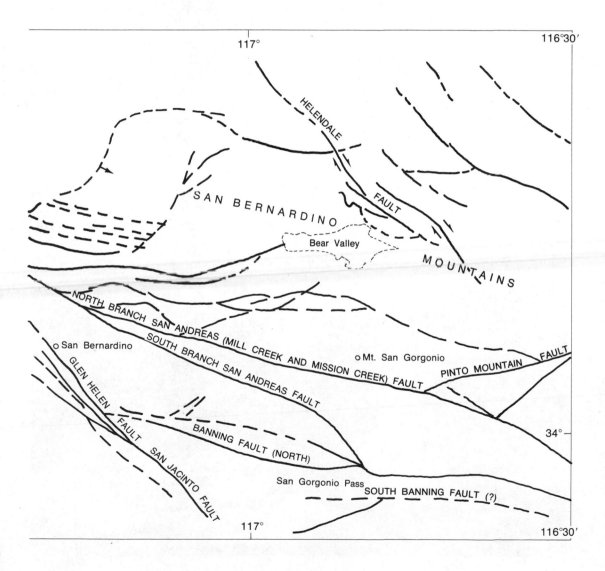

south flank of the range along the deep canyon of Mill Creek, passes south of Mount San Gorgonio, and enters the Coachella Valley near Desert Hot Springs. To the southeast, near Indio, this branch joins the south branch of the San Andreas fault, which follows the north side of San Gorgonio Pass. This south branch of the San Andreas is also known as the Banning (or South Banning) fault. The Banning fault probably caused the precipitous south face of the San Bernardino Mountains along San Gorgonio Pass; for a few miles east of Cabazon it has been mapped as a thrust.

Like the San Gabriel chain to the west, the San Bernardino Mountains are basically a horstlike block uplifted along bounding faults (Figure 8-12). Many smaller faults are present within the range. Several prominent Mojave Desert fault systems die away in the range or are lost where they enter the crystalline rocks.

The distribution of rock types in the San Bernardino Mountains is as varied as in the San Gabriel chain, but differs in some important respects. Anorthosites are absent in the San Bernardino Mountains, but Precambrian gneisses and schists are equally abundant in both ranges. These rocks are particularly prominent on the south flank of the San Bernardino Mountains and in Morongo Valley. Smaller patches of Precambrian rocks occur in the central and northeastern parts of the range, and Precambrian and Cambrian

Figure 8-12. *Looking east at Mount San Gorgonio and the San Bernardino Mountains. The San Andreas fault zone is at the base of the mountains. (Photo by Spence Air Photos, courtesy of Department of Geography, University of California, Los Angeles)*

marine sedimentary rocks have been reported from west of the Cushenbury grade.

Along the north flank and east and southeast of Big Bear Lake are extensive exposures of fossiliferous marine late Paleozoic rocks. These consist mostly of the Pennsylvanian Furnace limestone, but include other limestones and quartzites of Mississippian, Pennsylvanian, Permian, and possibly earlier ages. Total thickness is about 10,000 feet (3000 m).

At least 70 or 75 percent of the San Bernardino chain is composed of a light-colored quartz monzonite dated between 70 and 85 million years (late Cretaceous), called the Cactus granite. About 11 miles (18 km) southeast of Old Woman Springs, another quartz monzonite has been cut by pegmatitic dikes tentatively dated as middle Jurassic (150 million years old), although the true age may be somewhat less. Thus the granitic rocks in this range date from Jurassic to Cretaceous and also reflect several intrusive episodes. No Mesozoic sedimentary rocks have yet been recognized.

The San Bernardino Mountains contain few Cenozoic deposits, although some Pliocene basaltic flows are exposed in the eastern end of the range. There are also some Quaternary deposits including lake beds, alluvial gravels, stream deposits in larger canyons, and alluvial fans around the north and east flanks.

EASTERN BOUNDARY RANGES

Slightly southeast of the higher San Bernardino Mountains, the Little San Bernardino Mountains continue the east-west trend characteristic of the Transverse Ranges. The Little San Bernardino and adjoining Pinto and Eagle mountains are desert ranges and mark the eastern end of the Transverse Range province, although as noted previously some authorities include the Orocopia Mountains to the southeast. The northern edge of these eastern ranges lies along the Pinto Mountain fault zone, which merges on the west with the Morongo Valley fault. The southern margin of the Little San Bernardino Mountains is the Mission Creek fault, part of the San Andreas system. Eastward into the Eagle Mountains, the southern boundary is less clear, any faults present being concealed under large alluvial fans. The Mission Creek–San Andreas system turns south away from the ranges and strikes into the Salton Trough. The Pinto and Eagle mountains are separated by the Pinto Basin, a large upland alluviated valley.

Rocks of these three ranges are about equally divided between supposed Precambrian schists and gneisses and the granitic rock exposed on either side of Morongo Valley. The granitic rocks probably correlate with the Cretaceous Cactus granite of the San Bernardino Mountains. It is these granitic rocks that provide the spectacular boulder piles seen in Joshua Tree National Monument. Minor amounts of dolomite and other sedimentary rocks of presumably early Paleozoic age occur in both the Pinto and Eagle mountains.

The Pinto and Eagle mountains both contain important iron ore bodies, though the principal mines are in the Eagle Mountains. The ores are located in contact metamorphic zones between the plutonic rocks and the older Paleozoic dolomitic limestones. The ore minerals are chiefly hematite and magnetite and carry from 50 to 53 percent iron. Since 1948, the Eagle Mountain mine has been the main source of ore for the Fontana steel plant.

STRUCTURAL HISTORY

The Transverse Range province is dominated by east-west trending folds, often faulted, and by faults that have east-west or northwest-southeast trends. Even the San Andreas fault is deflected by the province and assumes a more east-west direction as it slices obliquely across the Transverse Ranges. The San Andreas, however, has offset the central axis of the ranges at least 25 miles (40 km) at Cajon Pass. The western ranges and basins are primarily folded structures, although they often are bounded by steep faults that show both vertical and horizontal components of movement. The faults of the western Transverse Ranges are shown in Figure 8-13. The eastern ranges are chiefly elevated fault blocks; many of their bounding structures are reverse faults.

Big Pine Fault

The Big Pine fault has long been recognized as one of California's major left-slip faults. According to some interpretations, it forms the northern boundary of the Transverse Range province in Ventura County. The fault is believed to extend about 50 miles (80 km) from Frazier Mountain west into Santa Barbara County, to a belt of Franciscan rocks where it apparently merges with the northwest-trending faults of the southern Coast Ranges. Based on presumed lineations revealed in high-altitude photographs, it has been suggested that the Big Pine fault is somewhat offset to the north and continues west from the Franciscan belt to the coast south of Point Sal. Field work provides little support for this idea, but some earthquake epicenters have been located along the supposed extension. If this extension is confirmed, it may be more logical to consider the Big Pine fault as the northern boundary of the western Transverse Range, rather than the Santa Ynez fault—which is the practice in this book.

Because both the Big Pine and Garlock faults show prominent left slip, it has been inferred that they are part of the same system.

As such, they may be subordinate to and older than the San Andreas, along which the western end of the Garlock is separated by about 7 miles (11 km) from the eastern end of the Big Pine. It may be, however, that the Big Pine fault is more logically related to faults in the Soledad Basin, from which it has been offset by the San Gabriel fault.

Earthquakes of 1852 are attributed to movement on the eastern segment of the Big Pine fault, when ground breakage was observed in Lockwood Valley. Investigations on the central and western parts of the fault suggest Quaternary horizontal stream offsets up to 3000 feet (900 m), and one fan may have been offset a mile (1.6 km). Much greater displacements are indicated since Miocene time, possibly 4 to 10 miles (6.4–16 km).

Santa Ynez Fault

The Santa Ynez fault is generally considered the northern boundary of the Transverse Ranges west of the Santa Barbara–Ventura county line. It is a relatively long fault, extending 75 miles (120 km) west from the Agua Blanca thrust in central Ventura County to the coast (Figure 8-3). At Gaviota Pass, the fault divides into a south branch, which enters the Santa Barbara Channel, and a north branch, which continues west for about 12 miles (19 km) and includes the Pacifico fault.

Matching the rock section north of the fault with that exposed to the south suggests a vertical separation of 9500 feet (2900 m). The differences in Tertiary rocks on opposite sides of the fault also have been interpreted as evidence of considerable horizontal displacement. Investigators disagree regarding direction of offset, however, although most support left slip. Furthermore, amount of slip is undetermined; post-Eocene estimates range from 1 or 2 miles (1.6–3.2 km) to as much as 37 miles (60 km).

Topographic characteristics measured between 1957 and 1971 (including offset streams, sag ponds, offset terrace deposits, and vertical changes in elevation) all suggest that the Santa Ynez fault is currently active. A strong 1927 earthquake, with epicenter west of Point Arguello, has been attributed to movement on a presumed western extension of the fault.

Santa Cruz Island and Santa Rosa Island Faults

Santa Cruz and Santa Rosa islands are each cut by a prominent east-west trending fault. Whether these two faults are related is uncertain, but many geologists project them into the channel between the islands and also assume left-lateral offset on a younger, northwest-trending associated fault.

The dissimilarity of rocks on either side of the two faults im-

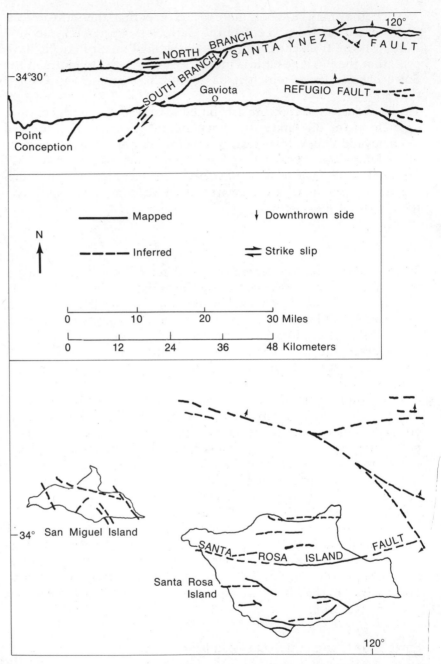

Figure 8-13. *Some known and inferred faults of the western Transverse Ranges. (Sources: California Division of Mines and Geology and U.S. Geological Survey)*

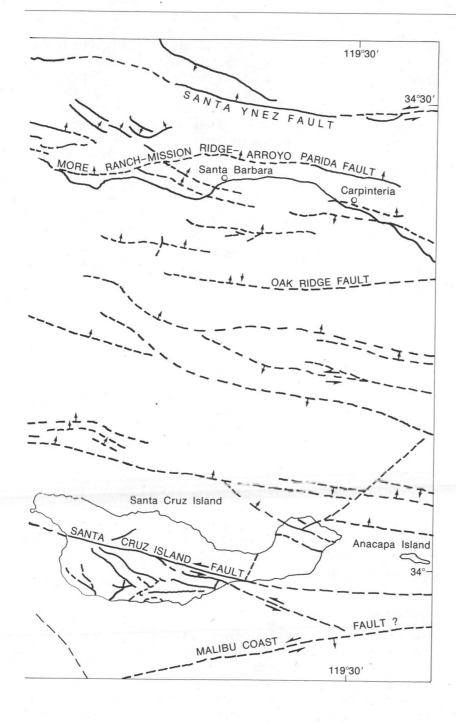

plies considerable displacement. For the Santa Cruz Island fault, vertical displacement of 7500 feet (2300 m) and horizontal separation of about 5000 feet (1500 m) has been proposed; for the Santa Rosa Island fault, a maximum of 10 miles (16 km) of horizontal displacement has been suggested. Perhaps more important from a practical point of view is the recency of movement on the two faults. Streams and terraces have been displaced left laterally in both instances, showing that the faults have been active during the Quarternary.

Santa Clara Valley Faults

Two faults approximately outline the valley of the Santa Clara River, and their movement appears to be closing the valley as if it were caught in the jaws of an enormous vise. The river and its tributaries, of course, keep the valley open by erosion, and in historical terms the "jaws" are unlikely to close very soon. Earthquakes are quite possible, however.

On the valley's north side is the San Cayetano thrust, with a sinuous trace from near Ojai to east of Piru. Its dip varies from 15° to 50° north, which explains the irregular surface trace. Dip-slip displacement may be as much as 20,000 feet (6100 m), with the Topa Topa mountain block riding up the fault plane in a southerly direction. East of Fillmore, the fault lies close to the base of these mountains, and to the west it is in their lower foothills. The fault can be seen fairly easily where it has brought Eocene rocks with bold outcrops and steep faces over softer, more gently dipping Miocene and Pliocene rocks.

On the south side of the valley is the poorly exposed Oak Ridge fault. This fault has been traced, through oil wells, from west of Piru to the mouth of the Santa Clara River and offshore into the Santa Barbara Channel. On land, it is a high-angle reverse fault, with south dips of 60° or more. The anticlinal folds of South Mountain and Oak Ridge have moved upward on the fault, which also has offset Quaternary terraces.

San Gabriel Fault Zone

About 90 miles (144 km) long, the San Gabriel fault is mapped as part of the San Andreas system and like the San Andreas shows predominantly right slip. Moreover, it somewhat defies the general east-west trend of Transverse Range structure, as it strikes southeast from Frazier Mountain and enters the San Gabriel Mountains at their western end. North of Mount Wilson, the fault does become nearly east-west, however. It appears to be offset in San Antonio Canyon by north-south trending faults, with the eastern segment terminating against the San Jacinto fault. The San Gabriel's wide crushed zone has strongly affected topography and drainage, prompting the East and West forks of the San Gabriel River to follow the fault for most of their lengths.

The San Gabriel fault has undergone about 21 miles (34 km) of right slip since the Miocene. This figure has been deduced from displacement of middle Miocene anorthosite-bearing conglomerates southwest of the fault from their only available source area, the western San Gabriel Mountains across the fault. The San Gabriel fault also has experienced varying vertical displacement, with the north side up in some places and down in others. Along the southwest of the Ridge Basin, vertical displacement is as much as 14,000 feet (4250 m). Apparently the San Gabriel fault has not been active recently. Some Pleistocene deposits and stream courses were affected by it, but there is no disturbance of late Quaternary alluvial deposits.

In Big Tujunga Canyon, the San Gabriel fault bifurcates into the main east-trending San Gabriel fault and a south branch that becomes the Sierra Madre fault. The Sierra Madre and its eastern counterpart, the Cucamonga, are the steep, north-dipping range-front faults along which has occurred most uplift of the San Gabriel Mountains. Activity is quite recent, because the Cucamonga fault has offset Quaternary alluvium. In a broad sense, the Sierra Madre fault zone comprises all the range-front thrust and reverse faults from east of Arcadia to the arcuate, supposedly discontinuous faults in the west San Gabriel Mountains and the northern foothills facing the San Fernando Valley. The latter include the Santa Susana thrust, which cuts across the Santa Susana Mountains to join the Oak Ridge fault.

Seismologists have shown that the 1971 San Fernando earthquake defined a north-dipping reverse fault that corresponded to the surface breaks observed along segments of the Sierra Madre fault zone. Main motion below the epicenter was pinpointed at a depth of about 7 miles (11 km), with no ground breakage above. A plane dipping northeast under the range at about 45° would include both the focus of the quake and the surface offsets observed near San Fernando.

Soledad and Vincent Thrust Faults

The Soledad fault can be seen in Soledad Canyon, where it brings Precambrian crystalline rocks into contact with the nonmarine middle and late Tertiary rocks of the Soledad Basin to the north. It is a normal fault.

On the north side of Mount San Antonio is the oldest major fault known in the San Gabriel Mountains. This is the Vincent thrust, a fault that dips south and west with an extremely irregular trace. As noted before, the Vincent thrust seems to have caused northeast movement of plutonic rocks over the Pelona schist. Although it has influenced modern topography, the thrust is an old feature.

Malibu Coast–Santa Monica–Raymond Hill Fault System

The southern frontal faults of the Santa Monica and San Gabriel mountains constitute a fault system of substantial magnitude, with inferred left slip of 40 miles (64 km) or more. The Malibu Coast fault extends from Arroyo Sequit at the south front of the Santa Monica chain to near the city of Santa Monica. It continues east as the Santa Monica fault, largely buried beneath the alluvium at the base of the Santa Monica Mountains. Near Glendale the Santa Monica fault seems to merge with the Raymond Hill fault, which has offset Quaternary terrace deposits and is traced from South Pasadena to near Monrovia.

The trend of this fault system aligns with the Santa Cruz Island fault, which it may well join. Such a connection is frequently assumed when accounting for the peculiar distribution of rocks like the Eocene Poway conglomerates of the San Diego coastline and the middle Miocene San Onofre breccia.

ORIGIN OF THE TRANSVERSE RANGES

The San Andreas fault and its history are of major importance in understanding the origin of the Transverse Ranges. Although the San Andreas is probably older in other areas, in this part of California the fault system apparently developed no earlier than late Oligocene. Since then, movement has been nearly continuous, usually reflecting right slip with perhaps a substantial vertical component. Amount of displacement is disputed, but 130 to 160 miles (210–260 km) is frequently suggested.

Most interpretations of plate tectonic theory identify the San Andreas as a transform (fault) and as the shear boundary between a northwest-moving Pacific plate and the American plate. If this is correct, the Transverse Ranges must be relatively modern—at least post-Oligocene—irrespective of total offset on the fault. As far east as the San Gabriel Mountains, the province definitely belongs to the Pacific plate, necessarily sharing in the northward migration. The apparent eastward continuation of the province, which incorporates only about 25 miles (40 km) of offset, must be just a coincidence, if this interpretation of plate history is accepted.

A related point is that the enigmatic rock distribution of the Transverse and Peninsular ranges can be most satisfactorily explained by substantial strike-slip movement on the San Andreas and east-west Transverse Range faults. Present geography alone cannot account for the puzzling distribution of the widely scattered

Poway-like conglomerates and San Onofre breccia. This is another indication that in former times little existed of the modern Transverse Range province.

Another view has been presented by A. O. Woodford and his colleagues. These investigators have called attention to substantial data to support the contention that the Transverse Ranges have persisted as a unified block of east-west trend since *before* emplacement of Cretaceous batholithic rocks. In particular, they cite east-west trends shown by the general petrology, crystalline rock patterns, chemical affinities, and structural features.

There is a lesson inherent in this puzzling situation. Scientists must continually reexamine their assumptions when developing a new hypothesis and must constantly check observed facts to see what they require of the hypothesis. Plate tectonic theory has been remarkable in providing a coherent explanation for a wide range of geologic phenomena. It is risky, however, to ignore geologic features that seemingly do not agree with theory. When a theory has been as successful and revolutionary as that of plate tectonics and sea-floor spreading, there is a tendency to force a recalcitrant fact to fit the grand plan, rather than to reexamine the assumptions underlying the theory in the first place. In this regard, no application of plate tectonic theory to the west coast of North America can be wholly valid unless it accords with the geologic realities of the Transverse Ranges.

SUBORDINATE FEATURES

Streams

The Transverse Range province has few sizable permanent streams. Small as most of them are, however, the streams are disproportionately important when one considers their modest discharges. This seeming paradox results from the combination of dry climate and the presence of about 40 percent of the state's population along the southern fringes of the Transverse Ranges.

The largest drainage system wholly within the province is that of the Santa Clara River (Figure 8-14), which drains most of central and southern Ventura County, the northwestern San Gabriel Mountains, Liebre Mountain, the Sierra Pelona, and the Soledad Basin. The river is more than 75 miles (120 km) long, and although its surface flow is normally small, it provided almost all the domestic and agricultural water for its basin until the arrival of Feather River water. In addition, the river bed has been a valuable source of sand and gravel for the Los Angeles and Santa Barbara–Ventura areas.

In the west, the Santa Ynez River, about 60 miles (96 km) long, is the most important stream (Figure 8-15). It drains the northern

Santa Ynez Mountains and the southernmost Coast Ranges and approximately follows the Santa Ynez fault toward the sea. This river was dammed to provide water for the south coast of Santa Barbara County.

The San Gabriel and San Bernardino mountains are drained by several permanent streams that have been crucial to communities of the region. On the western slope of the San Gabriel Mountains, streams drain into the San Fernando Valley and the Los Angeles River. The Los Angeles River is about 65 miles (105 km) long and has a total drainage area of about 650 square miles (2710 km²). Moreover, although the butt of many a local joke, it actually furnishes the city of Los Angeles with about 15 percent of its water supply from underflow and groundwater.

Important streams to the east include the San Gabriel River, which drains the central San Gabriel Mountains. The San Gabriel branches near Duarte into the San Gabriel River proper and the Rio Hondo, which joins the Los Angeles River near Downey so that the two drainages intermingle. The Los Angeles River reaches the sea near Long Beach, and the San Gabriel discharges near Seal Beach.

Figure 8-14. *Looking east over the valley of the Santa Clara River. The even skyline of the San Gabriel Mountains forms the background, and the Santa Susana Mountains are on the right. The town of Fillmore is in the center. Sespe Creek joins the Santa Clara River in the foreground. (Photo by Spence Air Photos, courtesy of Department of Geography, University of California, Los Angeles)*

Figure 8-15.
Santa Ynez River and its headwater drainage, north of the city of Santa Barbara. Gibraltar Reservoir is in the foreground. The east-west linearity of the ridge-valley topography is controlled primarily by faults. (Photo by Spence Air Photos, courtesy of Department of Geography, University of California, Los Angeles)

The Santa Ana River has the largest drainage basin in southern California. It drains both the high San Bernardino and the eastern San Gabriel mountains, plus the northern Peninsular Ranges. In the late 1930s, its limited supplies were augmented by Colorado River water.

Floods

Despite their normally modest discharges, like water courses in other arid and semiarid regions, the Transverse Range streams fluctuate greatly in volume. Every 20 or 25 years, heavy winter rains in the mountains cause widespread flooding in the lowlands. Owing to the high intensity of rainfall in such storms, small streams can accomplish tremendous erosion and transportation in short periods. In January 1934, a storm swept boulders weighing 100 tons (80 metric tons) from the San Gabriel Mountains into the town of Montrose—with most destructive results. One creek was deepened 16 feet (5.2 m) in a few hours.

The Los Angeles River has repeatedly caused severe flooding, with the result that more than three-quarters of its channel has been lined with concrete. In several instances, the river even changed its course. Prior to 1815, the Los Angeles River drained into the Long

Beach area, as it does today. The flood of 1815 caused the river to change its course westward, to join with Ballona Creek and empty into Santa Monica Bay. In 1825, flooding forced the river to abandon the new course and cut the first well-defined channel south toward the Long Beach region. The flood of 1862 reportedly converted most of the coastal plain into a lake; at one place the Santa Ana River widened to 3 miles (4.8 km). In 1884, all bridges but one over the Los Angeles River were washed out—the wettest season ever recorded in Los Angeles.

The flood of 1938 produced a discharge in excess of any previously recorded, although annual rainfall totals had been higher. Unfortunately, the ground had already been saturated by a series of heavy rains before the big storm arrived in late February. With rainfall of nearly 30 inches (760 mm) in the mountains and 10 to 15 inches (250–380 mm) in the lowlands in just four days, the streams in the Los Angeles region simply were unable to accommodate the sudden volume of water. Large areas of the San Fernando Valley and the Los Angeles coastal plain were inundated (Figure 8-16). The raging Los Angeles River damaged more than 100 bridges, caused the loss of 43 lives, and resulted in perhaps $40 million in property damage. Total loss of life in southern California was 87, and property damage was about $78 million.

Palos Verdes Marine Terraces

Between Redondo Beach and San Pedro, the Palos Verdes Hills form a prominent headland that rises to 1480 feet (451 m). From late Pleistocene until recently, the peninsula probably was an island, because land on the northeast is low and was very swampy until altered by man. An 80-foot (25 m) rise in sea level would make it an island again. The hills form cliffs and terraces along their seaward margin with some steps as high as 300 feet (90 m), although the average is 100 to 150 feet (30–45 m). Near San Pedro on the south and Redondo Beach on the north, cliff heights decrease to 50 feet (15 m) or less.

Palos Verdes Peninsula is an anticline uplifted as a horst between faults on the northeast and southwest. The exposed rocks belong mostly to the Miocene Monterey formation, which covers a core of Franciscan Catalina schist. Some Pliocene marine rocks are present in the northeast, and the marine terraces contain thin late Pleistocene deposits.

The terraces are the landscape's most striking feature and can be seen best on the seaward side of the peninsula. Thirteen have been recognized, with elevations ranging from 100 to 1300 feet (30–400 m), and it is likely that the subdued topography above 1300 feet (400 m) indicates complete late Pleistocene submergence. Relative ages of the terraces have not been determined, but the lower ones

*Figure 8-16.
Flooding in the Los
Angeles Basin,
March 1938. View
of the San Gabriel
River on its flood
plain near Seal
Beach. (Photo by
Spence Air Photos,
courtesy of De-
partment of Geog-
raphy, University of
California, Los
Angeles)*

appear to be younger. Although many contain shelly marine sand, they cannot be distinguished paleontologically from one another. Elevations show that most of the terraces have been minimally deformed since they were cut, but near San Pedro the Gaffey anticline has gently warped the youngest terraces. Apparently deformation is continuing there, because recent alluvium is deformed.

It should be remembered that the Palos Verdes and other southern California terraces are products of both local uplift and Pleistocene eustatic sea-level fluctuations that reflect worldwide cycles of glaciation. These two independent processes have so complicated local histories that correlation of the Palos Verdes terraces with any other terraces is difficult and uncertain.

SPECIAL INTEREST FEATURES

Landslides

Several Transverse Range landslides have achieved considerable notoriety, not because they are remarkable geologically, but because they have severely affected urbanized regions. Regrettably, some slides were prompted by man's own activities.

Portuguese Bend Near Portuguese Bend, in the southwest Palos Verdes Peninsula, are a recently active large landslide and two smaller slides at higher elevations. Another slide, of moderate size, lies near the coast west of White's Point. The large Portuguese Bend slide attained prominence in 1956, when it experienced renewed activity. Although the slide had been clearly mapped in a 1946 published report, building proceeded and by 1956 many homes existed in the area. Several of these were destroyed or seriously damaged when the slide began to move, sending porches downhill, developing faults in driveways and sags and cracks in walls, and jamming doors and windows.

Roads were patched repeatedly, and eventually many were abandoned as segments moved downhill. Scarps 20 feet (6 m) or more in height developed. Buried utility lines were excavated and laid on the ground; overhead wires required constant adjustment as distances between poles changed. Deep, undrained hollows and ominous bulges and buckles appeared within a few months. Near the toe of the slide, just offshore, a buckle arched a pier upward at its midpoint. The head of the slide was characterized by arcuate scarps and cracks, giving a steplike appearance.

Only the southeastern portion of this large slide moved during the most recent episode. Larger and larger portions were involved, however, until the area affected was roughly three times as extensive as the area first deformed. Monitoring of movement showed that slippage was more rapid after winter rains, with a lag time of a month or two. Rates of movement varied. A station near the head of the slide moved downward 12 feet (3.6 m) in 600 days, a centrally located site subsided about 5 feet (1.5 m) in a similar period, and upward movement near the slide's base was almost 4 feet (1.2 m) in 388 days.

Causes of the Portuguese Bend landslide have been investigated thoroughly, with complex and controversial results. Apparently it was triggered by erosion of the seaward side of a synclinal fold, thus removing support from the dipping rocks. In addition, the area's rocks are tuffaceous and contain numerous beds of an altered volcanic ash that becomes plastic when wet. So the combination of structure and lithology was initially unfavorable even under normal moisture conditions. In the few years preceding renewed movement, about 150 homes were built in the slide area and none had municipal sewer service. This dependence on septic tanks and cesspools introduced huge volumes of water into the slide, intensified by garden watering estimated at a minimum of 40,000 gallons (152,000 l) per day. Another factor, alleged by some to be the chief cause, was emplacement of highway fill near the slide's head. Prevailing geologic opinion, however, was that introduction of water into the old slide was the critical factor, though the addition of load certainly exacerbated the situation.

Efforts to check movement were largely unsuccessful and consisted initially of installing concrete caissons in the slide. These were 4 feet (1.2 m) in diameter and 20 feet (6 m) long and extended 10 feet (3 m) into the presumably stable material below the slide. Twenty-five of these "pins" were installed, but no appreciable change in slide movement resulted. Another plan entailed stabilizing the slide by placing a large fill on the toe, but this was never done. As homes were abandoned, the volume of water introduced into the slide decreased, and rate of movement consequently slowed. Nonetheless, by 1968 the slide had caused nearly $10 million in damage to homes, utility lines, and roads.

Point Fermin   In 1926, a landslide at the southeastern corner of the Palos Verdes Hills made a small segment of Point Fermin slip seaward. The landslide crack intersected the main street of San Pedro, rupturing streetcar tracks and damaging a few houses. There was no geologic evidence that the slide was the reactivation of an older one, as in the Portuguese Bend case. This was an original slide, one of hundreds discernible along the coast of the Transverse Range province. The causes of the Point Fermin slide were similar to those of the Portuguese Bend slide, however: seaward-dipping beds undercut by wave action and garden irrigation that contributed to lubrication of expansible clay and ash layers of the bedrock.

Blackhawk   Certainly the largest slide in the Transverse Range province is the Blackhawk, on the north slope of the San Bernardino Mountains. This prehistoric slide is one of the largest in the world. It was studied in detail by R. L. Shreve, who showed that the slide moved to its resting place on a cushion of compressed air. (This mechanism has since been recognized as applicable to other slides.) The end of the slide can be seen from State Highway 247 about 10 miles (16 km) east of Lucerne Valley; the only satisfactory way to see the entire slide is from the air (Figure 8-17).

The Blackhawk slide is 5 miles (8 km) long, about 2 miles (3.2 km) wide, and 30 to 100 feet (10–30 m) thick. It is a tonguelike sheet of brecciated Pennsylvanian Furnace limestone derived from the Blackhawk Mountains about 4000 feet (1200 m) above. In the source area, the Furnace limestone has been thrust northward over uncemented sandstone and weathered gneiss that subsequently were eroded away, leaving a precipitous slope. Initial slippage was probably in late Pliocene or early Pleistocene, when movement on thrust faults occurred, followed by erosion of the weaker supporting rocks.

Once the softer rocks were undermined, a mass of limestone breccia collapsed and slipped rapidly into upper Blackhawk Canyon, forming a stream of rubble about 2000 feet (600 m) wide and 300 to 400 feet (90–120 m) deep. As the slide moved down the canyon (at about 170 mph or 274 kmh), it passed over a resistant gneissic ridge that crosses the canyon and so was launched into the air—a geologic version of a flying carpet. Calculations indicate that the sheet of

Figure 8-17. *Blackhawk slide in the northern San Bernardino Mountains, with Mount San Gorgonio in the background. View to the south from the Mojave Desert. (Photo by John S. Shelton)*

moving breccia was probably as much as 400 feet (120 m) above the canyon floor immediately after becoming airborne, but that it settled quickly, compressing the air trapped beneath to a frictionless blanket only 1 or 2 feet (0.3–0.7 m) thick. While airborne, the slide possibly attained velocities of 270 mph (435 kmh), and the entire distance from launching point to resting place was covered in about 80 seconds. These values are based on a consideration of local geometry and are consistent with the behavior of similar slides observed during formation. As the slide spread over the desert floor, the air cushion became thinner, permitting the slide to settle.

A characteristic of such slides is the presence of large blocks that, although badly shattered, have fragments that retain their original orientation to one another—much like a jigsaw puzzle with pieces pulled slightly apart. This feature supports the view that carpetlike sheets of rock can be moved intact on cushions of compressed air.

Earthquakes

The Transverse Range province has sustained a number of severe earthquakes. Some are among the strongest observed in California, although at present the central Coast Ranges and the Imperial Valley are experiencing earthquakes more frequently. The following earthquakes of magnitude 6.0 or greater have originated in the Transverse Ranges since 1800: Santa Barbara Channel (7.5?) in 1812, Big Pine (7.0) in 1852, Fort Tejon (8.0?) in 1857, Santa Barbara (6.3) in 1925, Point Arguello (7.5) in 1927, and San Fernando (6.4) in 1971.

Seismologists have determined that a correlation exists between the length of a fault and the severity of earthquakes it generates. Unfortunately, fault length cannot always be measured accurately because part may be concealed. Until the fault is revealed (by ground breakage or drilling, for example), mapped segments are often erroneously regarded as separate faults.

With its possible 1000-mile (1600 km) length, the San Andreas fault is the most likely source of major earthquakes in the province, a probability confirmed by the historical record. The severest earthquake originating in the Transverse Ranges since Spanish settlement is the Fort Tejon earthquake, which occurred in 1857. Its magnitude is estimated at 8.0 ±0.5 on the Richter scale, and about 200 miles (320 km) of the San Andreas trace experienced ground breakage. Extensive rupture was accompanied by the greatest right-lateral offset yet observed on the San Andreas system, approximately 20 feet (6 m). According to one account, a circular sheep corral astride the fault in eastern San Luis Obispo County was broken across the middle and converted to an open S-shaped figure. The earthquake was strongly felt on the southern California coast, severely damaging Ventura and Santa Barbara missions and reportedly throwing the Los Angeles River from its channel.

During the past 40 years, repeated surveys have been made across the portion of the San Andreas affected in 1857. They have revealed no sign of creep, nor have any earthquakes been attributed to movement on this segment since 1857. Today this part of the San Andreas is frequently cited as locked into position by its bend into the Transverse Ranges—while presumably strain energy accumulates for the next all-but-inevitable strong earthquake.

Certainly the 1971 San Fernando earthquake was the most damaging in the Transverse Ranges so far this century. Consequences were severe even though the quake was of moderate magnitude (Richter scale 6.4) because the epicenter lay near the densely populated San Fernando Valley. The quake caused 64 deaths and property losses between $500 and $1000 million. The disaster would have been even worse had the shock been longer or stronger, for the Van Norman Dam was perilously close to failure when the 60-

second earthquake ended. Had the dam failed, a large residential area in San Fernando Valley would have been inundated. This quake was accompanied by the greatest ground accelerations ever recorded, mostly in the 0.5 to 0.75 g range but including a few peaks of more than 1.0 g.

On the positive side, many people survived because at the time of the shock (about 6 A.M.) they were in single-story, wood-frame homes that, owing to their flexibility, withstand earthquakes notably well. Furthermore, this shock was measured and studied more thoroughly than any previous quake, which will contribute substantially to understanding earthquake forces and promoting the most suitable safeguards.

St. Francis Dam Disaster

A tributary of the Santa Clara River, San Francisquito Creek, was the site of the St. Francis Dam failure of 12 March 1928, which killed about 600 people and destroyed bridges, several miles of highway, several hundred homes, and more than 10,000 acres (4000 ha) of field crops. Completed in May 1926, the dam had been built by the city of Los Angeles to be a reservoir for water from the Owens River and a source of electricity. It was 200 feet (60 m) high and 700 feet (210 m) long.

Unfortunately, the dam was planned without geologic advice. It was constructed in a narrow part of the canyon where the east wall is composed of thin-layered Pelona schist dipping down toward and underlying the canyon floor. This schist is fragile and readily breaks into small flakes. On the canyon's west wall the schist is separated from the younger Sespe conglomerate by the San Francisquito fault and a gouge zone of a few inches to more than 5 feet (1.5 m) thick. In this locality, the Sespe is not only badly sheared and fractured, but also has a dry crushing strength of 520 pounds per square inch (14.6 kg/cm²). It was discovered, moreover, that a sample of this conglomerate placed in water almost immediately disintegrated to an incoherent pile of gravel and mud easily stirred with the finger! Even the pebbles broke up along tiny fractures. This simple test, regrettably not performed prior to construction of the dam, showed that the conglomerate was cemented with only thin films of clay.

Ironically, upon learning of the proposed dam site and being familiar with the area's rocks and the location of the San Francisquito fault, several geologists suggested that the dam not be built. Their proffered advice was ignored, however.

The reservoir was first filled on 5 March 1928, and shortly afterwards seepage was observed in the conglomerate. Later analyses of the seepage water showed a marked increase in dissolved calcium sulfate, which was derived from solution of the gypsum in the conglomerate. The significance of the seepage was not fully appreciated by those on duty at the dam, and in the middle of the night of 12 March, a week after the reservoir had been filled to a depth of 185 feet (56 m), the dam failed.

It is probable that seepage increased rapidly in volume shortly before the dam's failure, washing out the soft conglomerate, undermining the west abutment of the dam, and permitting the concrete to crack into large blocks. Blocks up to 10,000 tons (9100 metric tons) were swept down the canyon as much as a half-mile (0.8 km). The intense swirling of the water probably undercut the schist on the east wall, allowing a huge block of concrete to slide down the dip of the rocks into the canyon. The torrent of water swept every shred of vegetation and loose rock from the canyon to a height of 50 feet (15 m) near the dam. The flood poured into the Santa Clara River near Castaic, sweeping people, houses, groves of trees, and bridges seaward. Some victims were not missed until their remains turned up during excavations for sand and gravel.

Rancho La Brea

In west Los Angeles, near the old Salt Lake oil field, are the tar seeps of Rancho La Brea. From this locality has come one of the most famous collections of Plcistocene animals in the world (Figure 8-18). The site consists of pools of viscous tar that oozed to the surface from deep petroleum reservoirs. In moving toward the surface, crude oil tends to lose its more volatile constituents, becoming viscous and asphaltic. Normally the tar pools were covered with thin sheets of slightly salty water, providing effective traps for the Pleistocene animals. As they waded into the pools seeking salt and water, the animals became mired in the sticky tar below.

Both extinct and living species have been collected. Included are nearly 50 species of mammals, 110 species of birds, a few snakes, turtles, and toads, and an assortment of land mollusks and arthropods. A few plant remains are preserved too, showing that pine, oak, cypress, and manzanita grew in the area. Even man himself was trapped by the tar, for at least one human skeleton has been recovered. Many of these plants and animals are displayed in the Los Angeles County Museum.

Oil and Gas Fields

The Transverse Ranges contain more than 40 separate oil and gas fields, some long since abandoned and others still producing. The oldest producing field in California is in Pico Canyon, near Newhall. Oil was collected from seeps here as early as 1850 and 1869, and several spring pole wells were sunk although regular production did not begin until 1875. Another pioneering "well" was a tunnel completed in 1866 on Sulphur Mountain near Santa Paula. This tunnel

Figure 8-18. Rancho La Brea, from the west. The tar pits are preserved as a park, seen in the bottom right. (Photo by Spence Air Photos, courtesy of Department of Geography, University of California, Los Angeles)

was only 80 feet (25 m) long, and oil flowed by gravity to the tunnel's entrance. Later tunnels dug in the same area were as long as 1600 feet (490 m). Production could be 60 barrels (9576 l) a day, but usually less. Oil seeps in this area are still active; north of Santa Paula a seep produces a tarry oil that flows down the hillside and across State Highway 150.

Ventura Avenue The Ventura Avenue field, just north of the city of Ventura, is the province's most productive field and ranks among the top producers in the state. Gas was first produced here in 1903. Early drilling was done with cable tools, which precluded control of the unusually high pressures encountered. As a result, great difficulties were experienced with cave-ins, large flows of water, and blowouts. Despite the field's area (about 2500 ac or 1000 ha), its oil-producing sands are so thick, sometimes more than 1000 feet (300 m), that the field had produced nearly three-quarters of a billion barrels (100 million metric tons) of oil by 1969.

Conejo One of the smallest fields in the province is perhaps the most interesting. This is the tiny Conejo field east of Camarillo at the base of the Conejo Grade. The field was unique in California because oil came from the fractured Miocene Conejo volcanics rather than from sedimentary rocks, which normally provide oil and gas reservoirs.

First discovered in 1892, the oil was at shallow depth, generally between 60 and 80 feet (18–25 m). For many years the wells were pumped in groups by little donkey engines that jerked cables fanned out on pulleys to the wells. Yield was never great, but the oil produced was highly desirable for lubricating oil stock at refineries.

The field was abandoned in the late 1940s, but an enterprising machinist subsequently leased the property and decided to let nature do the pumping for him. He set some old windmills to pumping six or eight of the wells (Figure 8-19). To his dismay, he got about 99 barrels (16,000 l) of water for every barrel of oil. He then fitted each windmill with endless loops of chain that were turned by the windmill. Oil clung to the chains and water dropped off. At the well-head the chain passed a wiper, and the oil dripped into a collecting trough. At best this operation recovered only about 5 barrels (800 l) of oil a week, so the field was again and finally abandoned.

Santa Barbara Channel   The Santa Barbara Channel has more than 20 oil fields, several produced from the shore. The Summerland field, discovered in 1896, was the first offshore field developed in North America and at one time had several hundred wells producing from piers extending up to 700 feet (210 m) from shore.

The most famous offshore field in this region is the Dos Cuadras field, first produced from a platform in 1968. It is a major field by American standards and is unusual in several respects. The highest point in the reservoir rock is only 300 feet (90 m) from the sea floor, in contrast to the normal pattern of giant oil fields where 1000 feet (300 m) of rock may lie above the reservoir. Furthermore, at Dos Cuadras these intervening rocks are surprisingly permeable sandstones, siltstones, and clays. It was from Platform B of this field that the notorious Santa Barbara oil spill of 28 January 1969 occurred.

During drilling of the well that blew out, the casing had been cemented only about 230 feet (70 m) below the sea floor when higher

Figure 8-19.
Windmills pumping
oil, Conejo oil field.
(Photo by Robert
M. Norris)

than expected pressures were encountered and a gaseous mist erupted from the well. The blowout preventer was closed, but the gas and oil were under high pressure and able to move up the mostly uncemented hole into lower-pressure reservoirs that failed to contain them. In a few minutes the gaseous fluid boiled up from the sea floor, and within 24 hours oil and gas were exuding from fractures along a zone nearly 1300 feet (400 m) long. Not until 8 February was the flow checked, and it was months before the rate was reduced to about 10 barrels (1600 l) per day. In the meantime, beaches along the Santa Barbara coast were blackened with heavy oil, and many birds and sea mammals were killed.

Though the long-term effects of the oil spill are disputed, there is little evidence of the event today. Both the beaches and marine plants and animals have recovered substantially, perhaps partly because oil seeps have been part of the channel's natural environment for millennia.

Wilmington Field Subsidence   Subsidence of land over the Wilmington field was first observed in 1937, a year after the field's discovery. By 1941, an elliptical area near the east end of Terminal Island had sunk 1.2 feet (0.36 m). By 1958, subsidence had increased to 25 feet (7.5 m) and subsequently has reached about 30 feet (10 m). After much litigation and geologic study, it was established that removal of oil and gas had allowed the rock and mineral grains in the reservoir rocks to pack together more closely, reducing thickness of the beds and encouraging subsidence. In a geologically active area, of course, it is always possible that subsidence is unrelated to extraction of oil and gas. In the Wilmington case, however, the pattern of subsidence plotted on a map outlined the form of the oil reservoir almost perfectly, with maximum subsidence near the center of the field.

The subsidence caused severe problems in this heavily industrialized district. Bridges were jacked up and straightened; high dikes were built to protect a large power plant from encroachment by the ocean; the U.S. Navy drydock was endangered by flooding; and numerous buildings were damaged or had to be raised. About the only advantage was the deepening of ship channels without dredging. Since 1958, the *rate* of subsidence has been greatly reduced by pumping salt water into the reservoirs to replace the oil and gas withdrawn. This nearly stabilizes the fluid pressure in the system, helps flush out more oil and gas, and reduces subsidence.

Water flooding itself may engender serious consequences, however. For example, it was the principal cause of renewed movements on small faults that ultimately provoked failure of the Baldwin Hills Dam in 1963. Two hundred and fifty million gallons (950 million l) of water were released into a residential neighborhood, killing 5 people, damaging 277 homes, and effecting property loss of $12 million.

*Figure 8-20.
Home threatened
by active seacliff re-
treat west of Santa
Barbara. (Photo by
Robert M. Norris)*

Seacliff Retreat at Santa Barbara

Though nearly all seacliffs are eroding landward, in only a few places in California has the rate of erosion been established. Santa Barbara is one of these places, and here rates vary from 3 to 11 inches (7.5 −28 cm) annually.

Santa Barbara cliff erosion is primarily attributable to the following five processes, all well-known to geologists but hitherto less appreciated by homeowners, builders, and government officials.

1. Undercutting of the base of the cliff by direct wave attack.

2. Weathering resulting in disintegration of rocks making up the cliff face.

3. Emergence of underground water at the cliff face, weakening the rocks.

4. Rainwash on the face of the cliff.

5. Various kinds of landsliding.

Because most of these processes are slow, they generally are not appreciated by the public until some property is threatened or lost.

Development of coastal bluff property has been relatively intense in the Santa Barbara area, with the result that examples of all processes can be observed. Figure 8-20 shows a house being threatened by cliff retreat caused chiefly by processes 1, 2, and 4.

REFERENCES

General

Dibblee, T. W., Jr., 1950. Geology of Southwestern Santa Barbara County. Calif. Div. Mines and Geology Bull. 150.

———, 1966. Geology of Central Santa Ynez Mountains, Santa Barbara County, California. Calif. Div. Mines and Geology Bull. 186.

———, 1970. Geology of the Transverse Ranges. Mineral Information Service (now California Geology), v. 23, pp. 35–37.

Elders, Wilfred A., ed., 1971. Geological Excursions in Southern California. Univ. Calif. Riverside Campus Museum Cont. no. 1.

Hoots, H. W., 1931. Geology of the Eastern Part of the Santa Monica Mountains, Los Angeles County, California. U.S. Geological Survey Prof. Paper 165C, pp. 83–134.

Rogers, John, 1961. Igneous and Metamorphic Rocks of the Western Portion of Joshua Tree National Monument, Riverside and San Bernardino Counties, California. Calif. Div. Mines and Geology Spec. Rept. 68.

Vedder, J. G., and others, 1969. Geology, Petroleum Development and Seismicity of the Santa Barbara Channel Region, California. U.S. Geological Survey Prof. Paper 679.

Yerkes, R. F., and others, 1965. Geology of the Los Angeles Basin, California—An Introduction. U.S. Geological Survey Prof. Paper 420A.

Special

Clements, Thomas, 1966. St. Francis Dam Failure of 1928. Assoc. Eng. Geol. Spec. Pub., pp. 90–91.

Crowell, John C., ed., 1975. San Andreas Fault in Southern California. Calif. Div. Mines and Geology Spec. Rept. 118.

Hanna, G. Dallas, 1969. Diatoms and Diatomite. Mineral Information Service (now California Geology), v. 22, pp. 111–118.

Hill, Mary, ed., 1971. San Fernando Earthquake. California Geology, v. 24, pp. 59–85.

Kiessling, Edmund, 1963. A Field Trip to Palos Verdes Hill. Mineral Information Service (now California Geology), v. 16, no. 11, pp. 9–14.

Morton, Douglas M., and Robert Streitz, 1967. Landslides. Mineral Information Service (now California Geology), v. 20, pp. 123–129, 135–140.

Norris, Robert M., 1968. Seacliff Retreat at Santa Barbara, California. Mineral Information Service (now California Geology), v. 21, pp. 87–91.

Ransome, F. L., 1928. Geology of the St. Francis Damsite. Econ. Geol., v. 23, pp. 553–563.

Shreve, Ronald L., 1968. Geology of the Blackhawk Slide. Geol. Soc. Amer. Spec. Paper 108.

U.S. Geological Survey, 1971. San Fernando Earthquake of Feb. 9, 1971. Prof. Paper 733.

Weaver, Donald W., and others, 1969. Geology of the Northern Channel Islands. Amer. Assoc. Petrol. Geol. and Soc. Econ. Paleon. Miner. (Pacific Sections) Spec. Pub.

Chapter Nine

Coast Ranges

*False facts are highly injurious to the progress
of science, for they often endure long; but
false views, if supported by some evidence do
little harm, for everyone takes a salutary
pleasure in proving their falseness.*

Charles Darwin

Interpretations of Coast Range geology have been greatly affected by the theory of plate tectonics. As a result, much of the geologic history previously accepted for the province has now been substantially revised. In addition, studies of Coast Range rock relationships have contributed significantly to plate tectonic theory in general. Once mainly an enigma of local concern, the Coast Ranges are now a reference area of world importance.

The Coast Ranges stretch about 600 miles (960 km) from the Oregon border to the Santa Ynez River and fall into two subprovinces: the ranges north of San Francisco Bay and those from the bay south to Santa Barbara County. This division is really one of convenience rather than geologic distinction, for the ranges have more in common than they have differences. The differences that do exist probably occur because the northern ranges lie east of the San Andreas fault zone, whereas most of the southern ranges are to the west. Moreover, the southern ranges are better known because of their oil and gas resources, easier accessibility, more intensive land development, and clearer rock exposures due to sparser vegetation and lower rainfall.

GEOGRAPHY

From south to north, the more important Coast Range geographic units are as follows.

1. The San Rafael Mountains, which adjoin the Transverse Range province in Santa Barbara County. The San Rafael Mountains merge to the northwest with the Sierra Madre, but are separated from them by the Nacimiento fault.

2. The Temblor Range, facing the San Joaquin Valley. This range extends northwest and divides near Cholame into an eastern branch, the Diablo Range, and a western branch, the Cholame Hills. The eastern foothills of the Temblor Range form several tiers, which become lower toward the east until they are buried completely by valley alluvium.

3. The Carrizo Plain, a broad, flat desert valley in southeastern San Luis Obispo County.

4. The Caliente and La Panza ranges, also located in southeastern San Luis Obispo County.

5. The hill country of northern San Luis Obispo and southern Monterey counties, which becomes the Gabilan Mesa on the east side of the Salinas Valley. The eastern and highest part forms the Cholame Hills.

6. The Santa Lucia Range, a major mountain block extending from the Cuyama River to Monterey.

7. The Gabilan Range, east of Salinas Valley, extending from the Gabilan Mesa to the Pajaro River.

8. The Salinas Valley.

9. The Diablo Range, extending from northern Kern County to central Contra Costa County.

10. The Santa Cruz Mountains and the San Francisco Peninsula.

11. San Francisco and adjacent bays.

12. The Mendocino Range, extending along the coast from San Francisco Bay to Humboldt Bay. This is the dominating block of the northern Coast Ranges.

13. The eastern ranges, from Clear Lake to the Oregon border. From Clear Lake south, named units include Petaluma-Cotati and Sonoma valleys, and the Mayacmas, Sonoma, Howell, and Vaca mountains.

Figures 9-1 and 9-2 show the locations of the main features of the Coast Ranges.

The province contains many elongate ranges and narrow valleys that are approximately parallel to the coast, although the coast usually shows a more exact northerly trend than do the ridges and valleys. Thus some valleys intersect the shore at acute angles and some mountains terminate abruptly at the sea. Only minor streams enter the sea at right angles to the shore; most major streams flow many miles through inland valleys that roughly parallel the coast. Except at San Francisco Bay, where a pronounced gap separates the northern and southern Coast Ranges, travel in any direction in the province involves crossing range after range.

Although elevations are moderate, relief is sometimes considerable. For example, within a mile (1.6 km) of the ocean are several peaks of the Santa Lucia Range that are more than 2500 feet (760 m) high; Cone Peak (5155 ft or 1572 m) is only 4 miles (6.4 km) from the ocean. Travelers along the coastal highway from San Simeon to Monterey are almost always impressed with the precipitous seaward face of the Santa Lucia Range, for the road snakes along cliffs often 500 or 600 feet (160–180 m) above the water. Highest elevation in the southern Coast Ranges is Big Pine Mountain (6828 ft or 2083 m) in the San Rafael Mountains. The northern Coast Ranges are higher, particularly in southern Trinity County, where Solomon Peak rises to 7581 feet (2312 m), the highest point anywhere in the Coast Range province.

DRAINAGE

Drainage is primarily controlled by structure. The large streams in the northern ranges, such as the lower courses of the Klamath, Mad, Eel, and Russian rivers, all follow the structural grain of faults or folds for much of their lengths. Of these, the Russian River has the most noteworthy drainage. This river flows south in a normal way more than 40 miles (64 km) to near Healdsburg where it abruptly turns west, crosses the Mendocino Range by a gorge up to 1000 feet (300 m) deep, and reaches the coast near Jenner. Because it would be simpler geologically for the river to flow south through the valley occupied by Santa Rosa and Petaluma and then into San Francisco Bay, geologists have long speculated about the origin of the river's lower course. Studies now indicate that the river established its initial channel on a more subdued Pliocene terrain. The river then flowed seaward across a blanket of sedimentary deposits, through which it subsequently cut downward into underlying Franciscan rocks. By the time the Santa Rosa–Petaluma lowland began sinking to its present elevation, the Russian River had established the course that it retained even though the valley from Healdsburg to Santa Rosa and Petaluma continued to sink and the Coast Ranges continued to rise.

The southern ranges also contain drainages strongly controlled by faults and synclinal folds. The best example of a structurally controlled stream is the Salinas River, which lies in a synclinal trough for most of its course. Some faulting is involved in the lower valley, but even here folding seems to be the dominant structural control. Before joining the Salinas River, the San Antonio and Nacimiento rivers also follow linear systems of folds and faults.

On the other hand, streams such as the Pajaro and the Santa Maria–Cuyama drain broad inland valleys and then follow gorges across intervening ranges before emptying into the sea. The Pajaro occupies a deep gorge between the Santa Cruz and Gabilan chains and, with its tributary the San Benito River, drains the southern Santa Clara Valley. The Santa Maria River and a primary tributary, the Cuyama, have cut a zigzag course across the trend of the south-

248

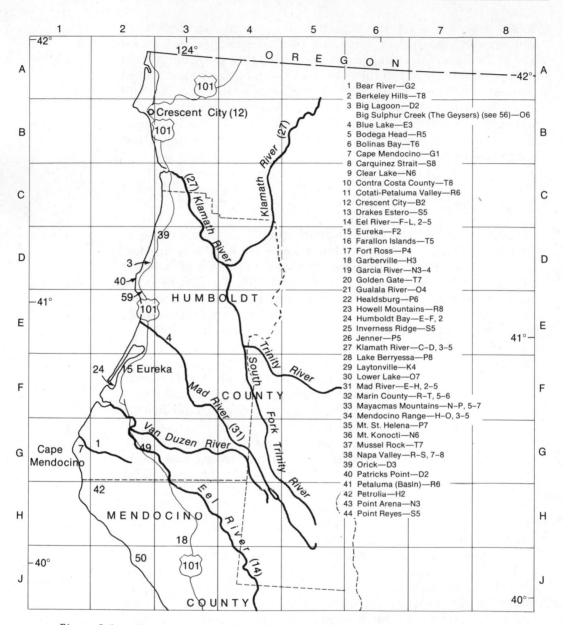

1 Bear River—G2
2 Berkeley Hills—T8
3 Big Lagoon—D2
   Big Sulphur Creek (The Geysers) (see 56)—O6
4 Blue Lake—E3
5 Bodega Head—R5
6 Bolinas Bay—T6
7 Cape Mendocino—G1
8 Carquinez Strait—S8
9 Clear Lake—N6
10 Contra Costa County—T8
11 Cotati-Petaluma Valley—R6
12 Crescent City—B2
13 Drakes Estero—S5
14 Eel River—F-L, 2-5
15 Eureka—F2
16 Farallon Islands—T5
17 Fort Ross—P4
18 Garberville—H3
19 Garcia River—N3-4
20 Golden Gate—T7
21 Gualala River—O4
22 Healdsburg—P6
23 Howell Mountains—R8
24 Humboldt Bay—E-F, 2
25 Inverness Ridge—S5
26 Jenner—P5
27 Klamath River—C-D, 3-5
28 Lake Berryessa—P8
29 Laytonville—K4
30 Lower Lake—O7
31 Mad River—E-H, 2-5
32 Marin County—R-T, 5-6
33 Mayacmas Mountains—N-P, 5-7
34 Mendocino Range—H-O, 3-5
35 Mt. St. Helena—P7
36 Mt. Konocti—N6
37 Mussel Rock—T7
38 Napa Valley—R-S, 7-8
39 Orick—D3
40 Patricks Point—D2
41 Petaluma (Basin)—R6
42 Petrolia—H2
43 Point Arena—N3
44 Point Reyes—S5

Figure 9-1.   Place names: northern Coast Ranges.

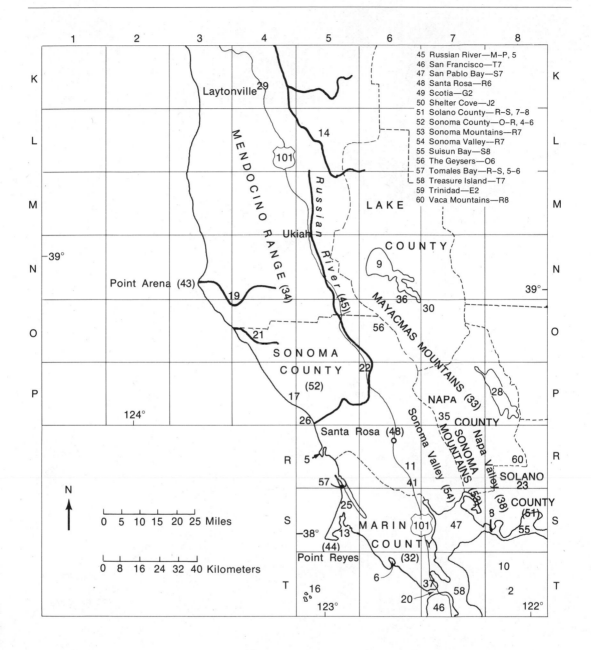

45 Russian River—M–P, 5
46 San Francisco—T7
47 San Pablo Bay—S7
48 Santa Rosa—R6
49 Scotia—G2
50 Shelter Cove—J2
51 Solano County—R–S, 7–8
52 Sonoma County—O–R, 4–6
53 Sonoma Mountains—R7
54 Sonoma Valley—R7
55 Suisun Bay—S8
56 The Geysers—O6
57 Tomales Bay—R–S, 5–6
58 Treasure Island—T7
59 Trinidad—E2
60 Vaca Mountains—R8

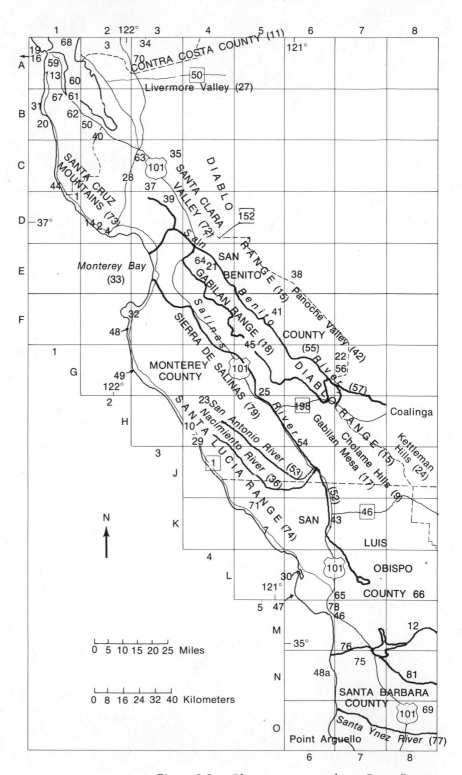

Figure 9-2. Place names: southern Coast Ranges.

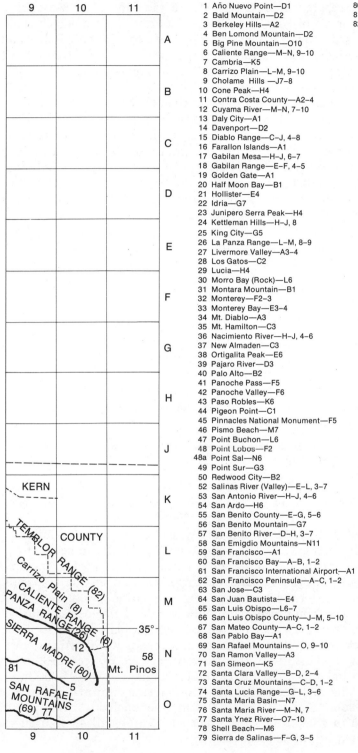

1 Año Nuevo Point—D1
2 Bald Mountain—D2
3 Berkeley Hills—A2
4 Ben Lomond Mountain—D2
5 Big Pine Mountain—O10
6 Caliente Range—M–N, 9–10
7 Cambria—K5
8 Carrizo Plain—L–M, 9–10
9 Cholame Hills—J7–8
10 Cone Peak—H4
11 Contra Costa County—A2–4
12 Cuyama River—M–N, 7–10
13 Daly City—A1
14 Davenport—D2
15 Diablo Range—C–J, 4–8
16 Farallon Islands—A1
17 Gabilan Mesa—H–J, 6–7
18 Gabilan Range—E–F, 4–5
19 Golden Gate—A1
20 Half Moon Bay—B1
21 Hollister—E4
22 Idria—G7
23 Junipero Serra Peak—H4
24 Kettleman Hills—H–J, 8
25 King City—G5
26 La Panza Range—L–M, 8–9
27 Livermore Valley—A3–4
28 Los Gatos—C2
29 Lucia—H4
30 Morro Bay (Rock)—L6
31 Montara Mountain—B1
32 Monterey—F2–3
33 Monterey Bay—E3–4
34 Mt. Diablo—A3
35 Mt. Hamilton—C3
36 Nacimiento River—H–J, 4–6
37 New Almaden—C3
38 Ortigalita Peak—E6
39 Pajaro River—D3
40 Palo Alto—B2
41 Panoche Pass—F5
42 Panoche Valley—F6
43 Paso Robles—K6
44 Pigeon Point—C1
45 Pinnacles National Monument—F5
46 Pismo Beach—M7
47 Point Buchon—L6
48 Point Lobos—F2
48a Point Sal—N6
49 Point Sur—G3
50 Redwood City—B2
52 Salinas River (Valley)—E–L, 3–7
53 San Antonio River—H–J, 4–6
54 San Ardo—H6
55 San Benito County—E–G, 5–6
56 San Benito Mountain—G7
57 San Benito River—D–H, 3–7
58 San Emigdio Mountains—N11
59 San Francisco—A1
60 San Francisco Bay—A–B, 1–2
61 San Francisco International Airport—A1
62 San Francisco Peninsula—A–C, 1–2
63 San Jose—C3
64 San Juan Bautista—E4
65 San Luis Obispo—L6–7
66 San Luis Obispo County—J–M, 5–10
67 San Mateo County—A–C, 1–2
68 San Pablo Bay—A1
69 San Rafael Mountains— O, 9–10
70 San Ramon Valley—A3
71 San Simeon—K5
72 Santa Clara Valley—B–D, 2–4
73 Santa Cruz Mountains—C–D, 1–2
74 Santa Lucia Range—G–L, 3–6
75 Santa Maria Basin—N7
76 Santa Maria River—M–N, 7
77 Santa Ynez River—O7–10
78 Shell Beach—M6
79 Sierra de Salinas—F–G, 3–5

80 Sierra Madre—M–N, 8–10
81 Sisquoc River—N8–10
82 Temblor Range—K–M, 9–10

ern ranges, separating the Sierra Madre and San Rafael Mountains from the ranges of San Luis Obispo County. The upper Cuyama drains a broad structural depression between the Caliente Range and the Sierra Madre. The Sisquoc River, another principal tributary of the Santa Maria, follows the trend of faults and folds within the San Rafael Mountains.

Why the Pajaro and Santa Maria rivers behave as they do is not fully known, but there are several possibilities. Both rivers may be antecedent, with their present courses set before folding and faulting outlined modern Coast Range structure; they may have been superimposed on sedimentary rocks that previously covered older Coast Range basement; streams eroding headward may have breached mountain barriers and captured upland drainages; or some combination of these events may have occurred. As with the Russian River, a complex explanation is likely, probably involving both antecedency and superimposition.

ROCKS

Since the 1914 work of A. C. Lawson, it has been recognized that the Coast Ranges have two dissimilar core complexes (basement rocks) in contact along major longitudinal faults: a Franciscan complex of eugeosynclinal and basic intrusive rocks and a granitic-metamorphic complex that includes the Sur series. Present distribution of these rocks is not fully understood, although many hypotheses have been offered.

Franciscan Basement

The Franciscan eugeosynclinal rocks have been variously labeled a *series*, a *formation*, or an *assemblage*. Some portions have been called a *melange*, a tectonic unit produced by fragmenting and mixing several rock types. Lithologically, the Franciscan is dominated by grayish green graywackes (sandstones), generally in beds 1 to 10 feet (0.3–3 m) thick. These graywackes were derived from rapid erosion of a volcanic highland and deposited in deep marine basins, usually by turbidity currents or submarine mudflows. The graywackes are composed mainly of quartz and plagioclase feldspar, with a chlorite-mica matrix that confers the dark greenish color. These rocks have immense volume and constitute 90 percent of the Franciscan. It is estimated that they average 25,000 feet (7600 m) in thickness and are exposed over 75,000 square miles (194,000 km^2), on both land and the sea floor. This gives a minimal volume of 350,000 cubic miles (1,500,000 km^3)—enough to cover all of California to a depth of 10,000 feet (3000 m) or all 48 contiguous states to a depth of 600 feet (180 m).

The graywackes are interbedded with lesser amounts of dark shale and even occasional limestone. Sometimes associated are thick accumulations of reddish radiolarian cherts, thought to represent organic deposition in marine waters possibly 10,000 feet (3000 m) deep. One limestone, the Calera, occurs discontinuously along the east side of the San Andreas fault from San Francisco to near Hollister. Another is the red Laytonville limestone found north of San Francisco Bay almost to Eureka; it has fewer and smaller outcrops than the Calera. Previously both limestones were thought to be chemical precipitates, because of their fine grain and lack of obvious fossils. Studies have shown, however, that these rocks contain siliceous radiolarians and planktonic calcareous foraminifers, with bulk composition resembling modern deep-sea oozes. Furthermore, the dark color and bituminous character of some of the Calera suggest that parts were deposited in stagnant basins. Within this array of sedimentary rocks are some altered submarine volcanics, now mostly greenstones, and other metamorphic rocks such as the unusual blue glaucophane schist and the more common green chlorite schist.

All these Franciscan rocks have been intruded by ultrabasic igneous rocks, now serpentinized peridotite (serpentinite). Sometimes the serpentinites have been injected as normal molten intrusives, but in other instances they occur in sill-like sheets that lack the thermal alteration of enclosing rocks expected in most sills. In still other cases, these plastic serpentinites have squeezed up through the overlying rocks as plugs or diapirs. The prevailing view is that these serpentinized peridotites are altered masses derived from the upper mantle and transferred tectonically to the earth's surface.

Although the Franciscan is more than 50,000 feet (15,000 m) thick, no recognizable top or bottom has been observed. This is surprising, because many rocks in adjacent provinces are much older. This curious record—like a book missing its first and last pages—has prompted the suggestion that Franciscan sediments were deposited in a deep oceanic trench directly on mantle material or on a thin oceanic crust overlying the mantle. Supporting this contention is the presence in the Franciscan of ophiolites, distinctive assemblages of ultramafic and mafic rocks thought to represent typical oceanic crust. A complete ophiolite sequence includes: (top) pillow lavas often containing pockets of radiolarian chert; a mass of basaltic dikes and sills (the sheeted complex); gabbro and diorite; and (bottom) ultramafic complex including serpentinites and dunites.

Such sequences, or parts of them, have been observed in the Coast Ranges, the Sierra Nevada, and elsewhere in the world, always associated with eugeosynclinal sedimentary rocks similar to the Franciscan. Ophiolites are interpreted as masses of oceanic crust because of: their lithologic similarity to samples dredged from oceanic fracture zones; their bulk chemical composition; their close association with pillow lavas and radiolarian cherts, indicating deep-sea volcanic extrusion; and similarity of seismic characteristics measured in both ophiolites and oceanic crust.

Glaucophane, which gives blue schist its characteristic color, jadeite, and lawsonite occur in some Franciscan graywackes. These minerals are thought to form under low temperature (not over 300°C) and high pressure (about 70,000 ft or 21,300 m of burial). Consequently, some geologists have proposed that Franciscan rocks were carried rapidly down from their depositional site on the deep-sea floor along a subduction zone beneath the edge of the continent, where they were subjected to high pressure. Before becoming thoroughly heated, they were forced back up with equal rapidity, thus producing minerals that reflect both high pressure and low temperature. Other geologists think these minerals were generated nearer the surface by high pressure involved in folding and faulting.

Franciscan sedimentary rocks contain few fossils, but widely scattered localities have yielded specimens ranging from late Jurassic to late Cretaceous (one report claims Paleocene and Eocene). Radiometric dating of several associated ophiolite sequences has given a late Jurassic age of 155 million years.

Crystalline Basement

The second type of basement underlying the Coast Ranges occurs between the Nacimiento and San Andreas fault zones in the southern ranges and west of the San Andreas in the northern ranges. Known as the Salinian block, this basement consists of metamorphic rocks and granitic plutons, a common association in California.

The Salinian metamorphic rocks present in the Santa Lucia Range are known as the Sur series and include gneiss, schist, quartzite, and marble. Lesser amounts of similar metamorphics occur in the Santa Cruz, Gabilan, and La Panza chains. The age of these rocks is not firmly established, but poorly preserved fossils from the Gabilan Range suggest at least partial Paleozoic age.

The granitic rocks intruding these metamorphics are widespread and probably underlie much of the Salinian block, although they frequently are concealed by younger sediments. They vary from granodiorite and quartz monzonite to quartz diorite, compositions much like the plutonic rocks of the Sierra Nevada and the Peninsular Ranges. In contrast to the poorly dated metamorphic rocks, good radiometric dates exist for Salinian plutonics—from 69 to 110 million years (late Cretaceous). Generally the dates are younger on the block's western side, suggesting that the western plutons cooled later, probably because they were deeper in the batholith. This in turn indicates that the western part of the Salinian block has been uplifted more than the eastern.

The Salinian granitic plutons are definitely younger than at least some of the Franciscan rocks. Furthermore, no contact metamorphism exists where Salinian granitics are in contact with the Franciscan. This suggests that present proximity of the two basements results from large displacement along the San Andreas and Nacimiento faults.

Cretaceous Sedimentary Sequences

Besides the Franciscan formation and the Sur series, the only major pre-Cenozoic sediments in the Coast Ranges are in the Great Valley sequence. This is an enormous thickness of miogeosynclinal late Jurassic to late Cretaceous (some claim Paleocene) shale, sandstone, and conglomerate, generally quite unlike the contemporary Franciscan assemblage.

The lower part of the Great Valley sequence is the late Jurassic Knoxville formation, a dark shale found mainly in a belt 110 miles (177 km) long and 16,000 feet (4900 m) thick along the western Sacramento Valley. Similar rocks occur as far south as Kern County east of the San Andreas fault and in the San Rafael Mountains west of the Nacimiento fault. The Knoxville contains some graywacke and even some pillow lava and basalt-derived sandstone reminiscent of the Franciscan, but it is chiefly a dark, rhythmically bedded shale with minor sandstone beds.

Above the Knoxville are lower Cretaceous sandstones formerly called the Shasta series. These are as much as 34,000 feet (10,400 m) thick and are associated with minor conglomerate and other sedimentary rocks. The lower Cretaceous appears only occasionally in the southern Coast Ranges. Miogeosynclinal upper Cretaceous rocks (formerly designated the Chico group) are widespread both east and west of the San Andreas fault, however, and make a nearly continuous belt from northern California to Kern County. (Although the terms Shasta series, Chico group, and Jurassic Knoxville have been applied throughout California, recent studies tend to discredit their use because they do not correspond to mappable units.) The thickest section of late Cretaceous rocks occurs in the northern Coast Ranges, in the eastern Mendocino Range (15,000 ft or 4550 m) and in the eastern Diablo Range (28,000 ft or 8500 m).

So far there is no indication that material from the Salinian block contributed to the lower part of the Great Valley sequence. Instead, all materials seemingly came from the east. This absence of any materials derived from the now adjacent Salinian block is evidence that the block was not there at all during the late Jurassic and early Cretaceous. This is consistent with the late Cretaceous dates for the block's intrusives. Had the Knoxville and Shasta rocks been deposited on the Salinian block, they would have been intruded and metamorphosed by the younger granitic rocks and subsequently incorporated into the Sur series.

Latest Cretaceous marine sedimentary rocks are present on the Salinian block and on the Great Valley basement to the east. The two occurrences differ, however, and do not represent a continuous sedimentary sheet deposited across discrete blocks situated in modern position. Latest Cretaceous beds east of the San Andreas fault and the Salinian block are the final members of the thick Great Valley sequence. On the other hand, latest Cretaceous rocks on the Salinian block, west of the San Andreas, rest unconformably on granitic and metamorphic basement. These Salinian beds are almost 6000 feet (1800 m) thick. It seems evident that the San Andreas fault forms a major break between these two late Cretaceous sequences.

Cenozoic Sedimentation

Correlation Problems By Cenozoic time, the sediments being deposited in the Coast Ranges were primarily of continental shelf origin. Although every Cenozoic epoch is represented, nowhere is there a complete section (this supports claims of repeated but localized tectonism). Since the gaps in the Cenozoic record occur at different points and in different localities, severe correlation problems exist. Ben Page points out that correlating these Cenozoic units requires the incorporation of multiple approaches simultaneously, since any single method may succeed in one area only to fail in another. He suggests five approaches to the problem.

1. Study of megafossils, particularly mollusks.
2. Analysis of heavy mineral detrital grains.
3. Use of microfossils, chiefly foraminifers.
4. Evaluation of palynomorphs (spores and pollen grains).
5. Study of the relationship of critical floras and faunas to radiometric dates to avoid the confusion caused by the tendency of plant and animal groups to form communities in response to environmental constraints.

Correlation of *pre*-Cenozoic Coast Range rocks across structural boundaries has not yet been accomplished. The reason presumably is that movement in most fault zones has been great enough to bring very different pre-Cenozoic materials into contact with one another, sequences like the Franciscan and Great Valley units. On the other hand, Cenozoic rocks, being younger and less extensive originally, are not so widely separated across structural breaks and so have been less affected by Coast Range tectonism. Moreover, they are less affected by intrusion and metamorphism than the older rocks. Nevertheless, it is unlikely that a detailed Cenozoic history of the Coast Ranges will be available very soon; our understanding of the relevant lithologies and stratigraphic relationships is not yet clear enough to permit unequivocal interpretations.

For many years the differences in Tertiary beds on either side of the San Andreas were explained by vertical uplift on the fault: first one side up and then the other. According to this view, a depressed block would receive thick deposits that might then be partially or totally removed when the block was reversed and became high standing. In such interpretations, opposing Tertiary sequences could be matched even though they involved notable differences in thickness or had unmatched gaps or extra units on one side of the fault. For example, if a basaltic flow west of the fault was absent in the corresponding eastern sequence, it was presumed that the eastern flow had been eroded when the eastern block stood higher than the western block. Complicated explanations often were necessary to justify all observed differences, and yet large-scale horizontal slip was not generally accepted until about 1953.

Detailed mapping on the San Andreas zone has now established that several Cenozoic sequences on opposite sides of the fault match amazingly well, though today they are separated by many miles. Mason L. Hill and T. W. Dibblee located an Eocene sequence near Palo Alto that they believed closely matched one in the San Emigdio Mountains, on the opposite side of the San Andreas. This implied that post-Eocene separation on the San Andreas could amount to about 225 miles (362 km). Hill and Dibblee also matched a Miocene section across the fault and suggested that post-Miocene displacement amounted to 175 miles (282 km). Skeptics point out, however, that the Eocene (Butano) sandstone, which was used for part of Hill and Dibblee's evidence, occurs on both sides of the fault near Palo Alto. Although these occurrences do differ in rock facies and stratigraphy, similar degrees of change exist within other pertinent formations when no faulting is involved.

Miocene: The Monterey Formation Miocene rocks are more widely distributed in the Coast Ranges than any other Cenozoic deposit. Most of this Miocene is marine and is characterized by organic deposits and silicic phosphatic members. In addition, in the southern part of the province Miocene rocks reflect more volcanic activity than any other Tertiary series.

The Monterey formation is perhaps the province's most distinctive Miocene sedimentary unit. Although it is preeminently a Coast Range rock, as noted previously it extends south into neighboring provinces. The Monterey is characterized by abundant silica that occurs organically as diatomites and inorganically as silicic ash beds. In many places the formation is a half-mile to a mile (0.8–1.6 km) thick. M. N. Bramlette observes that the volume of sediment involved in the Monterey amounts to thousands of cubic miles and notes that although its thickness and lithology vary, its silicification distinguishes it from other California rock units. The Monterey is easy to recognize in outcrop; it is pale buff to white, occurs in thin to very thin beds, and is often cherty where silicified and punky where rich in diatoms. It typically weathers to a dark, adobelike, clay-rich soil that normally supports grass rather than trees.

California phosphatic deposits are most common in Miocene rocks, notably in the Monterey formation where phosphatic shale, pelletal sandstone, and phosphatic mudstone occur. The formation's phosphate and abundant diatoms both reflect deposition in a marine environment in which organic productivity was unusually high. It has been suggested that fluctuating temperatures played an important role. When sea water was cool, diatoms and silica were deposited; when temperatures rose, phosphates were deposited instead.

Pliocene and Pleistocene   At first Pliocene rocks were chiefly marine and more restricted areally than Miocene units. By the end of the epoch, however, the sea had withdrawn and widespread stream gravels and sands appeared. Pliocene deposits are especially prominent on valley floors, as are Pleistocene deposits. In addition, the Pliocene often records continuing tectonic activity. Typically the deposits make thick, conformable sequences with older valley rocks; thicknesses are often so great that continuing subsidence is indicated. Around valley margins there has been important deformation, as shown by numerous unconformities that occur between late Cenozoic units.

Except along the coast and in the Sonoma and Clear Lake regions, Pliocene and Pleistocene strata are alluvial, with some lake-bed deposits. In several areas, these lake beds attest to the former presence of water bodies such as Lakes Merced and San Benito in Santa Clara Valley.

In the Sonoma area, volcanic activity began in the Pliocene and shifted north into the Clear Lake district during the Pleistocene. Chiefly lava flows and pyroclastics, the Sonoma volcanics are well displayed in the Sonoma Range and on either side of the Napa Valley. Mount St. Helena is a prominent peak composed mainly of fragmental Pliocene volcanic rocks. Near Clear Lake the volcanics are younger, with some of late Quaternary age, and include dacite, basalt, and obsidian. Mount Konocti is a nearly perfect, almost uneroded, dacitic strato-volcano that rises 2700 feet (820 m) above the western shore of Clear Lake.

STRUCTURE

Certainly the dominant characteristic of the Coast Ranges is its division into elongate topographic and lithologic strips underlain by discrete basement rocks that are separated by profound structural boundaries (Figure 9-3). The pattern extends east, and probably also west onto the sea floor. On the east, concealed beneath the Central Valley, is the enigmatic boundary between Sierra Nevada basement and the Coast Range Franciscan. Westward, the next major boundary is the San Andreas fault zone, which separates Franciscan basement from the granitic-metamorphic basement of the Salinian

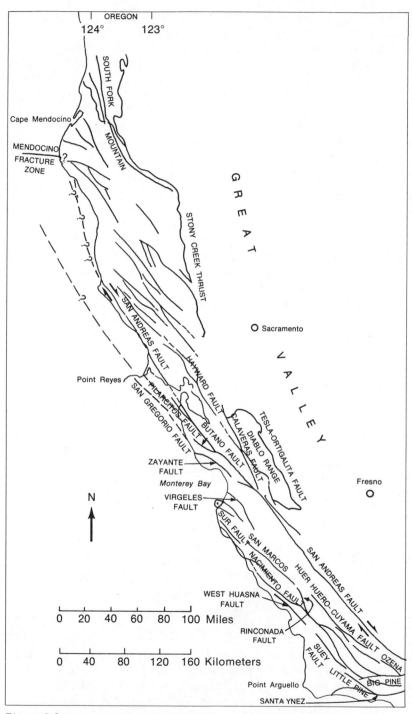

Figure 9-3.

Some known and inferred faults of the Coast Ranges. (Sources: California Division of Mines and Geology and U.S. Geological Survey)

block. South of Monterey, the Nacimiento-Sur fault zone separates Salinian rocks from more Franciscan basement to the southwest. Another boundary should occur farther to the west, offshore, where Franciscan basement is replaced by normal oceanic crust.

South Fork Mountain Thrust Fault

Most of the boundary between the Sierran and Franciscan basements lies beneath thousands of feet of late Mesozoic and Cenozoic sedimentary rocks in the San Joaquin and southern Sacramento valleys. North of Red Bluff, the boundary emerges as the South Fork Mountain thrust, separating the Klamath Mountains from the Coast Ranges.

Several unsolved problems are associated with the Sierran-Franciscan boundary. One is the apparent depositional overlap of the South Fork Mountain thrust fault by lower Cretaceous Great Valley strata west of Red Bluff. Here the thrust appears to cut Franciscan rocks that some geologists think are actually *younger* than the overlapping Great Valley beds. No completely satisfactory explanation for this situation has yet been offered, although it has been postulated that several stages of late Jurassic to early Tertiary faulting are involved. The South Fork Mountain thrust also appears to be essentially continuous with the Stony Creek thrust to the south, along which Great Valley shelf deposits have been thrust west over the eugeosynclinal Franciscan.

San Andreas Fault

The San Andreas fault, the next major boundary to the west in the Coast Range province, brings into contact such contrasting basement rocks that large-scale movement is necessarily involved. The fault cannot be the simple demarcation between continental and oceanic crusts because the Salinian block possesses granitic, continental crust and lies seaward of the Franciscan basement, which presumably includes some oceanic crust. Any analysis of Coast Range structure must incorporate the role of the San Andreas fault. The fault itself is considered more fully later, in Chapter 11.

Nacimiento-Sur Fault

Disagreement exists concerning the name of this fault zone. Some call the northern end the Jolon-Rinconada fault; others assign the name Nacimiento to several parallel faults in Santa Barbara and southern San Luis Obispo counties. For our purposes, however, the Nacimiento-Sur fault zone forms the western boundary of the Salinian basement and is mapped from Point Sur southeastward. The fault approximately parallels the San Andreas south to the Big Pine fault, which truncates or offsets the Nacimiento. Like the mysterious Sierran boundary fault under the Central Valley, the Nacimiento brings granitic basement on the east into contact with Franciscan and ocean basement on the west. This has prompted several intriguing hypotheses.

1. The Sierran-Franciscan boundary fault and the Nacimiento fault are parts of the same structure, which was broken and offset by right slip on the San Andreas.

2. The Franciscan rocks of the Santa Lucia Range west of the Nacimiento fault represent a huge remnant of the upper plate of a large thrust carried west across the Salinian block from east of the San Andreas.

3. Both the Nacimiento and the Sierran-Franciscan boundary are continental margin tectonic features that have been affected by large differential movements.

Unfortunately, the correct explanation for the puzzling basement pattern in the Coast Ranges is far from established. Further study of the ages of pertinent faults and a better understanding of Franciscan rocks and their relationship to the Great Valley sequence are still needed.

Hayward and Calaveras Faults

The Hayward and Calaveras faults are important parts of the San Andreas system in the San Francisco Bay region. The Hayward is currently active, but much of the Calaveras seems to be dormant. The Calaveras does offset some Plio-Pleistocene rocks, however, and a section of the fault is presumably responsible for offset curbs and sidewalks in Hollister.

The Hayward fault branches off the San Andreas south of Hollister. The junction itself is poorly exposed, and part of the Hayward is concealed beneath Quaternary alluvium. The fault extends north along the western foot of the Berkeley Hills to San Pablo Bay, although some authorities trace it as far as Petaluma.

The Calaveras fault branches off the Hayward fault east of San Jose, and their junction is not exposed at all. Some investigators extend the Calaveras fault south to Hollister and the San Andreas fault, but the U.S. Geological Survey now assigns this southern portion to the Hayward fault. Regardless of exact placement of its southern end, most geologists trace the Calaveras north as the Calaveras-Sunol fault, which disappears in the eastern Berkeley Hills.

Stony Creek Thrust Fault

As noted earlier, the Stony Creek thrust fault separates the Great Valley sequence from the Franciscan. The fault extends from near

Red Bluff south to the latitude of Clear Lake and perhaps as far as Sacramento. In fact, some workers have proposed that the Stony Creek is the northern portion of a major thrust system that extends south to eastern San Luis Obispo County and includes the Tesla-Ortigalita fault. Justification for grouping these faults together as the Coast Range thrust is that they all form a boundary along which Great Valley rocks have been forced up and over Franciscan rocks to the west. Proponents of this hypothesis point out that the thrust system is characterized nearly everywhere by long narrow outcrops of serpentinite. According to this interpretation, it follows that the Coast Range thrust must represent the late Mesozoic–early Cenozoic continental margin along which subduction occurred. This theory is strongly disputed. It has been maintained, for instance, that the Tesla-Ortigalita fault cannot be the same structure as the Stony Creek thrust because the serpentinite found along the Stony Creek thrust is virtually missing along the Tesla-Ortigalita fault.

Whatever the case, faults like the Stony Creek and Tesla-Ortigalita do separate Great Valley beds from the Franciscan and no depositional contact between the units has yet been found. Furthermore, the faulting that separates these two major sequences also seems to conceal the base of the Great Valley sequence. The Great Valley sequence has been thrust as much as 50 miles (80 km) westward along these faults. Age of this large-scale thrusting appears to be Paleocene or Eocene, and it certainly must have accompanied a major orogenic event.

Folding

Folding in the Coast Ranges is intense and widespread and involves all rocks. Fold axes tend to parallel faults, usually striking between N40°W and N60°W. Like the faults, the fold axes frequently intersect the coastline at acute angles.

Orogenic activity has affected the province repeatedly, beginning in earliest Cretaceous time or perhaps before. Older structures have been refaulted and refolded, often with concomitant uplift and erosion, and unconformities are numerous and localized. This pattern has jumbled the geologic record, forcing investigators to work backward and figuratively "peel off" younger rocks and "iron out" structures so that the province's older features can be determined.

Often younger rocks resting on granitic basement are less intensely folded than rocks of similar age resting on Franciscan basement. It seems, therefore, that the granitic rocks tend to form tectonically resistant cores, such as that in the Gabilan Range, which protect younger Tertiary sediments from excessive deformation.

Some ranges, like the Diablo, are crudely anticlinal, with younger strata dipping outward from the axis of the range. In the case of the Diablo Range, the archlike structure is traced for more than 90 miles (144 km), an unusually long distance for any Coast Range fold.

Among the intriguing structures of the Coast Ranges are the diapirs (cold intrusions) composed of Franciscan rocks, particularly serpentinite. These intrusive bodies usually occur in anticlinal arches through which they have been forced upward as massive, elongate plugs, a mechanism of emplacement first proposed by William Dickinson. A well-known example occurs near Idria in the Diablo Range; it is 13 miles (20 km) long and 5 miles (8 km) wide and consists chiefly of serpentinite. Another example is Mount Diablo, a plug of mixed Franciscan rocks including serpentinites and some sediments and volcanics. A few of these features may have been extruded as viscous surficial flows, but most have been unroofed by erosion. Examination of the surrounding younger sedimentary beds normally permits establishing the time of the intrusive body's first exposure. If the surrounding sedimentary rocks include little serpentinite detritus, the diapirs probably were covered when the sedimentary beds were deposited. If the sediments contain large amounts of clastic serpentinite, the plugs probably had breached the surface and contributed the detrital serpentinite.

GEOLOGIC HISTORY

Four main orogenic events have been tentatively proposed by Ben Page for the history of the Coast Ranges. (1) Early(?) Cretaceous orogeny was accompanied by granitic intrusion in the Salinian block and by metamorphism of older rocks to form the Sur series. (2) Early Tertiary (Paleocene and Eocene) thrusting of the Great Valley beds over Franciscan rocks probably occurred, followed by (3) prolonged simultaneous Cenozoic strike-slip faulting on the San Andreas and related faults and (4) late Pliocene and Pleistocene orogeny (the Coast Range orogeny). These points make a useful frame of reference when considering the following discussion of Coast Range history.

Pre-Cenozoic

No Precambrian or early Paleozoic rocks have been found in the Coast Ranges, although Paleozoic rocks occur in the Klamath Mountains and Transverse Ranges. Consequently, there is little knowledge of Coast Range history prior to deposition of the sedimentary beds metamorphosed into the Sur series. Confined to the Salinian block, Sur series rocks are composed of gneisses, marbles, schists, quartzites, and granulites. They presumably were a marine shelf sequence originally, a frequently metamorphosed and folded miogeosynclinal assemblage. This multiple metamorphism probably explains the scarcity of fossils, which are poorly preserved but do suggest late Paleozoic age for their enclosing rocks. The Sur series may also include early Paleozoic and even Mesozoic rocks, but it is reasonably certain that it does not include any extensively altered Franciscan material.

The Sur series has undergone several plutonic intrusions, each imprinting folding and metamorphism. It is difficult to assign ages to these plutonic events, despite the availability of potassium-argon dates. Some dates give *latest* Cretaceous ages for plutonic rocks, but these rocks are sometimes overlain depositionally by *late*, not latest, Cretaceous sediments. The aberrant potassium-argon dates may reflect some postintrusive event, perhaps further deformation.

At least partly contemporary with these plutonic events was the deposition of Franciscan and Great Valley sediments. Locations of the depositional basins have not yet been precisely determined, however, in relation either to one another or to the site in which the Salinian intrusives were forming.

During this same time interval, deep-water deposits were accumulating on the continental slope and in an offshore trench, where possibly the deposits rested on oceanic crust. Mixed with land-derived materials delivered to the offshore sites were radiolarian oozes and submarine extrusive and intrusive volcanics. This assemblage subsequently became the Franciscan formation. Deposition was followed and accompanied by rapid downward transport, with underthrusting of both sedimentary and volcanic materials into a subduction zone beneath the continental margin. During this underthrusting, Franciscan beds were mixed tectonically and intruded by ultrabasic peridotitic rocks from the upper mantle. Subduction ceased rather abruptly, allowing the Franciscan assemblage to rise rapidly to the surface—as indicated by its high-pressure, low-temperature mineralogy. Contemporaneously, the Stony Creek and Tesla-Ortigalita thrusts carried Great Valley deposits tens of miles westward over Franciscan terrain. This zone of thrusting may have marked the boundary between the oceanic and continental plates during subduction.

Early Tertiary

Paleocene and Eocene rocks rest on both Salinian and Franciscan materials. The Paleocene and Eocene rocks generally follow patterns of late Cretaceous marine deposition, but were more restricted as seaways became smaller and shallower. Paleocene sandstones, conglomerates, and shales rest on the granites at Point Lobos, on upper Cretaceous in the San Francisco Peninsula, on upper Cretaceous and Sur series in the Santa Lucia Range, and on upper Cretaceous in the Diablo Range. Reconciliation of large-scale subduction with continued continental shelf deposition from latest Cretaceous through the early Eocene is a major problem. Although available evidence indicates that these events were contemporary, the actual geography still eludes us.

The presence of coal beds, deeply weathered clays, and quartzose sandstones all suggest nearly tropical conditions during the Eocene. Minor unconformities between Eocene rock units testify to continued crustal unrest, but not to major orogenic activity.

From late Eocene to early Miocene, the sea withdrew from much of the Coast Ranges. The north particularly seems to have remained above sea level. In the south, shallow seas covered much of the Santa Cruz, Santa Lucia, and Diablo mountain units, but most of the extreme southern ranges remained above the sea.

Miocene

In early Miocene time, the sea again spread over the Coast Ranges, forming numerous bays, straits, islands, and inlets. Deposits are characterized by notable facies and thickness changes over short distances, particularly in early Miocene beds, which were laid down in a complex archipelago. By the middle of the epoch, land areas were reduced, and deep elongate basins had developed. Volcanism, much of it submarine, became prominent. The diatomaceous shales of the Monterey formation were deposited at this time, perhaps in a setting like the modern Gulf of California where deep elongate basins are receiving diatomaceous deposits.

By the close of the Miocene in the northern Coast Ranges, orogenic activity was again underway. The lower Eel River valley became deeply depressed and flooded, a condition that persisted for most of the Cenozoic. Other embayments developed near Petrolia, along the Bear River north of Cape Mendocino, along the Mad River near Eureka, and at Crescent City. These downwarps and the intervening anticlinal arches all have northwesterly axial trends. Much of the marine Miocene and Pliocene in the northern Coast Ranges has been removed by late Cenozoic erosion. Small, down-faulted inliers still preserve isolated blocks of such rocks some miles inland, however, showing that marine rocks were once more extensive.

Middle Miocene seas spread inland across the San Andreas fault into the San Joaquin Valley, but near the end of the epoch, orogeny was renewed and seaways became more restricted and the sediments coarser. In the southern ranges, diatomaceous shales were replaced by the clean quartz sands of the Santa Margarita formation, probably derived from Sierran granitic rocks to the east.

Pliocene

The Pliocene sea persisted chiefly in the seaward ends of what are now prominent valleys such as the Santa Maria Valley and in Half Moon and Bolinas bays. In addition, a long Pliocene embayment extended inland across the southern Gabilan Range and along both sides of the San Andreas fault from Kettleman Hills to the Diablo Range.

The orogenic episode begun in the late Miocene continued into the Pliocene, and by the end of that epoch most of the present ranges were dry land. In the southern ranges particularly, thick blankets of stream gravels were being deposited in the valleys. These sheets of gravel ultimately formed extensive, low-relief surfaces that almost covered some of the ranges. They generally rest by angular unconformity on the older rocks near the range margins, but are conformable with valley floor deposits. Practically all the Cholame Hills and the northern Temblor Range from Paso Robles east are covered with these gravels. Patches of similar materials occur as far north as Eureka, but are generally absent in the northern Coast Ranges.

The Pliocene gravels have various local names. In the southern ranges they are known as either the Paso Robles or the Tulare formation. In the central ranges, equivalent units are the Santa Clara formation, the Livermore gravels, the Merced formation, and the San Benito gravels. Near Eureka, the Packwood gravels are correlative.

Quaternary

Tectonism increased during the Quaternary and culminated, according to most views, in the middle Pleistocene Coast Range orogeny that produced today's topography. Much evidence for the orogeny comes from the southern ranges, where late Quaternary deposits often are unconformable on Pleistocene and older strata. These late Quaternary beds include some marine horizons and thus record a fluctuating sea level. In the San Francisco Bay region and in the central Salinas Valley, however, similar deposits are often conformable on Pleistocene beds.

This pattern of conformable relations between successive deposits in the valleys and unconformable relations between the same units in the foothills and ranges has persisted from the Miocene to the present. The pattern shows that since the middle Tertiary the basins and valleys have been intermittently depressed and nearby mountains and hills have been correspondingly uplifted. In several cases, sinking of the valleys has carried land-laid alluvial and near-shore deposits thousands of feet below their former elevations. In the Sacramento Valley, some land-laid deposits now lie 3000 feet (900 m) below sea level, and in the Santa Clara Valley some freshwater deposits are 300 feet (90 m) below sea level. The northern Coast Range record is less complete, owing to scarcity of Pliocene and Quaternary deposits; where these are present, a history similar to that of the southern ranges is indicated.

Perhaps the most striking examples of Quaternary Coast Range tectonism are young faults and their accompanying belts of intensely deformed Quaternary strata. Movements on the major strike-slip faults have been substantial, although opinions differ regarding cumulative Quaternary slip on the San Andreas particularly. Some suggestions for slip on the San Andreas fault since the middle Pliocene are listed below.

N. E. A. Hinds (1952)	less than a mile (1.6 km)
C. G. Higgins (1961)	4 to 10 miles (6.4–16 km)
L. F. Noble (1954)	20 miles (32 km)
T. W. Dibblee (1966)	20 to 40 miles (32–64 km)

Elevated marine terraces along the shoreline also reflect Quaternary tectonism. Inevitably, however, eustatic changes associated with continental glaciation compound the picture. Because maximum Quaternary sea level probably was not more than a few feet above the present level, virtually all elevated coastal terraces involve tectonic activity. Typically, higher terraces show greater deformation than younger, lower terraces. Further evidence of tectonism is provided by the disparity in elevation from one locality to another.

In the Coast Ranges, elevated terraces have been recognized as high as 900 feet (275 m) above sea level. The terraces identified so far seem no older than Pleistocene, but there is considerable disagreement about whether they are middle or late Pleistocene. For a few localities, there is little argument. Near the mouth of the Santa Ynez River, for example, marine terraces were cut into folded early and middle Pleistocene strata. These terraces are now more than 700 feet (210 m) above sea level, and some have been dated radiometrically at about 100,000 years.

SUBORDINATE FEATURES

Santa Lucia Range

For coastal scenery, the Santa Lucia Range has few equals in the 48 coterminous states. At several places along the Monterey coast, these mountains rise from the sea to significant heights less than 4 miles (6.4 km) from shore (Figure 9-4). The range is about 140 miles (225 km) long and extends from Monterey to the Cuyama River. For much of this distance, it is 20 to 25 miles (32–40 km) wide and so rugged that north of Cambria it is crossed by only one minor road. The highest point is Junipero Serra Peak (5862 ft or 1788 m), west of King City.

The Salinian granitic rocks of the Santa Lucia Range are crucial to an understanding of the history of the Coast Range province and

Figure 9-4. Looking north at the bold coastline of the Santa Lucia Range. (Photo by Robert M. Norris and David Doerner)

its prominent faults. Unlike the extensively metamorphosed Sur series, the granitic rocks can be dated, presumably precisely, by radiometric methods. Most dates from Santa Lucia granites vary from 81 to 92 million years, making them younger than some Franciscan rocks. In the eastern part of the Salinian block, Franciscan sometimes is in fault contact with the granitic basement. Some geologists interpret this fault relationship as evidence for major strike-slip displacement of the granitic rocks. The younger granitic rocks have not yet been found in intrusive contact with the Franciscan.

Although we like to consider radiometric ages valid, actually metamorphic or structural events occurring after plutonic rocks cool can reset radiometric clocks, subsequently yielding ages too young for the original magmatic crystallization. Robert R. Compton has found Santa Lucia granitic rocks with radiometric dates of 69.6 and 75 million years overlain in *normal depositional contact* by unmetamorphosed fossiliferous marine Cretaceous beds of 80 to 85 million years age. This dilemma has not been completely resolved, but Compton has suggested that the anomalous dates reflect post-granitic deformation. In reality, then, the granitic rocks are older than their radiometric dates.

Apart from the Franciscan formation, no pre-Cretaceous un-metamorphosed sedimentary rocks are known in the Santa Lucia Range. The thick section of Great Valley beds seen in the eastern ranges is missing here and apparently was never deposited. Any late Jurassic or early Cretaceous beds once deposited on the Salinian block would undoubtedly be part of the Sur series anyway.

The important Miocene Monterey formation takes its name from extensive exposures south and east of the city of Monterey. The formation dominates the eastern half of the Santa Lucia Range, including the ridge adjoining the Salinas Valley. The prominent ridge northeast of San Luis Obispo, with its oaks, grassy lower slopes, and pine-covered summits, is carved mainly from the Monterey, as is the ridge southwest of San Luis Obispo from Point Buchon to Pismo Beach.

Salinas Valley

The structural history of the Salinas Valley is difficult to interpret. Stream-produced features are readily visible, but apparently folding has been more important in development of the valley than either stream erosion or faulting.

The basement rocks exposed in the adjacent Santa Lucia and Gabilan ranges are either missing or concealed in the Salinas Valley. Although the basement surface may once have had considerable relief (which might account for the present valley), probably it was only slightly irregular when latest Cretaceous seas spread over the land and covered the granitic rocks. There are several reasons for inferring this.

First, geophysical studies indicate that the basement's surface lies more than 5000 feet (1500 m) below the valley floor in many places and as much as 10,000 feet (3000 m) in a few. The only logical explanation is that the basement under the valley has been either folded sharply downward or dropped downward on faults, or both. Second, detailed geologic mapping of the Salinas Valley has demonstrated that sedimentary rocks exposed on the flanks of the Gabilan and Santa Lucia ranges dip down toward the valley. Individual rock units thicken in the same direction, suggesting downward folding as the valley floor sank and sedimentary materials accumulated. Third, faulting is evident along the west. Moreover, the rocks have a pattern that would be expected where a valley block has been down-dropped with respect to adjacent highlands.

Although the generally synclinal character of the Salinas Valley is not favorable for oil accumulation, there are several oil fields near San Ardo. Most of the oil is trapped by variations in rock permeability rather than by anticlinal or fault traps characteristic of most oil fields, and for this reason the San Ardo field was not discovered until 1947.

Diablo Range

The Diablo Range is a well-defined topographic feature 130 miles (210 km) long and as much as 30 miles (48 km) wide. San Benito Mountain (5238 ft or 1598 m) is its highest point, but Mount Hamilton (4209 ft or 1284 m), site of Lick Observatory, and Mount Diablo (3849 ft or 1174 m) are better known. Some authorities include the Berkeley Hills in the Diablo Range, although the Berkeley Hills are separated from the main range by the San Ramon Valley and the Calaveras-Sunol fault zone.

The southern part of the Great Valley sequence is well developed in the Diablo Range and rests on Franciscan basement along the west San Joaquin Valley. The late Jurassic Knoxville formation accounts for only a small part of this sequence in the range, but the late Cretaceous portion is thick and widely distributed. These late Cretaceous marine rocks are usually divided into the Panoche group (mostly sandstone, shale, and minor conglomerate) and the overlying Moreno shale. They are notably developed in the Panoche Pass area, where 28,000 feet (8500 m) are exposed.

In a broad sense, the southern Great Valley units, dominated by late Cretaceous beds, form large collars around the elliptical masses of Franciscan rocks that are the core of the Diablo Range. The Cretaceous rocks do not contain distinctive Franciscan fragments, but do incorporate some granitic materials and numerous clasts of a dark, fine-grained porphyritic rock of unknown source. It is suggested that both the dark rock and the granitic detritus came from the Sierra Nevada. The absence of Franciscan clasts implies that the extensive Franciscan exposures seen today were covered during late Cretaceous time and consequently were unavailable for deposition in the Diablo Range.

Younger sedimentary rocks vary from Paleocene to Pleistocene. They are widely distributed around the margins of the range and in some intermontane areas such as the Livermore, San Ramon, and Panoche valleys. Rocks of every Tertiary epoch appear in the Diablo Range, but nowhere is there a complete section.

Some particularly distinctive Cenozoic rocks are the middle Eocene nonmarine beds east of Mount Diablo, which contain clean, washed quartz sands and coal beds and clays suitable for ceramic use. These and similar beds in California are thought to indicate almost tropical conditions, because the clays and quartz sands appear to be the end results of intense weathering.

Miocene volcanic rocks occur at several places within the Diablo Range, but the best examples are the Quien Sabe volcanics east of Hollister. Flows, dikes, plugs, and agglomerates are included, some of which may be of submarine origin. Flows range from basalt through andesite to dacite; plugs are mostly andesites or rhyolites. The lavas and agglomerates often produce steep cliffs and block-strewn surfaces that contrast sharply with the gentler topography developed on the underlying sedimentary rocks. In some places, the scenery developed on the agglomerates resembles that of Pinnacles National Monument in the Gabilan Range. The volcanics in both regions are of similar age, but different composition.

The general structural pattern of the Diablo Range shows large anticlinal folds with Franciscan cores arranged en echelon and separated by synclinal folds containing younger rocks. Sometimes the crudely anticlinal features (antiforms) are diapirs composed of a mixture of serpentinite and other Franciscan volcanic and sedimentary rocks that have been forced up along faults into and even through the younger rocks.

A large antiform with typical Franciscan core is the one dominating the range from the Panoche Valley northwest to the Livermore Valley. It is about 90 miles (144 km) long and averages 15 miles (24 km) in width, with a maximum of more than 20 miles (32 km) east of San Jose. Two notable intermontane synclinal folds are the Panoche Valley and a syncline between Panoche Valley and the New Idria mining district. Rings of Paleocene, Eocene, and Miocene rocks surround these valleys, which have floors filled by Pliocene and Pleistocene gravels.

Santa Cruz Mountains and the San Francisco Peninsula

The Santa Cruz Mountains and the San Francisco Peninsula are aspects of the same topographic unit. As in some of the Coast Ranges to the south, parts of the Salinian block and parts of the Franciscan block east of the San Andreas are incorporated within the Santa Cruz Mountains. The western Franciscan block, last exposed in the southern Santa Lucia Range, presumably lies far offshore.

The Santa Cruz Mountains extend from the San Francisco Peninsula about 80 miles (130 km) southeast to the Pajaro River, where they merge with the Gabilan Range. Generally less than 10 miles (16 km) wide, between Santa Cruz and San Jose they widen to nearly 20 miles (32 km). The chain inclines to more modest elevations than other ranges in the province. Maximum is about 3800 feet (1160 m) near New Almaden, and average summit elevation is only 2500 feet (760 m).

Most of the range is developed on the Salinian block, but the southern portion is about equally divided between the Salinian block and the Franciscan block. Between Los Gatos and Redwood City, most of the range has Salinian basement, which is well exposed at Ben Lomond Mountain and Montara Mountain. From Shelter Cove north, the entire San Francisco Peninsula has Franciscan basement.

Numerous faults (most probably parts of the San Andreas system) parallel the range or slice obliquely across it. Near Redwood City, the Pilarcitos branches off the San Andreas and follows a westerly course along Montara Mountain to Shelter Cove, where it strikes out to sea. This fault is apparently the eastern demarcation of the Salinian block, for it separates the granitic rocks exposed in Montara Mountain from the Franciscan assemblage of the peninsula. Some geologists have suggested that the Pilarcitos fault represents an older course of the San Andreas that was eventually abandoned in favor of the present trace. This view is corroborated by the Quaternary inactivity of the Pilarcitos.

The major crustal blocks of the Santa Cruz Mountains are subdivided into elongate slices separated by high-angle faults of substantial vertical movement. One of these faults, the Zayante, extends southeast from Ben Lomond Mountain and was the middle Tertiary boundary between a depressed block on the northeast and an elevated block on the southwest. (Elevation and depression refer here to the basement surface and not necessarily to the topographic surface. Even today, there is considerable difference in basement elevation on either side of the Zayante fault, but little topographic difference.) On the southwest, Salinian granitic rocks are exposed at the surface in and around Ben Lomond Mountain or are present at modest depth. In contrast, geophysical studies show that granitic rocks lie 6000 to 9000 feet (1800–2750 m) below the surface northeast of the Zayante fault.

Other major faults include the Seal Cove–San Gregorio fault zone. This extends from Half Moon Bay across the bay but parallel to the coast, returning to shore near San Gregorio and extending south to Año Nuevo Point where it again goes out to sea. Thick Cretaceous sedimentary rocks occur in the narrow coastal block west of this fault, indicating that this slice of seacoast has been elevated with respect to the area on the east.

The San Andreas fault has long been active in the Santa Cruz Mountains. Normally it is marked by a distinct, elongate valley that has been used for reservoir sites north of Redwood City (for example, San Andreas Lake and Crystal Springs). Unfortunately the valley also has been used for intensive housing developments, especially in the Daly City area. This has happened despite displacement of this segment of the fault up to 10 feet (3 m) during the 1906 San Francisco earthquake. Virtually all authorities regard this portion of the San Andreas as active today.

Oldest known rocks in the Santa Cruz Mountains are undated, probably Paleozoic, metasediments of the Ben Lomond district. They form small roof pendants in the granitic rocks and are doubtless equivalent to the Sur series of the Santa Lucia Range.

Santa Cruz Mountain granitic compositions range from gabbro to granite, with most samples being quartz diorites. Well records have shown that these granitic rocks are continuous beneath the surface with those exposed in the Gabilan Range. In addition, quartz diorites almost identical with those of Montara Mountain occur on the bleak Farallon Islands, 28 miles (45 km) west of San Francisco, and along the shelf edge north of the islands for about 30 miles (48 km).

Northeast of the San Andreas and Pilarcitos faults, the San Francisco Peninsula has Franciscan basement with a typical array of graywackes, volcanic sills and dikes, deep-water red cherts, and ultrabasic intrusives with serpentine derivatives. The type locality for the Franciscan formation is the northern San Francisco Peninsula and was first described by A. C. Lawson in 1895. The hills that dot the city of San Francisco are all exposures of various Franciscan rocks.

On the Salinian block, the only occurrences of Cretaceous marine sediments are those west of the San Gregorio fault, in the Pigeon Point area, and possibly north of Montara Mountain in the highly sheared rocks along the Pilarcitos fault. No Cretaceous marine rocks are known from well records in the central Santa Cruz Mountains. East of the San Andreas fault, late Cretaceous marine sedimentary rocks are exposed, primarily near the crest of the range. These rocks are in fault contact with the Franciscan, but are overlain by as much as 16,000 feet (4900 m) of clastic, mostly marine, Cenozoic sediments.

Tertiary marine strata are present throughout the range and possess an aggregate thickness of more than 22,000 feet (6700 m). Considerable tectonic activity is suggested by the nature and distribution of the Tertiary, because some beds were deposited on land and others in deep marine waters. During the early Tertiary, deep-water conditions prevailed over much of the central Santa Cruz Mountain area. A small portion was uplifted during the Oligocene, however, forming a small island on which accumulated land-laid deposits assigned to the Zayante sandstone. Marine conditions subsequently prevailed until at least middle Pliocene.

San Francisco, San Pablo, and Suisun Bays

This system of bays, one of the world's finest harbors, is the only place where streams from interior California reach the sea. The system occupies a late Pliocene structural depression that has been flooded several times in response to Pleistocene glacial cycles.

The Pliocene date was determined in the San Francisco Bay area by stratigraphic studies. A thick Plio-Pleistocene deposit (the Merced formation) occurs here, of which the lower 4500 feet (1370 m) are marine and the upper 500 feet (160 m) are mostly nonmarine. The lower portion contains heavy minerals, indicating locally derived sediment. About 100 feet (30 m) above the marine-nonmarine transition, the mineral assemblage changes abruptly to one identical with that carried by the Sacramento River system and derived primarily from the Sierra Nevada. This mineral change reflects initial establishment of the present drainage pattern.

Figure 9-5.   *Carquinez Strait, with San Pablo Bay in the distance. (Photo by Robert M. Norris and David Doerner)*

San Francisco and San Pablo bays occupy part of the main structural depression that includes the Santa Clara Valley. This depression extends from south of Hollister and northward beyond the bays. In the north section, it divides into the Petaluma, Sonoma, and Napa valleys. Suisun Bay to the east is separated from San Pablo Bay by the narrow, winding Carquinez Strait, thought to be the channel of a superimposed stream cut into bedrock by the Sacramento River (Figure 9-5). The stream channel is now as much as 200 feet (60 m) below sea level. Its extension across the floors of San Pablo and San Francisco bays and out through the Golden Gate has not yet been traced, although the early river almost certainly followed such a course.

Most bays of the San Francisco system are shallow. About 85 percent of the water area is less than 30 feet (10 m) deep, sometimes less than 18 feet (5.5 m). Sediment entering from the Sacramento–San Joaquin system will fill the bays soon (geologically), unless tectonic activity or a rise in sea level intervenes. Another factor is that man's activities have accelerated the rate of fill. Both miners (hydraulic mining and gold dredging) and farmers (plowing and cultivating) have increased the sediment load in the Sacramento–San Joaquin system. Reclamation operations have developed such land additions as Treasure Island and San Francisco International Airport.

The San Francisco Bay area is of special geologic significance because it is the first place in California for which a geologic map was prepared (Figure 9-6). Drawn in 1826, this historic map was not

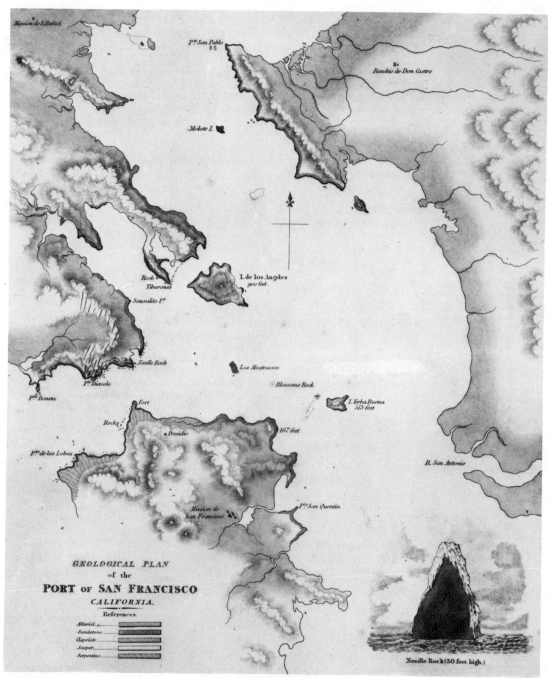

Figure 9-6.
Geologic map of San Francisco area, 1826. (Courtesy of Bancroft Library,
University of California, Berkeley)

only a first for California, but also was probably the second geologic map of any part of North America. It appeared only seven years after William Maclure's map of the eastern United States and only eleven years after William Smith in England drew the first geologic map for any region.

This pioneering map of San Francisco Bay was prepared by Edward Belcher, the surveyor, and Alex Collie, the surgeon, of H. M. S. Blossom, a British sloop sent to meet the explorers John Franklin and William Parry in the Bering Strait. Belcher and Collie accurately represented the form of the bay, its islands, and the Carquinez Strait and also closely calculated the elevation of Mount Diablo. Furthermore, they correlated main rock types with different colors, recorded the trend of rock units, and listed a variety of mineral species present.

Mendocino Range

The northern units of the Coast Ranges are not so clearly defined as the southern units. They also are less well known geologically, although general distribution of rock types is now understood. The Mendocino Range, the main unit, extends about 215 miles (350 km) from San Francisco Bay north to Humboldt Bay. It lies between the coast and the elongate depression occupied by the Russian and Eel rivers. Most peaks are about 3000 to 3500 feet (900–1050 m) high, except west of Laytonville where some are a little over 4200 feet (1280 m) in height. The northern Coast Ranges are rugged, heavily wooded, and crossed by fewer roads than the drier southern ranges.

The Mendocino Range lies almost entirely on the Franciscan block, and its rocks resemble those of the Diablo and Santa Cruz ranges to the south. North of the Golden Gate the San Andreas fault hugs the coast, so here only headlands, promontories, and long narrow slices of coast lie on the granitic Salinian block. Possibly these selvages of coastal northern California do not belong to the Mendocino Range at all, but it is convenient to treat them together.

The main trace of the San Andreas leaves the coast at Mussel Rock and crosses the Golden Gate to Bolinas Bay, where it comes ashore. It has been established that the Pilarcitos fault, not the San Andreas, separates the Salinian from the Franciscan in San Mateo County, but how the Pilarcitos and San Andreas join on the sea bottom off the Golden Gate is not known. In Marin County the San Andreas again forms the eastern edge of the Salinian block and granitic rocks are exposed west of the fault. The San Andreas occupies a well-defined valley from Bolinas to Tomales Bay, which is a long narrow depression opening to the northwest and lying between Bolinas and Inverness ridges (Figure 9-7).

*Figure 9-7.
Looking south along
the San Andreas
fault trace and To-
males Bay. Bolinas
Ridge is on the left
and Inverness Ridge
on the right. (Photo
by Robert E. Wal-
lace, courtesy of
U.S. Geological
Survey)*

A conspicuous geomorphic feature of this coastline is the Point Reyes headland. The landward side consists of granitic Inverness Ridge, almost 1500 feet (450 m) high. To seaward, the ridge faces a tapering lowland more than 8 miles (13 km) wide developed on synclinally folded Pliocene marine sedimentary rocks. Drakes Estero, a shallow lagoon open to the south, lies along the axis of this syncline. (It was here, in 1579, that Francis Drake formally took possession of the land for Queen Elizabeth of England.) Along the lowland's northern coast, northwesterly winds regularly drive beach sand ashore to form dunes that extend inland as much as three-quarters of a mile (1.6 km). The lowland terminates on the southwest at the granitic ridge called Point Reyes.

A superficially similar but smaller feature occurs a few miles north at Bodega Head, the most northerly exposure of definitely Salinian rock. This granitic eminence is separated from the Mendocino Range by the low sandy area surrounding Bodega harbor. Bodega Head was once an island, but has been connected to the mainland by sand deposited by longshore currents. Once the connecting spit was established, winds and sand-loving vegetation developed dunes that are prominently aligned with the northwest winds.

From near Fort Ross to Point Arena, the San Andreas fault zone is only a mile or two (1.6–3.2 km) inland, nearly paralleling the coast. The fault is again distinctively marked by a narrow, long, straight valley that is occupied by such streams as the Gualala and Garcia rivers. The slice of crust thus formed between the sea and the San Andreas is probably the northernmost land occurrence of the Salinian block. Neither granitic nor Franciscan basement is exposed in it, however, so there is some uncertainty about its affinities with Salinian rocks.

The bulk of the Mendocino Range is made up of Franciscan basement overlain by Cretaceous sedimentary rocks of the Great Valley sequence. North of the gorge through which the Russian River crosses the range, Cretaceous beds and typical Franciscan graywacke and shale constitute bedrock in about equal amounts. Fossils indicate that here, as in the southern ranges, the Franciscan is late Jurassic to late Cretaceous.

The relationship between the Great Valley and Franciscan sequences poses unresolved problems in the northern Coast Ranges as it does elsewhere in the province. Great Valley beds of shelf and nearshore origin (the miogeosynclinal suite) have the same ages as the Franciscan slope and deep-water beds (the eugeosynclinal suite). Great Valley beds sometimes rest on Franciscan, but this is not proof they were deposited there initially. In many cases, the two sequences are separated by faults. Even in areas that lack faulting, there are uncertainties; it is not unusual to find chert and greenstone, usually associated with eugeosynclinal Franciscan, interbedded with clay shales typical of the Great Valley sequence. In light of these problems, present geologic interpretation is quite likely to be considerably revised in the future.

The northern Mendocino Range includes two conspicuous down-warped basins, one near Garberville and another south of Cape Mendocino, where thick piles of Tertiary marine rocks are preserved. The range terminates near the mouth of the Eel River against a third, generally similar but larger basin that extends southeast from the coast for about 30 miles (48 km). This large synclinal structure is outlined by marine strata of probably late Miocene age and is filled with Pliocene and Quaternary beds. (The smaller basins lack the Pliocene and most of the Quaternary.) Along the coast, the basin extends from about Cape Mendocino north almost to Eureka. Its Miocene beds are equivalent to such southern units as the Rincon and Santa Margarita formations; the Pliocene is referred to the Wildcat group and is probably the age equivalent of the Pico and Repetto formations of the Los Angeles Basin. This Cenozoic section is more than 12,000 feet (3650 m) thick. It has been of particular interest for many years because of small oil seeps and evidence of natural gas, but only minor production has resulted.

Eastern and Northern Ranges

South of Clear Lake, the inland Coast Ranges are not only better known, but also seemingly more complex than the ranges north and east of the Mendocino Range and Clear Lake. The latter are dominated by Great Valley and Franciscan rocks with little else present.

The significant volcanism that occurred in the southeastern ranges produced the Sonoma volcanics, which dominate the Sonoma, Mayacmas, and Howell mountains. This complex array of lava flows and tuffs rests mostly on Franciscan or Great Valley rocks, Eocene sediments, or the Pliocene Petaluma formation. Because some Sonoma volcanics rest on Pliocene beds, the formation must be Pliocene or younger. The volcanics cover more than 350 square miles (900 km²), from the Petaluma-Cotati lowland east to the Howell Mountains and from Suisun Bay north to Mount St. Helena. They originally formed a nearly continuous blanket over the whole region, but have since been segmented by faulting, folding, and erosion. Andesites and andesitic tuffs are most common, but some basalts are present and near Mount St. Helena rhyolites are prominent.

A belt of younger volcanic rocks stretches from near the north end of Lake Berryessa to Clear Lake, the Clear Lake volcanic series. These rocks range from basaltic flows to rhyolites. On the southwestern shore of Clear Lake is Mount Konocti (4200 ft or 1280 m), a prominent multiple cone composed mainly of dacitic to andesitic lava flows and interbedded pyroclastics. Adjacent mountains either have similar composition or are rhyolitic.

The volcanic area south of Clear Lake contains silicic volcanic rocks including obsidian flows, layers of pumice, and numerous cinder cones. Ages range from Pleistocene to Recent, and residual volcanic activity persists as hot springs, steam vents, and a borax-rich lake on the southeastern shore of Clear Lake (Figure 9-8). Sulfur and quicksilver were mined during the nineteenth century in the Sulphur Bank area close to Borax Lake.

Clear Lake is the largest natural freshwater lake in the Coast Ranges; it is also the largest landslide lake in California. Its drainage is unusual because it has two outlets: north into the Russian River and south into the Sacramento River. According to one view, its original eastern outlet was dammed by a small lava flow that raised the lake so it spilled west into the Russian River. A few centuries ago, a landslide blocked this western exit, causing another rise in lake level. This resulted in reestablishment of the eastern drainage across the lava flow into the Sacramento River. Erosion then cut a gorge about 60 feet (18 m) deep into the lava flow, thereby reducing the lake level. Most of today's drainage is eastward, and only minor seepage occurs across the large landslide. In the 1920s, a control dam was built at the eastern outlet in order to regulate the lake level.

Coast Ranges

Figure 9-8.
Borax Lake, with
Clear Lake in the
distance. (Photo by
Robert M. Norris
ınd David Doerner)

Figure 9-9.   Strike ridges in Cretaceous rocks northwest of Williams. (Photo by Burt
Amundson)

Most of the Coast Range province north of Clear Lake is developed on Franciscan rocks that may be 50,000 feet (15,000 m) thick. The eastern edge, however, is underlain by a thick sequence of Jurassic and Cretaceous rock that constitutes the type locality for the Great Valley sequence. The oldest part is the 20,000-foot (6100 m) thick late Jurassic Knoxville formation, composed of dark shales, thin-bedded sandstone, and minor conglomerate. The Knoxville grades imperceptibly into overlying Cretaceous beds that aggregate at least 32,000 feet (9700 m) in thickness. The entire sequence is a continuum and can be subdivided satisfactorily only by fossil content, not lithology.

Great Valley rocks are steeply tilted toward the Sacramento Valley, and their regular bedding and dip have produced long, parallel hogback ridges. These mark the eastern side of the Coast Ranges from west of Red Bluff south to almost the latitude of Sacramento (Figure 9-9). The Franciscan rocks west of the Stony Creek thrust are less regular and do not form distinctive topography.

In the latitude of Red Bluff, the Coast Ranges narrow sharply, veer west, and wrap around the seaward side of the Klamath Mountains. The provinces are separated by the South Fork Mountain fault zone, which lies about 5 miles (8 km) east of the shoreline at Crescent City, the narrowest part of the northern Coast Ranges. Apart from scattered coastal Quaternary deposits, the northernmost part of the province is developed on Franciscan terrain.

SPECIAL INTEREST FEATURES

Tertiary Intrusive Plugs near San Luis Obispo

Southeast of the city of San Luis Obispo is the first of 14 volcanic plugs that dot the landscape for 18 miles northwest to Morro Rock. The plugs form a string of picturesque hills with bold, rocky upper slopes and grassy, oak-covered lower slopes. The most prominent are those between San Luis Obispo and Morro Bay. Morro Rock itself rises more than 500 feet (160 m) above sea level (Figure 9-10).

These features are interpreted as eroded Miocene volcanic necks, probably emplaced along the West Huasna fault. The volcanic edifices themselves have long since been eroded, leaving only stumps of the more resistant lavas that once congealed in the throats of the vanished volcanos. These volcanic centers are thought to be the sources of the Obispo tuff, a pyroclastic deposit widely distributed in the San Luis Obispo–Morro Bay region. Composition of rocks preserved in the volcanic necks varies from basalts and diabases to dacites. The larger peaks of the chain are andesites; Morro Rock is a dacite.

Pinnacles National Monument

The Pinnacles are located about 35 miles (56 km) south of Hollister, at the southern end of the Gabilan Range. This small area of rough and jagged topography is so conspicuous among the rolling hills characteristic of this part of the Coast Ranges that it was a popular attraction as early as the eighteenth century. In 1794, George Vancouver took time from his coastal explorations to endure a mule trip to the Pinnacles. The region was proclaimed a national monument by President Theodore Roosevelt in 1908, to a large degree because of the urging of Stanford University's naturalist and president, David Starr Jordan.

The distinctive topography at Pinnacles, with its towering spires, numerous caves, and jumble of fallen blocks, is the result of weathering and erosion of a rhyolitic breccia. Although the area includes other volcanics, the most striking topography is sculptured in the breccia. All the rocks have been preserved in a down-faulted block about 6 miles (9.6 km) long and 2.5 miles (4 km) wide.

Figure 9-10.   *Looking east at Morro Rock. The lagoon and sand spit are typical features of coastline topography in the Coast Ranges. (Photo by Spence Air Photos, courtesy of Department of Geography, University of California, Los Angeles)*

The eastern half of the volcanic area is composed mainly of rhyolitic flows varying from gray to red and green and including large amounts of obsidian. The obsidian is concentrated either near old volcanic vents or wherever the lavas came in contact with the granitic basement through which they intruded. Within the predominantly rhyolitic flows are some andesitic and basaltic flows. The western half is composed of the slightly younger, thick pyroclastic volcanic breccias from which the Pinnacles themselves are eroded. Total volume of volcanic material originally erupted is estimated at between 6 and 10 cubic miles (25–40 km³); all but 3 cubic miles (12.5 km³) has been removed by erosion. The volcanics are considered Miocene because the first influx of volcanic detrital material occurs in Miocene sedimentary rocks deposited nearby.

The main events in the geologic history of the Pinnacles are as follows.

1. Masses of rhyolitic lava were erupted from at least five vents as flows on a slightly irregular erosional surface developed on the granitic rocks of the Gabilan Range. During the eruptive cycle, magma composition varied from andesitic and basaltic to rhyolitic. The possibility exists, of course, that the rhyolitic flows are the result of remelting of the underlying granitic rocks by the rising, less silicic magma.

2. Toward the end of the eruptive cycle, some vents became plugged, and periodic steam eruptions blasted out solidified or nearly solidified material with new lava. This formed the breccias that subsequently were emplaced as avalanche deposits. There is no evidence to suggest that water transport was involved in forming these inclined, bedded rocks. They were deposited on steep slopes around the vents and have remained in nearly their original positions.

3. Faulting has preserved the volcanic rocks of the Pinnacles area by permitting the block on which they rest to be dropped down into the granitic rocks of the southern Gabilan Range. Most volcanic rocks that remained on the higher, unfaulted regions have been removed by erosion. The age of faulting usually appears to postdate volcanic activity, but it may well have begun before the eruptive cycle closed and perhaps was related to it.

4. Pliocene events are unrecorded, but by the Pleistocene the volcanics were being eroded into today's distinctive landscape.

The Geysers

This unusual area of hot springs and steam vents is located in the Mayacmas Mountains. It occupies the northwestern end of a graben about 5.5 miles (8.8 km) long and a mile (1.6 km) wide, drained by Big Sulphur Creek. The whole region lies within Franciscan rocks, though Clear Lake rhyolites are exposed about 3 miles (4.8 km) to the east and the subterranean heat source is probably attributable to residual volcanic heat at depth. The graben itself appears to be underlain mainly by Franciscan greenstones and serpentinites that have been faulted down into the surrounding graywackes.

The main natural thermal area is along Geyser Creek, a tributary of Big Sulphur Creek, and is only about 1300 feet (400 m) long and 600 feet (180 m) wide (Figure 9–11). Geyser Canyon contains numerous hot springs, with temperatures from 50°C to boiling. Several small steam vents occur here, with modern geothermal wells located just east of Geyser Creek.

This region is of particular interest because it is the only place in the United States where electric power is being produced from geothermal sources in commercial quantity. Power development began here, after a fashion, in 1921, when the first steam wells were drilled. By 1925, eight wells had been completed, but there was little market for the power and the project was abandoned. In 1955, more wells were drilled and generating equipment installed. Power was first produced in 1960, making The Geysers the third place in the world to use geothermal resources to produce commercial electric power. (Lardarello in Italy had been in operation since the early twentieth century and Wairakei in New Zealand since 1950.) The geothermal area at The Geysers is unusually favorable because it produces dry steam with few corrosive products apart from sulfur. This characteristic has greatly facilitated development, so that the region is now the world's largest producer of geothermally generated electric power.

Figure 9-11.   *The Geysers. (Photo by Robert M. Norris and David Doerner)*

*Figure 9-12.
Devil's slide, north
of Montara Moun-
tain. (Photo by
David Doerner)*

Landslides

Landsliding in California is certainly not restricted to the Coast
Ranges, but the phenomenon is so pervasive in the province that it
deserves special attention. Because Franciscan rocks are more
widely distributed in the Coast Ranges than any other group and
because their highly sheared serpentinites are so unstable, slides
involve Franciscan rocks more than any other formation (Figure
9-12). In the northern ranges, where Franciscan is the dominant
rock, slides are so prevalent that they may have accounted for more
downslope transport of material than any other process *including
streams.*

An extensive study of geologic hazards in the nine counties
around San Francisco Bay has pinpointed the locations of hundreds
of slides and provided an estimate of their direct and indirect costs.
Slides cause direct losses by damage to houses, buildings, utility
lines, and roads and indirect losses by reduced property values, tax
forfeiture, and need for emergency help from public agencies. In the
1968–1969 winter season alone, the nine-county area suffered minimal losses of $25 million; this included $9 million in direct property losses, $10 million in public property, and $6 million in miscellaneous costs.

In California, winter is the time of greatest slide activity, and wet winters promote more sliding than dry winters. Water is the crucial factor in activating slides, because water reduces internal coherence of earth materials and also it often represents considerable weight added to a potentially unstable mass. In addition, sliding is promoted by oversteepening of slopes caused by such activities as marine and stream erosion or earth moving and grading done by man. Shearing and crushing associated with faulting also make sliding more likely; for example, landslides are particularly common along traces of active faults like the San Andreas and Hayward. Table 9-1 shows the widespread occurrence of landsliding in the Coast Ranges, plus some of the rock types involved.

Marine Terraces

Marine terraces occur at numerous places along the Pacific shore, all reflecting changes in relative sea level in the recent past. Although only a few have been studied in detail (notably those at Santa Cruz), it is clear that terraces are discontinuous and do not necessarily correspond in elevation. The picture has been complicated by the

Table 9-1
Coast Range Landsliding and Associated Rocks

Area	Associated Rocks
Berkeley Hills area (Alameda and Contra Costa counties)	Orinda formation: weakly consolidated sands, silts, clays, and tuffs; Pliocene
Blue Lake area (Humboldt County)	Franciscan serpentinites; Falor formation: soft clays and sands, Pliocene
Eel River area (Humboldt County)	Rio Dell formation: mudstone; Yager formation: sandstone; Wildcat formation: mudstone; all Pliocene
Healdsburg area (Sonoma County)	Franciscan serpentinites
Lower Lake area (Lake County)	Franciscan serpentinites
Marine headlands (Marin County)	Franciscan cherts
Mount Hamilton area (Santa Clara County)	Alum Rock slide: Franciscan serpentinites; Oak Ridge area: mainly Franciscan serpentinites; Santa Clara formation: sands and gravels, Pliocene
Ortigalita Peak area (Merced County)	Franciscan serpentinites
Quien Sabe area (Merced and San Benito counties)	Large slide in Quien Sabe volcanics, Miocene; Franciscan shales and shaly sandstones
San Juan Bautista area (San Benito County)	Purisima formation, Pliocene
San Mateo coast area (San Mateo County)	Devils slide: faulted and sheared mass involving Franciscan, Cretaceous, and Paleocene rocks
Suisun Bay area, north side (Solano County)	Franciscan serpentinites; Sonoma volcanics, Pliocene; Petaluma sands and gravels, Pliocene

Pleistocene events that affected the California coast; it is difficult to distinguish eustatic changes in sea level from changes caused by diastrophism. The diastrophic record is both localized and complicated. Furthermore, there is disagreement regarding the magnitude and time scale of sea-level fluctuations.

No terrace has been dated earlier than Pleistocene, and as a rule the higher terraces are older and often show deformation not recorded in the lower, younger terraces. The principal areas of well-developed terraces along the Coast Range shoreline are, from north to south: Crescent City, Trinidad, Cape Mendocino, from near Mendocino to Point Arena, from Point Arena to Fort Ross, from Davenport to Santa Cruz, from Point Lobos to Point Sur, from Ragged Point to Cambria, from Morro Bay to Point Buchon, and the Shell Beach–Pismo Beach region.

At Santa Cruz, five conspicuous terraces have been recognized between elevations of 100 and 850 feet (30–260 m). The 100-foot (30 m) terrace is the lowest and most prominent and currently produces most of California's Brussels sprouts. It is cut in the Monterey formation and in some places is nearly 1.5 miles (2.4 km) wide. It generally slopes seaward at less than 1° and is typically covered with a thin layer of marine sand. Near the old shoreline angle on its landward side, the terrace and its thin veneer of marine sediment are often blanketed by thick nonmarine deposits. These have slumped and washed down from above in a pattern characteristic of marine terraces.

REFERENCES

General

Bailey, Edgar H., and others, 1964. Franciscan and Related Rocks and Their Significance in the Geology of Western California. Calif. Div. Mines and Geology Bull. 183.

Durham, David L., 1974. Geology of the Southern Salinas Valley Area, California. U.S. Geological Survey Prof. Paper 819.

Jenkins, Olaf P., ed., 1951. Geological Guidebook of the San Francisco Bay Counties. Calif. Div. Mines and Geology Bull. 154.

Oakeshott, Gordon B., 1960. Geologic Sketch of the Southern Coast Ranges. Mineral Information Service (now California Geology), v. 13, no. 1, pp. 1–13.

———, 1970. Geology of the California Coast Ranges. Mineral Information Service (now California Geology), v. 23, pp. 7–10.

Page, Ben M., 1966. Geology of the Coast Ranges of California. *In* Geology of Northern California. Calif. Div. Mines and Geology Bull. 190, pp. 255–276.

Rice, Salem J., 1961. Geologic Sketch of the Northern Coast Ranges. Mineral Information Service (now California Geology), v. 14, no. 1, pp. 1–9.

Special

Bedrossian, Trinda L., 1974a, Geology of the Marin Headlands. California Geology, v. 27, pp. 75–86.

———, 1974b. Fossils of the "Merced" Formation, Sebastopol Region. California Geology, v. 27, pp. 175–182.

Bolt, Bruce A., 1969. Earthquakes and the Structural Features of Northern California. Mineral Information Service (now California Geology), v. 22, pp. 51–53.

Bowen, Oliver E., 1965. Point Lobos. Mineral Information Service (now California Geology), v. 18, pp. 60–67.

Carlson, Paul R., and others, 1970. The Floor of Central San Francisco Bay. Mineral Information Service (now California Geology), v. 23, pp. 97–107.

Clague, John J., 1969. Landslides of Southern Point Reyes National Seashore. Mineral Information Service (now California Geology), v. 22, pp. 107–110.

Hinds, N. E. A., 1968. Pinnacles National Monument. Mineral Information Service (now California Geology), v. 21, pp. 119–121.

Jenkins, Olaf P., 1973. Pleistocene Lake San Benito. California Geology, v. 26, pp. 151–163.

Koenig, James B., 1963. The Geologic Setting of Bodega Head. Mineral Information Service (now California Geology), v. 16, no. 7, pp. 1–9.

———, 1969. The Geysers. Mineral Information Service (now California Geology), v. 22, pp. 123–128.

Morton, Douglas M., and Robert Streitz, 1967. Landslides. Mineral Information Service (now California Geology), v. 20, pp. 123–129, 135–140.

Oakeshott, Gordon B., 1951. Guide to the Geology of Pfeiffer Big Sur State Park, Monterey County, California. Calif. Div. Mines Spec. Rept. 11.

——— and Clyde Wahrhaftig, 1966. A Walker's Guide to the Geology of San Francisco. Mineral Information Service (now California Geology), v. 19 (supplement).

Pestrong, Raymond, 1972. San Francisco Bay Tidelands. California Geology, v. 25, pp. 27–40.

Rogers, Thomas H., 1969a. Where Does the Hayward Fault Go? Mineral Information Service (now California Geology), v. 22, pp. 55–60.

———, 1969b. A Trip to an Active Fault in the City of Hollister. Mineral Information Service (now California Geology), v. 22, p. 159.

———, 1972. Santa Cruz Mountain Study. California Geology, v. 25, pp. 131–134.

Saul, Richard B., 1967. The Calaveras Fault Zone in Contra Costa County, California. Mineral Information Service (now California Geology), v. 20, pp. 35–37.

Sullivan, Raymond, 1975. Geologic Hazards along the Coast South of San Francisco. California Geology, v. 28, pp. 27–36.

Chapter Ten

Great Valley

"Out of sight—out of mind." It must not be ever so.

Anonymous

Between the Sierra Nevada and the Coast Ranges is the elongate lowland known as the Great Valley. About 400 miles (640 km) long and 50 miles (80 km) wide, this lowland rises from slightly below sea level to about 400 feet (120 m) at its north and south ends. The valley is unusual for a lowland because it is a relatively undeformed basin bounded by highly deformed rock units. The Sacramento River drains the northern half, and the San Joaquin River the southern half. The lowland also is referred to as the Central Valley, its northern segment as the Sacramento Valley, and its southern segment as the San Joaquin Valley. The locations of some major features of the province are shown in Figure 10-1.

GEOGRAPHY

The Great Valley is monotonous geologically, representing primarily the alluvial, flood, and delta plains of its two major rivers and their tributaries (Figure 10-2). The region persisted as a lowland or shallow marine embayment during the entire Cenozoic and at least the later Mesozoic. In the late Cenozoic, much of the area was occupied by shallow brackish and freshwater lakes. This was particularly the case in the San Joaquin section, which has had internal drainage in its southern third since the Pliocene. Lake Corcoran (now extinct) spread over much of the northern San Joaquin Valley during the middle and late Pleistocene. Today the only outward drainage is through Carquinez Strait, into San Francisco Bay. The valley's most fertile lands lie at the head of this strait, where the deltas of the two main rivers converge.

Annual rainfall ranges from 5 to 20 inches (127–508 mm). The land is well watered, however, because its rivers are fed by the heavy rain and snowfall of the Sierra Nevada. In fact, during severe floods the valley's playas may fill and inundate thousands of acres of crops.

Only two topographic breaks occur on the flat lowland floor: Sutter (Marysville) Buttes in the Sacramento Valley and the Kettleman Hills and other anticlinal arches on the western side of the southern San Joaquin Valley. Sutter Buttes reach 2100 feet (640 m) in elevation, but have an area of only a few square miles (Figure 10-3). The Kettleman, Elk, and Buena Vista hills are outliers of the Coast Ranges and have elevations of about 1800 feet (550 m).

Figure 10-1.   Place names: Great Valley.

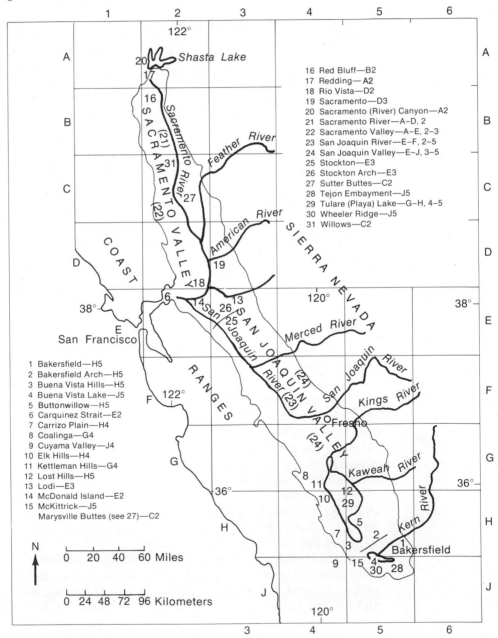

16 Red Bluff—B2
17 Redding—A2
18 Rio Vista—D2
19 Sacramento—D3
20 Sacramento (River) Canyon—A2
21 Sacramento River—A–D, 2
22 Sacramento Valley—A–E, 2–3
23 San Joaquin River—E–F, 2–5
24 San Joaquin Valley—E–J, 3–5
25 Stockton—E3
26 Stockton Arch—E3
27 Sutter Buttes—C2
28 Tejon Embayment—J5
29 Tulare (Playa) Lake—G–H, 4–5
30 Wheeler Ridge—J5
31 Willows—C2

1 Bakersfield—H5
2 Bakersfield Arch—H5
3 Buena Vista Hills—H5
4 Buena Vista Lake—J5
5 Buttonwillow—H5
6 Carquinez Strait—E2
7 Carrizo Plain—H4
8 Coalinga—G4
9 Cuyama Valley—J4
10 Elk Hills—H4
11 Kettleman Hills—G4
12 Lost Hills—H5
13 Lodi—E3
14 McDonald Island—E2
15 McKittrick—J5
    Marysville Buttes (see 27)—C2

Figure 10-2.

San Joaquin Valley and San Joaquin River. (Photo by Fairchild Aerial Surveys, courtesy of Department of Geography, University of California, Los Angeles)

ROCKS

The surface of the Central Valley is composed of unconsolidated Pleistocene and Recent sediments. Where streams have cut channels into these sediments, lake beds are occasionally exposed that include clay zones, diatomite, and other rocks that can be correlated and mapped. Generally, rock sequences must be inferred from well records and by extension of formations that are exposed on valley margins and then dip beneath the valley floor. Fortunately, information from the thousands of oil and gas wells drilled has permitted fairly confident reconstruction of the valley's geologic history. Thousands of water wells have been drilled also, but they can help with only the latest geologic record; water wells rarely extend deeper than 1500 to 2000 feet (450–600 m) and seldom penetrate any pre-Pliocene rocks.

The rock sequences of the Great Valley can be divided into two sections, a division supported by geophysical studies and well records. First is a belt along the west base of the Sierra; this includes minimally deformed alluvial fans and lake deposits that feather east

Great Valley

*Figure 10-3.
Sutter (Marysville)
Buttes in central
Sacramento Valley.
Note radial drain-
age and plug-dome
characteristics.
Northern Coast
Ranges are in the
distance. (Photo
courtesy of U.S. Air
Force and U.S.
Geological Survey)*

onto the Sierran basement. Second is a linear belt from 10 to 20 miles (16–32 km) wide at the east base of the Coast Ranges; this is composed of deformed Mesozoic and Cenozoic rocks that dip east beneath the valley. The sedimentary cover of gravels and sands is thinner in the western belt.

Although unconfirmed, the contact at depth between Sierran and Coast Range basements is presumed to be a fault. Nearer the Coast Range margin, the contact between the two basement units has been definitely located.

Establishing the relationships among Great Valley rock sequences has been a major geologic detective project. A fairly complete record is now available, however, as indicated by Figures 10-4, 10-5, and 10-6.

STRUCTURE

The Great Valley is a synclinal trough with its axis off center to the west. It is interrupted by two major surface cross structures that had developed by the beginning of the Cenozoic: the Stockton fault in the Stockton arch and the White Wolf fault in and south of the Bakersfield arch. Since the opening of the Miocene, both arches have been uplifted by movement on the Stockton and White Wolf faults. In the Bakersfield arch, a subsidiary southward extension known as the Tejon embayment crosses the White Wolf fault, which is the approximate southern boundary of the Bakersfield arch. The surface trace of the White Wolf fault was extended nearly 17 miles (27 km) by the 1952 Arvin-Tehachapi earthquake, although the subsurface existence of this segment had been documented previously by well records.

Many major folds trend parallel to the Coast Range–Great Valley boundary and include some of the most famous oil-producing domes in North America: Elk Hills, Lost Hills, Kettleman Hills, Buena Vista Hills, McKittrick, and Wheeler Ridge. Less well known are the more recently discovered gas-producing structures near Sutter Buttes, Willows, Dunnigan, Lodi, and Rio Vista.

GEOLOGIC HISTORY

Situated as it is between the Sierra with its two great orogenies and the presently active Coast Ranges, the Great Valley has a rather complex geologic history. Furthermore, in the Coast Ranges, both pre-Cenozoic and Cenozoic structural features suggest almost continuous deformation since the opening of the Mesozoic. Consequently, it is perhaps surprising that the Central Valley has persisted as a recognizable unit.

The valley's history begins with late Jurassic sedimentation from the east and northeast, derived from the rising Nevadan Mountains. Presumably restricted by the western shelf edge, the pattern of deposition was typically miogeosynclinal except for the absence of limestone. Concurrently, a deeper western basin received great thicknesses of eugeosynclinal deposits, including volcaniclastics, volcanic flows and intrusives, and other clastic materials. These are all represented today by the Franciscan formation of the Coast Ranges. Most geologists think these two geosynclinal units were deposited in contiguous areas. An alternate view is that during their formation the Franciscan and Great Valley units may have been separated by ophiolite sequences that overlay sutures between the North American and Pacific plates.

Deposition continued, with Cretaceous miogeosynclinal sediments accumulating from continued erosion of the Nevadan Mountains, especially in the northern regions. Cretaceous rocks are less evident in the south, possibly reflecting more distant or less well developed source areas. The Mesozoic sedimentary rocks usually show great continuity and uniformity and are often easily recognized in the field even when separated by several miles.

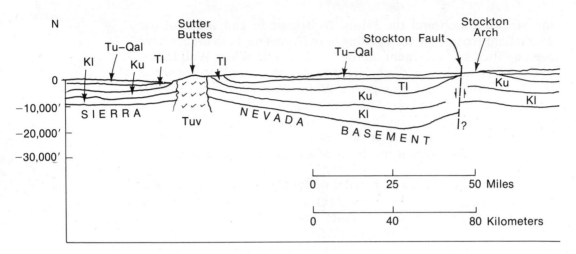

Figure 10-4. *North-south diagrammatic section of the Great Valley. (Sources: California Division of Mines and Geology and Sacramento Geological Society)*

Figure 10-5. *East-west diagrammatic section of the Sacramento Valley (latitude 39°N). (Sources: California Division of Mines and Geology and Sacramento Geological Society)*

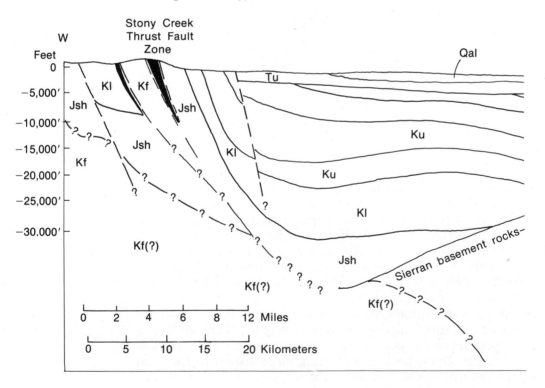

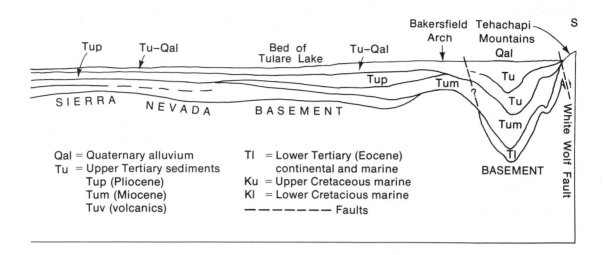

Qal = Quaternary alluvium
Tu = Upper Tertiary sediments
    Tup (Pliocene)
    Tum (Miocene)
    Tuv (volcanics)

Tl = Lower Tertiary (Eocene)
    continental and marine
Ku = Upper Cretaceous marine
Kl = Lower Cretacious marine
—————— Faults

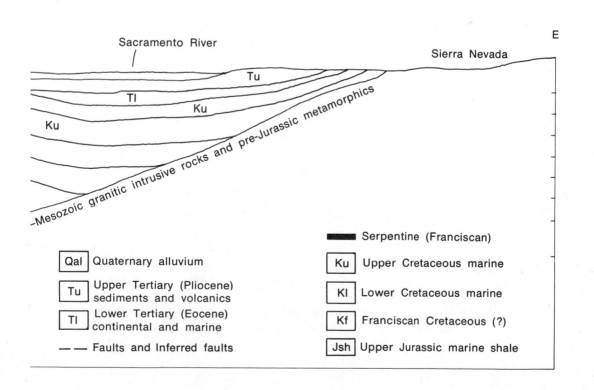

Qal	Quaternary alluvium
Tu	Upper Tertiary (Pliocene) sediments and volcanics
Tl	Lower Tertiary (Eocene) continental and marine

— — Faults and Inferred faults

    Serpentine (Franciscan)

Ku	Upper Cretaceous marine
Kl	Lower Cretaceous marine
Kf	Franciscan Cretaceous (?)
Jsh	Upper Jurassic marine shale

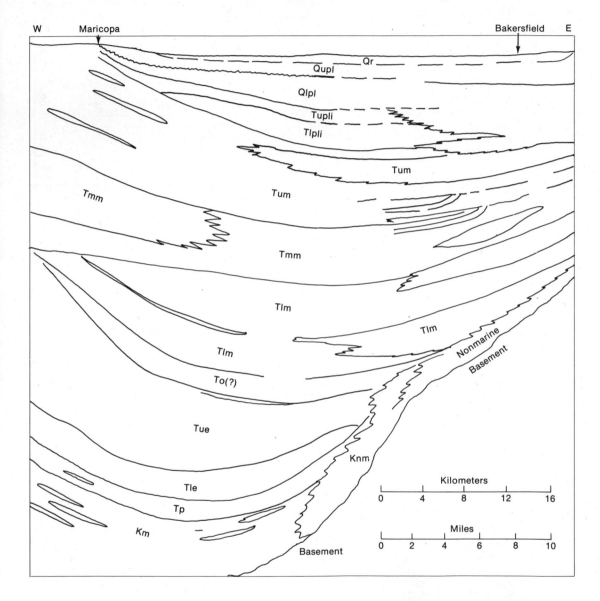

Figure 10-6. *East-west diagrammatic section of the southern San Joaquin Valley (latitude 35°30'N). (Source: California Division of Mines and Geology)*

QUATERNARY
Qr   = Recent
Qupl = Upper Pleistocene
Qlpl = Lower Pleistocene
TERTIARY
Tupli = Upper Pliocene
Tlpli = Lower Pliocene
Tum = Upper Miocene
Tmm = Middle Miocene
Tlm = Lower Miocene
To   = Oligocene
Tue = Upper Eocene
Tle   = Lower Eocene
Tp   = Paleocene
Km   = Cretaceous marine
Knm = Cretaceous nonmarine
Basement
Lenses in section
Interfingering contact, sometimes
with local unconformities

Age	Lithology	Maximum Thickness (feet)
Quaternary	Clay, sand, and conglomerate; buff to gray and greenish-gray color, poorly cemented and poorly sorted; almost entirely alluvial-fan and lacustrine material.	8000–10,000 (2440–3000 m)
Pliocene	Soft greenish-gray claystone and interbedded permeable sands; upper third nonmarine and marine; lower two-thirds marine, particularly in central basin areas; megafossil control.	8000–9000 (2440–2750 m)
Miocene	Brown and gray clay shale and hard siliceous shale, with numerous permeable sandstone and conglomeratic sandstone members; marine with foraminifers, diatoms, and megafossils, except uppermost and basal nonmarine members along eastern and southeastern borders; basaltic and andesitic flows and intrusions in lower part along southeastern borders.	12,000–13,000 (3650–4000 m)
"Oligocene" and Upper Eocene	Gray and brown shale and hard siliceous shale with some thin and thick permeable sands in local border areas; marine with foraminifers and megafossils, except for red and green nonmarine beds in "Oligocene" of eastern and southeastern border.	8000–9000– (2440–2750 m)
Lower Eocene and Paleocene	Gray shale with some sands that become thick and very permeable, particularly in Coalinga and southern border areas; marine, with foraminifers and megafossils.	5000–6000 (1500–1800 m)
Upper Cretaceous	Upper part weathered to purple, with dark gray siliceous and calcareous foraminiferal shale and clay shale with local sands; middle and lower parts massive thick concretionary sandstone, conglomerate, and dark shale with intercalated sandstone, marine.	25,000+ (7600+ m)

The Cenozoic was initiated by deformation that was regional in the north and localized in the south. Intense local deformation apparently continued throughout the San Joaquin region, where extremely thick marine sections accumulated during the Miocene. These sections were often highly localized, implying the presence of deep, narrow seaways (possibly block-faulted valleys) extending to the Pacific through embayments and straits across the site of the southern Coast Ranges. These linear structures may reflect extension of Basin Range block faulting into the Coast and Transverse ranges during the middle and late Cenozoic. Another possibility is that they were depressed slivers along major strike-slip fault zones, much like the Gulf of California.

The floor of the major basin sank and was concomitantly filled, although no significant deep-water areas ever developed. Nevertheless, localized nearly abyssal depths are indicated by some microfaunal associations in the southern San Joaquin Valley. Simultaneously, deformation continued until marine waters were expelled completely, presumably because of the increasing intensity of the Coast Range orogeny. By Pliocene time, most of the valley's seas were drained, via Carquinez Strait. Brackish and freshwater lakes replaced marine waters, and the Central Valley assumed its present form.

Figure 10-7 gives the inferred profile of Great Valley basement throughout the Tertiary, showing a continually sinking basin being filled with sediment. Increased deformation is related to accelerating strike-slip movement on the San Andreas fault and compressional forces from the rising Coast Ranges.

SUBORDINATE FEATURES

Sutter (Marysville) Buttes

A prominent monolithic cluster of steep-sided, pinnaclelike hills covering about 10 square miles (26 km²) breaks the topographic monotony of the Sacramento Valley near Marysville. Rising over 2000 feet (600 m) from the valley fill, this pile of volcanic rock is the only major igneous outcrop anywhere in the Great Valley (Figure 10-8).

What forces drilled these vents? Why this location? Why only one such volcanic center? The buttes are plug domes of andesitic porphyry, each with an inner core and an outer ring of intrusive rhyolite. In the earlier eruption, magma apparently was injected slowly and in pasty form, although there are some breccias and tuffs that suggest mildly explosive action. The andesite subsequently was intruded by the rhyolite. Late Cretaceous and Eocene sediments are included as xenoliths, and the entire structure is upwarped into the

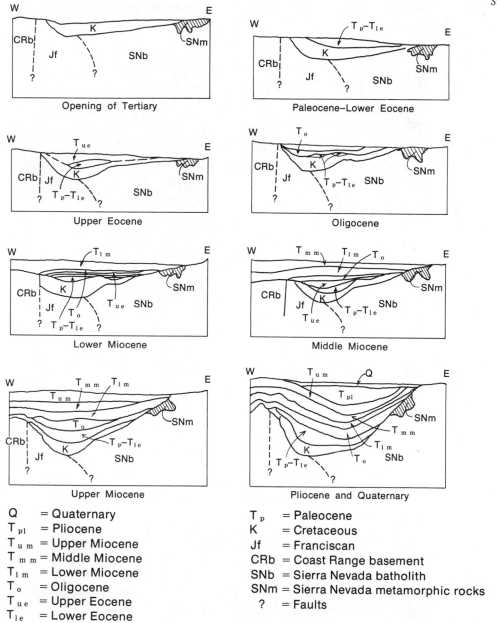

Q = Quaternary
T_{pl} = Pliocene
T_{um} = Upper Miocene
T_{mm} = Middle Miocene
T_{lm} = Lower Miocene
T_o = Oligocene
T_{ue} = Upper Eocene
T_{le} = Lower Eocene

T_p = Paleocene
K = Cretaceous
Jf = Franciscan
CRb = Coast Range basement
SNb = Sierra Nevada batholith
SNm = Sierra Nevada metamorphic rocks
? = Faults

Figure 10-7.

Evolution of sedimentation in southern San Joaquin Valley. Thicknesses of sediment increased as the Tertiary advanced, with depression of basement and concomitant development of east-west arches and domes. Although the sections are for the southern San Joaquin, they represent conditions throughout the Great Valley. Note that the contact between Franciscan and plutonic basement is not defined, although it is presumed to be an east-dipping fault. (Source: California Division of Mines and Geology)

Figure 10-8.
View to the south-west across Lassen Peak (center foreground), on the boundary of the Cascade–Sierra Nevada provinces. Lake Almanor is in the left center. The Great Valley is in the upper right, with Sutter Buttes showing as a circular protrusion in the extreme upper right. (Photo courtesy of U.S. Air Force and U.S. Geological Survey)

Plio-Pleistocene Tehama formation. Thus the eruption is dated, and related origins for the volcanics can be established.

The position of these intrusions has long intrigued geologists, but explanations have been inconclusive so far. The Tehama formation underlies the Tuscan formation, which is composed of extensive volcanic flows and volcaniclastics that connect the Great Valley and Sierra Nevada with the Cascade province. Are Sutter Buttes in fact the southernmost of the Cascade volcanos? Or are they unrelated minor pulsations that coincidentally broke the valley floor? The buttes take on added importance because their volcanic domes and plugs have deformed underlying sediments, producing anticlines that have trapped natural gas.

Kettleman Hills

The Kettleman Hills hug the western edge of the San Joaquin Valley, extending south from near Coalinga. They are usually considered an outlier of the southern Coast Ranges, but they also represent an interruption of the floor of the Great Valley. At their north end, they stand about 1500 feet (450 m) above the valley. From here they stretch south for 30 miles (48 km), descending in elevation and finally merging with the valley floor. Almost a perfect elongate dome, the Kettleman Hills are divided into the three anticlinal segments of North, Central, and South domes. The whole unit is about 5 miles (8 km) wide.

The Kettleman Hills lie in the driest part of the San Joaquin Valley. Consequently, since their surface rocks are weakly consolidated, drainage patterns have produced almost badland topography where the sparse vegetation has been destroyed. The hills are the largest of a series of folds in this area, all of which are important petroleum producers (Figure 10-9). Through 1964, North Dome had produced nearly half a billion barrels (70 million metric tons) of oil.

SPECIAL INTEREST FEATURES

Natural Gas Fields

The Great Valley has produced trillions of cubic feet of natural gas since gas was first consumed in Stockton in the 1850s. The Rio Vista gas field alone has produced over 2.25 trillion cubic feet (600 million m³) since its discovery in 1936.

Natural gas is classified in two categories, wet or dry, with most oil fields producing wet gas. After its more valuable components like ethane, butane, and propane have been extracted, wet gas is often pumped back into the wells to recharge pressure and increase petroleum yield. Most gas thus recycled reappears with oil produced subsequently from the wells. Dry gas is produced from porous and permeable rock reservoirs—like oil, but where oil is absent. Dry gas is primarily methane and usually is too low in other gases to make further treatment profitable, so it is piped directly to consumers.

Enormously important gas fields occur north of the Stockton arch and extend almost 200 miles (320 km) to the northern end of the valley (Figure 10-10). First discovery of commercial dry gas was in 1933 at Sutter Buttes, followed by discovery of the McDonald Island and Rio Vista fields in 1936. Dry gas was produced in the San Joaquin region as early as 1909, and some prolific fields were discovered at Elk Hills in 1919 and Buttonwillow in 1927. It is the Sacramento Valley, however, that dominates California in gas production.

The gas fields of the Sacramento differ from those of the San Joaquin by abnormally high reservoir pressures, almost complete absence of associated petroleum, and high production from Cretaceous formations. The reasons for these discrepancies are not yet known, although differences in parent organic material may be the answer.

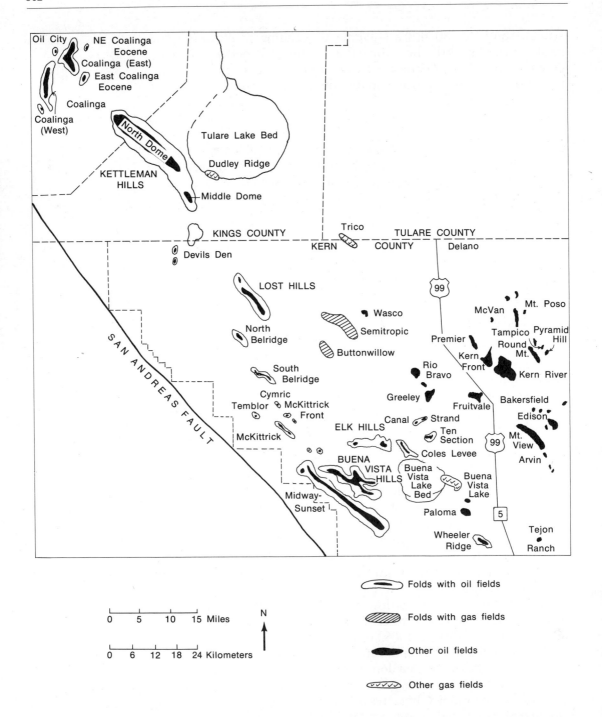

Figure 10-9. Some structures producing petroleum and natural gas in the San Joaquin Valley. (Source: California Division of Mines and Geology)

Subsidence in the Great Valley

Particularly since 1940, agricultural development of the arid San Joaquin Valley has considerably strained water resources. Although plentiful water actually enters the valley via Sierran rivers, inadequate distribution has necessitated several major irrigation projects. Each has been plagued with massive subsidence along its distributary canals, partially as a consequence of overpumping and overuse of groundwater.

Subsidence regions include the San Joaquin–Sacramento deltas where agricultural development has occurred in peat lands, the overpumped artesian basins in the valley's western and southern section, and places compacted by wetting of moisture-deficient ground by irrigation. Subsidence up to 23 feet (7 m) has occurred in some areas of overpumping, but 10 to 15 feet (3–5 m) is typical in the deltas. Recharge of the underground reservoirs has not produced rebound toward original levels.

Subsidence creates major problems; several areas 4 to 6 miles (6.4–9.6 km) wide and from 20 to 40 miles (32–64 km) long have been affected. A technique to help reduce subsidence has been developed, however. Suspect parts of an excavated canal are divided into small segments by earth-fill barriers, and the depressions are then filled with water that is maintained at a certain level until stability is achieved. The trench is then regraded and the canal developed. In an extreme case, subsidence up to 15 feet (5 m) occurred before stability was finally achieved 10 to 18 months later.

Pleistocene Lakes

The Pleistocene lakes of the San Joaquin Valley are all extinct and today form playas that are sometimes reclaimed temporarily for agriculture. The lakes were concentrated south of the Stockton arch and generally occupied the western side of the valley floor.

Lake Corcoran   Lake Corcoran occupied approximately the western half of the San Joaquin Valley, from the Stockton arch south to the area where the San Joaquin River turns abruptly north. The lake had its highstand about 600,000 years ago and has left lake-bed clays, diatomite, and other deposits. Sand dunes are preserved in places, marking the ancient shoreline. The extreme flatness of this part of the valley floor is due to the ancient lake bottom.

Lake Tulare   The floor of the valley south of the San Joaquin River is the playa of Lake Tulare. Were it not for irrigation diverting waters of the Kings and other rivers, in years of heavy Sierran rainfall the playa would become an extensive, shallow body of water because its basin is the normal terminus for drainage of these rivers. At present, the playa is cultivated for agriculture.

Great Valley

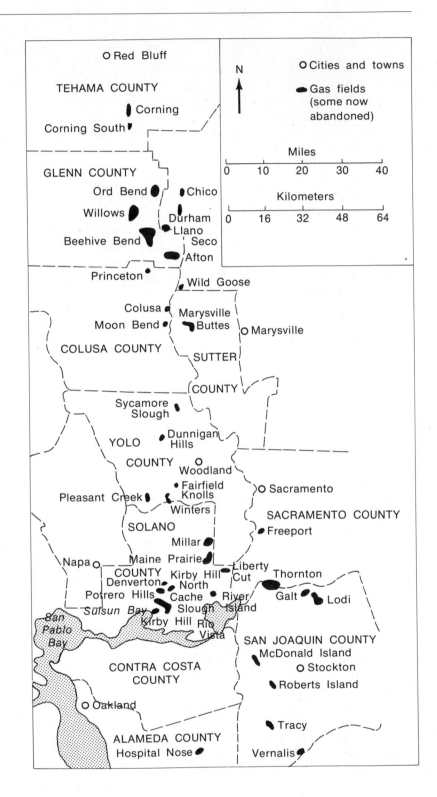

*Figure 10-10.
Gas fields of the
northern Great Val-
ley. (Source: Cali-
fornia Division of
Mines and Geology)*

Buena Vista and Kern Lakes   Buena Vista Lake was formerly the terminus of the Kern River. The Kern has one of the largest drainages of the Sierra, and the lake persisted as a permanent feature until the river was controlled by the Isabella Dam. This, plus irrigation diversion, dried Buena Vista Lake and its playa is now used for agriculture. Water rarely stands in the playa except in years of heavy local rainfall. The drainage divide between the Tulare and Buena Vista basins is low, and occasionally the two lakes have formed a single shallow water body. Evidence suggests that late Pleistocene drainage was south into Buena Vista Lake when glacial meltwater entered Lake Tulare from the Kings River. Buena Vista Lake overflowed into smaller Kern Lake, which extended southeast from Buena Vista into the southernmost San Joaquin Valley.

REFERENCES

General

Hackel, Otto, 1966. Summary of the Geology of the Great Valley. *In* Geology of Northern California. Calif. Div. Mines and Geology Bull. 190, pp. 217–238.

Poland, J. F., and R. E. Evenson, 1966. Hydrogeology and Land Subsidence, Great Central Valley, California. *In* Geology of Northern California. Calif. Div. Mines and Geology Bull. 190, pp. 239–247.

Smith, M. B., and F. J. Schambeck, 1966. Petroleum and Natural Gas. *In* Mineral and Water Resources of California, Part 1. Committee on Interior and Insular Affairs, Eighty-Ninth Congress, Second Session, pp. 291–328.

Woodring, W. P., and others, 1940. Geology of the Kettleman Hills Oil Field, California. U.S. Geological Survey Prof. Paper 195.

Special

Christian, Louis B., 1970. Ancient Windblown Terrains of Central California. Mineral Information Service (now California Geology), v. 23, pp. 175–179.

Manning, John C., 1973. Field Trip to Areas of Active Faulting and Shallow Subsidence in the Southern San Joaquin Valley. Far Western Section Nat. Assoc. Geology Teachers, Spring Meeting 1973.

Meehan, J. F., 1973. Earthquakes and Faults Affecting Sacramento. California Geology, v. 26, pp. 32–36.

Chapter Eleven

San Andreas Fault

Civilization exists by geological
consent . . . subject to change without notice.
                              Will Durant

With a total known length of over 1000 miles (1600 km), the San Andreas is the longest fault in California and one of the longest in North America (Figure 11-1). It extends 600 miles (960 km) through western California, cutting indiscriminately across rock and structural boundaries (Figure 11-2).

The San Andreas came to world attention initially because of the great 1906 San Francisco earthquake, which was caused by sudden right-slip movement on the fault of up to 16 feet (5 m). Subsequent investigation of the San Andreas established California as an internationally famous center for the study of structural geology and seismology. Since 1906, dozens of earthquakes have occurred on the San Andreas. Moreover, sizable tremors are continually being predicted, even though earthquake prediction is not yet possible in the sense in which such pronouncements are often interpreted.

The population explosion and concentration of cities on the coastal block of western California makes understanding of the San Andreas critically important. Furthermore, the San Andreas is just one of several faults with high potential for sudden slippage. The programs necessary to promote understanding of these faults must be implemented primarily at the state level, however. Federal legislators tend to be unimpressed by requests for substantial appropriations for earthquake study, especially when California and Alaska seem to be the principal beneficiaries. Nevertheless, a program was initiated after the 1964 Alaskan earthquake and is presently directed by the U.S. Geological Survey.

The San Andreas fault was not recognized as a continuous regional structure until after the 1906 San Francisco quake, although faulting as a cause of topographic expression in California had been recognized as early as 1891. The name San Andreas was apparently first employed by A. C. Lawson in 1895, after other geologists had used various names for discrete sections of the single feature now considered the San Andreas. The California Earthquake Commis-

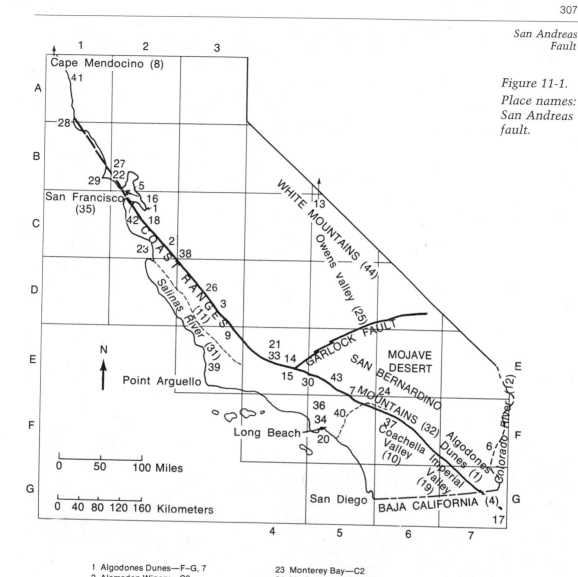

*Figure 11-1.
Place names:
San Andreas
fault.*

1 Algodones Dunes—F–G, 7
2 Alamaden Winery—C2
3 Avenal—D3
4 Baja California—G6–7
5 Berkeley—B2
6 Blythe—F7
7 Cajon Pass—E5
8 Cape Mendocino—A1
9 Carrizo Plain—E3
10 Coachella Valley—F6
11 Coast Ranges—A–E, 1–3
12 Colorado River—F–G, 7
13 Fish Lake Valley (Nevada)—B5
14 Fort Tejon—E4
15 Frazier Mountain—E4
16 Golden Gate—C2
17 Gulf of California—G7
18 Hollister—C2
19 Imperial Valley—G6–7
20 Long Beach—F5
21 Maricopa—E4
22 Marin Peninsula—B2

23 Monterey Bay—C2
24 Mt. San Gorgonio—E6
25 Owens Valley—C–D, 4–5
26 Parkfield—D3
27 Petaluma—B2
28 Point Arena—B1
29 Point Reyes—B2
30 Ridge Basin—E4–5
31 Salinas River—D–E, 2–3
32 San Bernardino Mountains—E–F, 5–6
33 San Emigdio Mountains—E4
34 San Fernando—F5
35 San Francisco—C2
36 San Gabriel Mountains—F5
37 San Gorgonio Pass—F6
38 San Juan Bautista—C3
39 San Luis Obispo—E3
40 Santa Ana River—F5
41 Shelter Cove—A1
42 Stanford University—C2
43 Valyermo—E5
44 White Mountains—C5

sion's 1908 report used the term San Andreas rift, and from this evolved the usage of San Andreas fault, San Andreas (fault) zone, and San Andreas system.

Generally, the San Andreas fault is defined as the principal surface rupture produced by recent movements within the San Andreas zone. This zone incorporates many nearly parallel fractures that are related but often 5 or 6 miles (8–9.6 km) apart, and many segments record movements older than those of the present fault. The system also contains faults that are subparallel, but not directly connected

Figure 11-2.   *Looking southeast at the San Andreas fault (center) from near Palmdale. The San Gabriel Mountains are in the right background. East of the fault, in the far left, are Mount San Gorgonio and the San Bernardino Mountains. The San Jacinto fault crosses the San Gabriel Mountains through the first notch on the skyline west (right) of the San Andreas notch. Palmdale Reservoir is in the center foreground, with the Sierra Highway between the reservoir and a smaller pond across the road. These are both sag ponds on the San Andreas, but the volume of the larger one has been increased artifically from its original size. (Photo by Spence Air Photos, courtesy of Department of Geography, University of California, Los Angeles)*

surficially to the San Andreas proper. Some examples are the San Jacinto and San Gabriel faults in southern California and the Calaveras and Hayward faults in northern California. *Rift* is normally used as a geomorphic term only, but it does encompass all features of a zone, both erosional and structural.

Although important earthquakes have been associated with sudden slippage on the San Andreas, the record is incomplete and too short historically to have much geologic significance. California's first recorded earthquake was reported in 1769 by Gaspar de Portola and was felt while he camped near the Santa Ana River. The state's first seismographs were installed at the University of California, Berkeley, in 1887, and the 1906 earthquake was the first on the San Andreas fault for which a seismogram is available. Earthquakes on the Hayward fault were reported on 10 June 1836, 8 October 1865, and 21 October 1868, when a maximum of 3 feet (1 m) of right slip apparently occurred. Earthquakes on the San Andreas were recorded in June 1838, on 24 April 1890, and on 18 April 1906 (the San Francisco disaster). The 1906 earthquake measured 8.25 on the Richter scale. (The Richter scale defines earthquakes of magnitude 7 to 7.75 as major and over 7.75 as great. Earthquakes of 2 are felt, and those of 4 to 5.5 cause local damage.) The San Francisco and Fort Tejon (1857) earthquakes were probably of equal magnitude. The 1872 Owens Valley earthquake (on the Sierra Nevada fault) may have been of higher magnitude, but no seismographic records were available then.

In the Fort Tejon earthquake, at least 250 miles (400 km) along the San Andreas fault showed surface rupture. Major surface breaks occurred as far as 100 miles (160 km) southeast and northwest of the Fort. No other significant quakes have been recorded on the southern California section of the San Andreas proper, but many have occurred on faults that may yet be established as parts of the San Andreas system.

On land the San Andreas fault extends south from Shelter Cove, across the Golden Gate under western San Francisco, and on through the Coast Ranges (where it is straight and narrowly confined) to the intersection with the Garlock fault at Frazier Mountain. Southeast of the Garlock, the San Andreas approximates the boundary between the Mojave Desert and Transverse Range provinces. The fault separates the San Gabriel and San Bernardino mountains in Cajon Pass, and as the southern boundary of the San Bernardino Mountains it forms the San Gorgonio Pass into the Coachella and Imperial valleys. Some authorities dispute the existence of a continuous trace from Cajon Pass south and suggest instead that the San Andreas proper terminates against a major east-west cross fault. The most recently active trace of the San Andreas is seldom more than a half-mile (0.8 km) wide and is often much less. An especially notable feature is the sharpness of the fault's definition (Figure 11-3).

GEOMETRY

Surface Geometry

Certain aspects of its plane geometry have been emphasized by describing the San Andreas as a seam, but the fault's planar features are actually far more complex. The San Andreas frequently consists of many closely spaced parallel fractures, all previously surfaces of movement that often had slip different from today's. Various sections of the fault have features that are strikingly dissimilar from one another. Sometimes this reflects the rocks involved, but often the patterns seem to disregard rocks and preexistent structure completely.

Structural Knot or Big Bend   The Frazier Mountain intersection of the Garlock, Big Pine, San Gabriel, and San Andreas faults has been described as the Big Bend or structural knot of southern California. When the gross tectonics of the southern San Andreas fault are eventually worked out, key ingredients certainly will be found here. Other areas, such as the Cajon–San Gorgonio in the San Andreas alignment and the Malibu Coast–Raymond Hill–Cucamonga alignment, also will figure prominently in the final structural interpretation of southern California.

The explanation of the Big Bend is related to left slip on the Big Pine and Garlock faults and to movement on the Frazier Mountain thrust and related faults. The Frazier thrust is a short fault that dips north and northwest beneath Frazier Mountain. It has forced Precambrian banded gneisses into a high, small massif that cuts across folded Pleistocene nonmarine sedimentary rocks. Frazier Mountain is interpreted as a squeeze-up block of previously buried Precambrian, a splinter that rose up and out from the San Andreas zone because of rotational compression. As the block moved south, it was thrust over the weak, underlying younger formation. Similar thrust slices occur elsewhere in the San Andreas zone.

Mendocino Escarpment   The Mendocino escarpment may terminate the San Andreas on the north, at Cape Mendocino. The Cape is at the eastern end of the east-west trending Mendocino fracture zone, which extends more than 1000 miles (1600 km). There is no evidence, however, of an eastward land extension of the Mendocino zone—a necessary condition in the opinion of some geologists if the San Andreas is displaced by the escarpment. Several other large east-west fracture zones off California apparently do extend landward.

One interpretation is that the Mendocino fracture zone terminates the San Andreas by offset, underthrusting the San Andreas into the subduction zone that appears to form the contact between the North American plate and the East Pacific rise. In this view, the San Andreas is a transform in the evolution of the North

*Figure 11-3.
San Andreas fault
on the Carrizo
Plain. Almost every
stream course in the
Carrizo Plain shows
right-lateral offset.
The dark line on the
left is made by
tumbleweeds col-
lected against a
fence. (Photo by
Robert E. Wallace,
courtesy of U.S.
Geological Survey)*

American–Pacific plate boundary. This is currently the favored interpretation.

Another hypothesis is that the San Andreas terminates at the intersection of the North American, Pacific, and Juan de Fuca (Gorda) plates. Some authorities, however, think that the San Andreas turns west to join the Gorda escarpment along the Mendocino fracture zone. Still others cite what they interpret as a pattern of earthquake epicenters across the shelf. They maintain that the San Andreas does not join the Mendocino fracture zone, but instead continues obliquely across the shelf and slope to the northwest. This last view is receiving less support as the San Andreas fault is interpreted as a major transform. Finally, even if the continuity of their structural trends is established, slip differences on the San Andreas and the Mendocino fracture zone must be explained. Based on offset magnetic anomalies, the Mendocino zone apparently has undergone 430 miles (700 km) of *left* slip whereas the San Andreas is characterized by *right* slip of possibly hundreds of miles.

Gulf of California The San Andreas fault disappears southward almost on the Riverside-Imperial county line and may be buried under the Algodones sand dunes. It is thought to reappear along the east side of the Gulf of California. The geology of the

Mexican section is still incomplete, but several pertinent theories have been presented. One idea is that spreading centers in the Gulf are offset, with the cumulative effect being the San Andreas. J. C. Crowell suggests that the San Andreas becomes involved in the Gulf's sea-floor spreading mechanisms, which may have separated Baja California from the North American mainland. In this view, the San Andreas is no longer a fault line with definable margins, but is a transform with an uncertain history over the past 10 million years. Another hypothesis postulates movement of the Baja California land mass north from the west coast of Mexico along an inferred southeastern extension of the San Andreas. This view emphasizes the fault's substantial right-slip displacement.

Vertical Geometry

Little is known about the San Andreas at depth. Only within the past 10 years have studies been initiated to define San Andreas zone fault patterns and their downward extensions. Defining the vertical geometry of any fault normally requires extensive drilling, maintenance of well logs, and collection of cores. Still, modern seismological instruments do permit inferences from records in areas where earthquake swarms occur.

A 1970 study in the Parkfield-Cholame region, along about 25 miles (40 km) of the San Andreas, shows the fault to be a vertical zone to a depth of about 9 miles (14 km). Accordingly, the San Andreas customarily is presented as a fault of vertical dip. (Exceptions are local thrust faults directly related to the San Andreas, like the Frazier Mountain thrust.) Furthermore, evidence for anything other than vertical dip is usually inconclusive—although positive evidence for such dip is often lacking. The Cenozoic history of the San Andreas, with its miles of strike-slip displacement, discourages the acceptance of any geometry but a nearly vertical dip for a tear fault of this magnitude. Moreover, it is hard to reconcile a transform fault and large-scale shear with any dip other than nearly vertical.

SEDIMENTATION

The sharp topographic rent along most of the San Andreas is the result of recent strike-slip movement. The generally accepted displacement of 130 miles (210 km) is difficult to establish throughout the fault if precisely the same evidence is demanded for all segments, however. Much evidence is obscured by sedimentation, and more has probably been removed by erosion.

In major basins, sedimentation along the fault is important. At the head of the Gulf of California, up to 20,000 feet (6100 m) of continental and deltaic sediments have accumulated. Other areas adjoining the San Andreas show similarly thick nonmarine and clastic deposits, all developed by erosion of linear highlands. The Ridge Basin, for example, contains over 33,000 feet (10,000 m) of marine and nonmarine rocks. Such basins commonly exist where irregularities have developed along the fault's trace. When the San Andreas became a plate margin, stretching, squeezing, and sagging occurred simultaneously in different segments. Structural irregularities and sedimentation thus mask the evidence needed to establish the nature of movement.

GEOMORPHOLOGY

At its two final land expressions, the San Andreas fault lies at sea level, but elevations as high as 6019 feet (2000 m) occur along the fault in the San Emigdio Mountains. Relief up to 1500 feet (500 m) is found where graben-horst fault slices are juxtaposed. Relief is often magnified by rapid erosion, which combines fault and fault-line features. Local, natural, closed depressions are common along the entire trace of the fault.

Most types of landforms produced by faulting or subsequent erosion are found in the San Andreas zone. Although unspectacular when compared with similar forms on Basin Range faults, facets and spurs abound on the San Andreas. Such features are usually considered evidence of vertical (dip-slip) movement, but this interpretation does not necessarily apply to their San Andreas occurrence. In strike-slip fault zones, localized but prominent facets and scarps are produced by horizontal shift where a hill or ridge is crossed by the fault. The lateral shift exposes offset facets that face each other. Such facets should not be confused with fault-line scarps, which face in opposite directions along the same fault and are produced by erosion along the fault rather than by fault movement.

R. P. Sharp divides geomorphic forms developed along faults into primary (movement) and secondary (erosion) features. Besides scarps, facets, and scarplets, primary features along the San Andreas include offset streams, fault-slice (squeeze-up) blocks and shutter-ridges, closed depressions, fault valleys and troughs, gaps, saddles, kernbuts, and kerncols. Secondary features include, besides fault-line scarps and valleys, drainage derangements and springs that often result from groundwater barriers incident to faulting.

Detailed analyses of surface stream derangements and fault patterns on segments of the San Andreas have been conducted by Robert E. Wallace. Figure 11-4 summarizes patterns in the Carrizo Plain, although similar patterns occur throughout the fault zone.

Geomorphic forms are often heavily influenced by the rock types involved. The San Andreas is special because practically all of California's rock types are found somewhere along the fault. They may occur in small quantities, and as granulated, mylonitized, brecciated, or simply unaltered material. In southern California, banded gneiss, anorthosite, and other Precambrian crystalline rocks are commingled with Pleistocene alluvium, lake beds, and Recent deposits. Slivers of early Cenozoic units, Mesozoic granites, and occasional Paleozoic metasediments are all found together. North of the structural knot, the rocks are less varied. They include Mesozoic Franciscan and granitic rocks and often thick sections of post-Jurassic sediments intimately associated with Pleistocene and Recent nonmarine and marine units.

Figure 11-4.   *Patterns of fault-related stream channels found in the Carrizo Plain area.* OFFSET CHANNELS: (a) *Misalignment of single channels directly related to amount of fault displacement and age of channel. No ridge on downslope side of fault. Beheading is common.* (b) *Paired stream channels misaligned.* COMBINATION OF OFFSET AND DEFLECTION: (c) *Compound offsets of ridge spurs and offset and deflection of channels. Both right and left deflection.* (d) *Trellis drainage produced by multiple fault strands, sliver ridges, and shutter ridges.* (e) *Offset plus deflection by shutter ridge may produce exaggerated or reversed apparent offset.* (f) *Capture by adjacent channel followed by right slip may produce "Z" pattern.* FALSE OFFSETS: (g) *Differential uplift may deflect streams to produce false offset.* (h) *En echelon fractures over fault zone followed by subsequent streams produce false offset.* (Source: California Division of Mines and Geology)

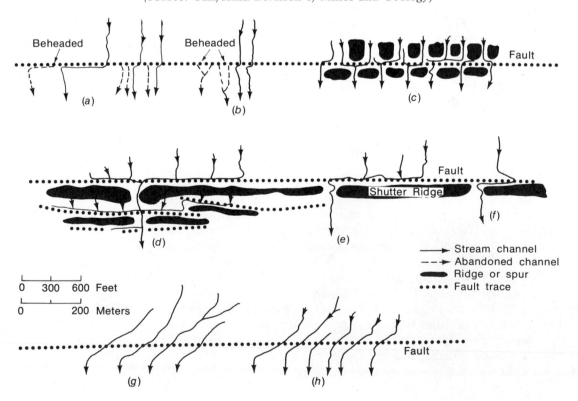

MOVEMENT

Recent ideas about movement on the San Andreas fault differ significantly from earlier views. Although no single interpretation is currently acceptable, consensus seems to exist on the following points.

1. The San Andreas has had many periods of activity followed by quiescence.

2. Different parts of the fault have been active at different times.

3. No single section of the zone has had a unique history.

4. Irregularity of movement over time has produced variety in rock deformation and geomorphism.

5. The conventional idea of a fault as an identifiable structure with similar characteristics throughout does not apply to the San Andreas.

Detailed study of portions of the San Andreas indicate that both compressional and extensional structures have been produced. These have resulted either directly from movement along the fault or from overriding deformation of broader regional origins. Often extensive folding, with axial trends at slight angles to the San Andreas, has involved major deformation of rock units. Subsequently these units have been displaced both vertically and horizontally.

A major redirection in study of the San Andreas occurred when Mason L. Hill and T. W. Dibblee advanced the view that the San Andreas had been active geologically since the Jurassic and that it had experienced 350 miles (560 km) of displacement since inception. Their hypothesis was based on suggested matching of similar rock types, fossil sequences, and structures on opposing sides of the fault. Moreover, it directed attention anew to the San Andreas problems of how much? when? how? and why? Although geologists had accepted 1 mile (1.6 km) of displacement, Hill and Dibblee's hypothesis was nothing less than outrageous to many investigators.

Vertical Components—Dip Slip

Prior to the advent of the Hill-Dibblee thesis, the San Andreas was assigned extensive vertical components of movement, particularly south of the Big Bend. Relative to the Mojave block, the San Gabriel Mountains have risen at least 4000 feet (1200 m) on faults that are subordinate but parallel to the San Andreas. The San Bernardino Mountains rise over 7000 feet (2150 m) above San Gorgonio Pass along faults in the San Andreas zone. The San Jacinto Mountains rise 10,000 feet (3000 m) relative to the Coachella Valley, another impressive evidence of vertical movement. In the San Gorgonio Pass, strands of the San Andreas appear to be the cause of the displacement. In the Coachella Valley proper, however, no evidence of a major fault is known except the extreme relief itself. Studies of comparative thicknesses of sediment in the central Coast Ranges suggest that the coastal block moved up a maximum of 6000 feet (1800 m) during the middle Cenozoic. In northern California, vertical displacement is less.

Apparently the San Andreas fault zone has experienced major vertical movement either at intervals between or concurrently with the horizontal displacement generally characterizing the fault. Although pictorially represented as a line, a shear zone like a transform is actually a wide belt in which vertical displacement may occur as the strike-slip shearing accrues horizontal displacement. Both regional basins of sedimentation and regional uplifted blocks have been produced in and along the splintered margins of the Pacific and North American plates. Concomitantly, where rotation of segments within the fault system occurs, squeeze may lift large units vertically. This often suggests a shift in tectonic pattern when in fact the vertical displacement is compatible with activity on the transform. Although attractive to some, this explanation for major vertical displacements along the San Andreas is not acceptable to all geologists. If such an interpretation is assumed, however, such vertical displacement is compatible with a long history of miles of strike slip. Furthermore, it does not require an important tectonic change to account for both major strike-slip and dip-slip movement at different times on the same structure.

Horizontal Components—Strike Slip

Right slip as the major movement on the San Andreas fault was deduced from topography. Horizontal slip, at least in the central Coast Ranges, gradually was deemphasized as a result of studies by N. L. Taliaferro. Taliaferro proposed that the San Andreas was a major structure of dominantly vertical movement, transgressed with a 10° to 15° strike variation by a younger strike-slip fault. This is the present-day rent, for which Taliaferro postulated up to a mile (1.6 km) of strike-slip displacement since its Pleistocene(?) inception.

The earlier suggestions of up to 30 miles (48 km) of right slip in southern California and 1 mile (1.6 km) in the Coast Ranges are quite consistent with today's studies. Geologists now acknowledge 130 to 180 miles (200–290 km) of displacement in southern California since the middle Cenozoic. This figure results when the Recent San Gabriel fault is interpreted as the Pliocene San Andreas.

In northern California, evidence of cumulative offset is less direct. Suggested amounts vary from 350 miles (560 km) in Mesozoic formations to 180 miles (290 km) since middle Tertiary. There is good evidence in the central section for up to 180 miles (290 km) of Miocene (or late Oligocene) offset and up to 50 miles (80 km) for the Pliocene. Determination of displacements north of the structural knot is complicated by a thick blanket of Cenozoic marine and nonmarine rocks, which precludes much field study of basement terrains. Nevertheless, evidence for displacement of magnitudes approximating those in southern California now seems almost conclusive for all northern sections of the fault.

Creep

Records of progressive horizontal movements along the San Andreas have been systematically maintained since 1930 through measurements of triangulation networks installed by the U.S. Coast and Geodetic Survey. Creep was first documented on the San Andreas fault at the Almaden Winery near San Juan Bautista, and then on the Calaveras fault at Hollister. In the 1960s, continuous creep was demonstrated along the Hayward fault in the San Francisco Bay region.

Rates of specific creep vary considerably, although 1966 reports show all movement to be horizontal. Average creep rate is summarized in Table 11-1. Very recent studies suggest that little or no creep is occurring on the San Andreas north of San Juan Bautista, but it has been documented for other Bay area faults besides the Calaveras and Hayward.

Creep has not yet been unequivocally related to the major causal mechanism of earthquakes, sudden slip. What are the relations? Are zones of creep on faults less prone to sudden slip, and adjacent areas therefore less prone to earthquakes? Or are the two processes unrelated? Is creep a subtle assurance of greater immunity from earthquakes of high Richter magnitude? Could the opposite be true?

ORIGIN

Fifty years ago, geologists were more confident that they understood the origin of faults than they are today. Faults were simply breaks where rocks had failed. They were thought to be major boundaries of mountain units, on which the mountains were lifted relative to the adjacent lowland. Movement was therefore dip slip. Strike-slip

Area	Years	Distance
Point Reyes to Petaluma	1930–38	2 cm/yr
	1938–60	1 cm/yr
Vicinity of Monterey Bay	1930–51	1.6 cm/yr
	1951–62	1.6 cm/yr
Vicinity of Hollister	1959–63	1.5 cm/yr
	1963–66	1.7 cm/yr
Salinas River Valley	1944–63	3 cm/yr
San Luis Obispo–Avenal	1932–51	2.5 cm/yr
Maricopa to Cajon Pass	To 1962	None
Imperial Valley	1935–41	Large (1940 earthquake)
	1941–54	3 cm/yr

Table 11-1
Creep on the
San Andreas Fault

Source: California Division of Mines and Geology.

movement was considered theoretically possible, but was recognized as subordinate to the vertical responsible for major uplifts.

Related to the standard dip slip expected on faults was the widely accepted (and still valid) theory of isostasy. According to this theory, rock failure on the margins of erosional-depositional boundaries results in fracturing and vertical movement. Isostasy produces greater instability on continental margins, which are usually sites of mountain ranges. Thus the San Andreas was a major fault of strong vertical displacement along the Pacific margin.

In global tectonic theory, however, the San Andreas is a lineament where the unstable edge of the Pacific plate (westernmost California) is drifting north beneath the westward-migrating, overriding continental plate. The conflicting drift directions segment the upper plate into irregular mobile units (Transverse Ranges), which act both as part of the whole plate and as separate features. Shear zones also develop, and the San Andreas transform is one result. Faults like the Garlock, San Jacinto, and San Gabriel also become major transforms. Although an explanation of the origin of the San Andreas is apparently crystallizing, major questions still exist. One critical problem is the fault's age. The fault may have influenced geologic history for the past 50 million years, with a profound influence in the past 15 million years, but these figures are not definitive.

Several questions fundamental to understanding the San Andreas were posed in 1967 by Gordon B. Oakeshott. They form a pertinent summary of problems concerning the San Andreas.

1. When did the San Andreas fault originate?
2. Should the late Quaternary and ancestral faults be regarded as different faults, developed by different stresses and with different characteristics?
3. Have the sense and direction of displacement been the same (right slip) throughout the history of the fault?
4. If there has been significant vertical movement, is it related to the present fault line, and in what manner?
5. What has been the post-Cretaceous rate of displacement?
6. Is cumulative displacement actually scores of miles?
7. To what depth does the faulting extend?
8. Is the San Andreas becoming more or less active?
9. Are frequent earthquakes an indication of high seismicity and earthquake probability?

TRANSFORM FAULTS AND GLOBAL TECTONICS

The relationships of California's many faults to one another have been considered throughout the discussions of the various geomorphic provinces. It seems apparent that most of the state's large faults and some of its smaller ones are controlled by forces associated with the San Andreas. Probably most right-slip faults are somehow related to the San Andreas system. Several left-slip faults also may result from forces generated by movements along the San Andreas. What, however, are the relationships of the San Andreas to major faults that are not obviously connected? The San Andreas is currently the major fault boundary in the western United States. Will other transforms develop as the North American plate continues riding west? It has been suggested that other transform boundaries existed earlier in California's Tertiary history.

It is appropriate to mention again two previously discussed geomorphic lineaments of California and western Nevada. First is the Furnace Creek fault, which cuts indiscriminately across rock and topographic boundaries and may be a major through-going transform like the San Andreas. Such a feature could extend from Fish Lake Valley southeast to the Colorado River near Blythe. Evidence is insufficient, however, to establish definitely such a strike-slip fracture subparallel to the San Andreas. A second trend appears from the southeast near Las Vegas, strikes northwest, and seems to extend about 200 miles (320 km). Known as the Walker Lane or the Las Vegas shear zone, this lineament is closely parallel to the San Andreas. In addition, it has been proposed that a lineament similar to the Walker and Las Vegas patterns extends over 450 miles (750 km) from central Oregon southeast into central Nevada. Such a structure might result from the same forces that initiated the San Andreas.

Faults similar to the San Andreas are known elsewhere in the world. An example is the major alignment of the Denali and Fairweather faults in Alaska. The Great Alpine fault, which cuts across the west side of New Zealand's South Island and is traced north by some geologists into two or three prominent North Island faults, may be nearly 1000 miles (1600 km) long. In addition to vertical displacement, the Alpine fault shows up to 350 miles (560 km) of strike-slip movement. In Asia Minor, major faults of the San Andreas type trend east-west to intersect similar northwest-southeast trending faults that extend several hundred miles south into the Jordan Valley and the Sinai Desert. The Red Sea and the Gulf of Suez are marked by large strike-slip structures. All these features are interpreted as transforms in areas of suturing between continental plates.

HUMAN INVOLVEMENT

In an unstable region with large population and continuing growth, maximum safety precautions to offset earthquake hazards are extremely important. In California, about 75 percent of the population resides in the areas of greatest instability. It is difficult to implement earthquake precautions, however. They are expensive, and since earthquakes may never affect the majority, appropriate precautions are often something we would like to dodge. Building for earthquake resistance may increase costs by as much as ten percent. In the past, incorporating earthquake safety into structures was hampered because geologic science could not provide adequate data for structural engineers, architects, construction firms, and so on. Today better information is available.

Establishing adequate safeguards assumes willingness to recognize earthquake hazards, expenditure of the large sums of money required, and counteracting the apathy that arises from the "it can't happen to me" philosophy. This last problem might be overcome quickly if many continuing, sufficiently strong but not disastrous earthquakes were to occur. Instead, there are infrequent, localized, but unfortunately sometimes severe tremors that affect comparatively few people at a time.

Engineering

It is now feasible to construct modern earthquake-resistant (*not* earthquake-proof) buildings. Since the locally disastrous 1933 Long Beach earthquake, the Field Act has required all California school buildings to be earthquake-resistant. Moreover, the uniform building codes adopted in the 1950s have applied Field Act standards to virtually all buildings. Yet damage sustained in the 1971 San Fernando quake shows that even these standards are not completely adequate. Nonetheless, dams, waterworks, highways, and utility structures may all be built with reasonable safety provided certain precautions are followed and provided they are not built *directly* on faults or on unstable ground subject to liquefaction or sliding.

Public lack of awareness and failure to insist on reasonable precautions are the main blocks to earthquake safety, although some hazards will probably always remain. In the 1971 San Fernando earthquake, gross damage to newly constructed freeways (built with reasonable safeguards) was tremendous. The quake's g factor led some seismologists to assert, however, that the San Fernando case can reasonably be expected to occur once every 5000 years. Should huge expenditures be allocated to safeguard structures, roads, and bridges that have an estimated life of 50 years? Reasonable precautions are required, but some acceptable risk must be involved when people choose to live in an unstable, earthquake-prone region.

Psychological Factors

The psychological discord produced in some people when an earthquake strikes can be extreme. One result is the tendency to accept rumors about earthquakes without regard for facts. The psychological consequences of other destructive natural phenomena like tornados, hurricanes, and floods are also emotionally traumatic, but they do not seem to be accompanied by false rumors to the same degree as earthquakes. Perhaps other natural disaster-producing phenomena are more acceptable to the psyche because they can be seen—seen to arrive and seen to leave—something rarely possible with an earthquake. As a matter of fact, the hazards of water, wind, and fire take far larger tolls of life and property in the United States than do earthquakes. Furthermore, safeguards against earthquakes are no more costly than those regularly required to prevent damage by other natural phenomena.

PREDICTION OF MOVEMENT

Prediction of earthquake-producing movement for the San Andreas fault is not new. As a rule, it has involved both geologic pronouncements and the ruminations of soothsayers, mediums, and clairvoyants.

In the geologic category, there are specific predictions like the one attributed to an eminent California geologist about 1920. This authority allegedly predicted that a great earthquake would occur in Los Angeles Basin from movement on the San Andreas fault. He theorized that the movement would occur closer to 5 years than 10, and closer to 15 years than 20. It is said that, when 1940 arrived, a national news service telephoned the geologist, noting that it was the twentieth anniversary of his prediction and that no earthquake had yet occurred from movement on the San Andreas. The reporter is said to have queried, "Have you any statement you would care to make?" "Oh, yes," responded the geologist, "it isn't every geologist who is privileged to live long enough to have his predictions disproved!" Regardless of its accuracy, this story makes a point: specific predictions about slippage on any fault can as yet be based only on statistical probabilities.

Studies in Japan, the United States, and the U.S.S.R. suggest, however, that definitive earthquake prediction may be near. In fact, a California quake was successfully predicted in 1974. In the U.S.S.R., it has been determined that ratios of the velocities of P and S waves measured in small shocks sometimes drop significantly, to be followed by a sharp increase in the ratios just before a major shock. Anomalous ground tilt in regions of active faulting may signal impending sudden movement. Porosity and dilatancy changes in critical areas adjacent to faults and shifts in electrical conductivity of rocks are being explored as possible avenues for prediction. The meaning of small shock patterns both before and after known major quakes is being restudied in conjunction with changes in rock properties when the quakes occur. It may be that before an earthquake is triggered, the rocks along faults undergo detectable compression. Apparently, the effects of compression in small earthquakes are measurable for a few days prior to the actual shocks; in the case of large earthquakes, it may be possible to make predictions three to four years in advance.

REFERENCES

Crowell, John C., 1962. Displacement along the San Andreas Fault, California. Geol. Soc. Amer. Spec. Paper 21.

———, ed., 1975. San Andreas Fault in Southern California. Calif. Div. Mines and Geology Spec. Rept. 118.

Dibblee, T. W. Jr., 1966. Evidence for Cumulative Offset on the San Andreas Fault in Central and Northern California. *In* Geology of Northern California. Calif. Div. Mines and Geology Bull. 190, pp. 375–384.

Dickinson, William R., and Arthur Grantz, eds., 1968. Proceedings of Conference on Geologic Problems of San Andreas Fault System. Stanford Univ. Pub. Geol. Sci., v. 11.

Kovach, Robert L., and Amos Nur, eds., 1973. Proceedings of the Conference on Tectonic Problems of the San Andreas Fault System. Stanford Univ. Pub. Geol. Sci., v. 13.

Lawson, A. C., and others, 1908. The California Earthquake of April 18, 1906. Rept. of State Earthquake Investigation Commission. Carnegie Institute, Washington, D.C., Pub. 87, v. 1, part I.

Chapter Twelve

Offshore

There is a path on the sea's azure floor,
No keel has ever ploughed that path before.
                    Percy Bysshe Shelley

Offshore California may be related to three discrete land provinces. Evaluating it as a separate entity, however, usually leads to the most accurate understanding of Offshore geology. One onshore counterpart is the Coast Range province, north of Point Arguello, where the Offshore includes continental shelf and slope that may be submerged extensions of the Coast Ranges. The other related land provinces are the Transverse Ranges and the Peninsular Ranges, south of Point Arguello. A distinctive geology occurs in the southern part of Offshore California that is quite unlike either the steep continental slopes or the gently outward sloping shelves seen elsewhere in the world. Francis P. Shepard named this section the Continental Borderland.

NORTHERN SHELF AND SLOPE

Usually the northern shelf is only a few miles wide, but off the Golden Gate it widens to about 30 miles (48 km). It widens again north of Cape Mendocino, probably in response to the westward bulge of the Klamath Mountains, which squeeze the dry land width of the northern Coast Ranges to only 8 miles (13 km).

Although the sedimentary cover on the northern shelf is reasonably well known, few bedrock exposures are available. Apart from rocky islets and sea stacks close to shore, the only exposures are on the Farallon Islands off San Francisco. Most information has been derived from geophysical studies, which include subbottom profiling and seismic surveys on deeper structures. These reveal some characteristics of the rocks immediately below the sedimentary blanket. Echo soundings made in the past 45 years, coupled with earlier wire soundings, have provided a fairly accurate and detailed picture of the shelf topography (Figure 12-1). All investigations have been hampered, however, by the persistently rough seas that pound the northern California coast.

The most striking topographic features of the northern shelf are the submarine canyons that cut both the shelf and the continental slope beyond. Several head within a mile (1.6 km) of the beaches and continue as distinctive topographic features to the deep ocean floor. By far the largest is Monterey Canyon and its branches. Not only is this the largest submarine canyon on the California coast, but also, so far as is known, it is the largest between Alaska and Cape Horn.

Monterey Canyon extends within a half-mile (0.8 km) of the coast and continues in recognizable form to depths of 12,000 feet (3650 m). It follows a winding course and occupies a V-shaped gorge, at least in its headward portion. Two main branches join the Monterey in its upper reaches: Soquel Canyon from the north and Carmel Canyon from the south. Although no bedrock is found in the

Figure 12-1. *Topographic features: northern California shelf. (Source: California Division of Mines and Geology)*

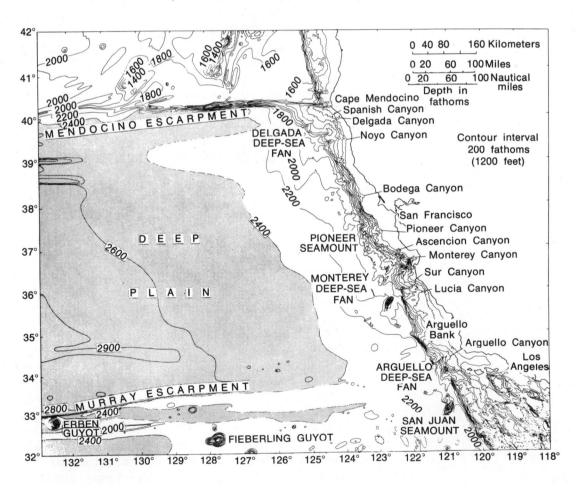

main Monterey Canyon until a depth of 1800 feet (550 m), borings along the beach show that a connecting canyon 5000 feet (1500 m) deep was once cut into bedrock but later filled with sediment. Although not so long, Monterey Canyon compares in depth and width with the Grand Canyon of the Colorado River. The Monterey's outer portion, below a depth of 5100 feet (1550 m), becomes less V-shaped and more troughlike with distinct levees up to 75 feet (24 m) high. These eventually disappear near the outer margin of the huge fan at the mouth of the canyon. This submarine delta is almost 100 times the volume of rock that the canyon above would have disgorged by itself, establishing that the canyon has long been a channelway conveying sediment from shelf and nearshore to the deep ocean. The submerged deltaic material is both mineral and organic, derived from beach and adjacent shallow-water areas.

Other submarine canyons include Delgada Canyon south of Cape Mendocino, Noyo Canyon about 40 miles (64 km) south of Delgada Canyon, Bodega Canyon off the mouth of Tomales Bay, Sur and Lucia canyons off the Monterey coast south of Carmel, and Arguello Canyon well off Point Arguello.

Rocks

The northern outer shelf and upper slope appear to be dominated by an elongate ridge of basement rock from near Cape Mendocino to about 60 miles (96 km) south of Monterey. Composition of this bedrock high is uncertain because geophysical measurements have not clearly distinguished between granitic (Salinian) and Franciscan basements. Nevertheless, available evidence does suggest that the ridge is chiefly granitic from Monterey north.

Although the surface of the northern shelf has a nearly regular outward slope, geophysical studies have shown that a prism of sedimentary rock lies between the present shoreline and the basement ridge. Both features are blanketed by recent sediments. It has been suggested that the bedrock ridge represents a slice of continental crust detached from the Coast Ranges, with the intervening elongate basin subsequently filled with sediments. This may be analogous to the Gulf of California, which some geologists regard as a pull-apart structure. Furthermore, gravity measurements on the shelf between the Golden Gate and the Farallon Islands have revealed a distinct negative anomaly that could reflect either a thick section of low-density sedimentary rocks or stretching and thinning of underlying crystalline crustal rocks.

Sediments   Much of the northern shelf is covered with sands and muddy sands. The few banks and higher places that exist usually are overlain by shelly materials often mixed with glauconite

(green sand) and phosphorite. (Both minerals are chemical precipitates or alteration products that accumulate where detrital sediments from land are scarce.) Sedimentary patterns were affected by
extensive exposure to wave action during lower stands of Pleistocene sea level. Several relict nearshore deposits are preserved well
below the reach of modern wave action.

An unusual sedimentary feature is the submerged lunate
sandbar off the mouth of Golden Gate. This sandbar has induced
many cases of sea sickness because its presence is often marked by a
patch of higher waves and rougher sea. It is formed from river sand
carried from the Golden Gate and spread as a low crescentic ridge on
the sea floor. A delicate balance is maintained with the tidal currents that sweep back and forth through the Golden Gate. Previously, the bar was somewhat farther offshore. As sediment from
mining and farming activities in interior California flowed into San
Francisco Bay, however, the bay became shallower and the volume
and velocity of tidal water moving back and forth through the Gold-
en Gate were reduced. This effect was augmented by man-made
fills added to the San Francisco Bay foreshore. Weakening of the
tidal currents subsequently allowed wave action in the open Pacific
to shift the bar hundreds of feet shoreward.

Structure

Knowledge of the structure of the northern shelf is incomplete. Although it is clear that many land structures extend onto the shelf,
important but puzzling changes seem to occur there. For example,
the San Andreas and Pilarcitos faults diverge near Redwood City and
continue onto the shelf as separate entities. These faults almost
certainly rejoin somewhere off the Golden Gate because the San
Andreas reappears on shore alone, but no details are yet established
about their offshore reunion. Investigations in the Monterey Bay
area, however, permit fairly confident extension of faults from the
Santa Cruz Mountains north of the bay across the shelf to the Santa
Lucia Mountains on the bay's southern shore.

Near Cape Mendocino, where the San Andreas fault finally
leaves the coast and curves west, the Mendocino fracture zone approaches the shelf. This major feature of the ocean floor has several
counterparts in the Pacific. All seem to reflect extensive strike slip,
as indicated by offset magnetic anomalies that disappear as they
approach the base of the continental slope. The Murray fracture
zone, for example, dies out on the sea floor west of Point Arguello
where the east-west trending Transverse Ranges begin. Many
geologists have tried to relate the 1900-mile (3050 km) Murray fracture zone to the Transverse Ranges because both have a strong
east-west trend. So far, no really persuasive evidence has been found
to corroborate such a connection.

CONTINENTAL BORDERLAND

The section of Offshore California south of Point Conception, the Continental Borderland, is far better known than the northern portion (Figure 12-2). In fact, it is among the most thoroughly investigated areas of sea floor anywhere in the world. Despite this, our knowledge of the area is still far from complete.

The Continental Borderland is composed of elevated blocks and ridges, sometimes with islands above marine datum and sometimes with deep, often closed basins. The seaward edge approximately connects Point Conception with Punta Banda in Baja California. To seaward, a typical continental slope with relief up to 13,000 feet (4000 m) descends to the deep floor of the Pacific. The landward margin is a narrow continental shelf seldom more than 5 miles (8 km) wide. In the Continental Borderland proper are basins and intervening ridges. Here maximum relief is about 8500 feet (2600 m), although the sedimentary fill in the basins suggests that previously it may have been much greater.

Shelves as wide or wider than the Continental Borderland are known throughout the world, but in only a few other localities do they depart from the typical pattern of almost flat, gently sloping platforms interrupted by low banks and occasional canyons or channels. The southern California–northern Baja California region is certainly not typical shelf or continental slope, although narrow shelves do extend along the mainland California coast and around the islands. Some steep declivities reminiscent of slopes occur also, but these normally lead into deep, closed basins. Only at the outer edge of this unusual region is there a well-defined slope leading into the deep sea (the Patton escarpment).

Topography

Apart from the Santa Barbara Basin and the Channel Islands, which follow Transverse Range trends, most of the basins and intervening banks and ridges trend distinctly northwest. Depths of basin floors vary somewhat systematically. Those closest to shore have the shallowest depths, flattest floors, and thickest sedimentary fillings. Partially filled inshore basins are Santa Monica, San Pedro, San Diego, and Catalina. As noted in Chapter 8, the Offshore region once included much of the Los Angeles area, but being closest to the copious sedimentary supply the basins were filled and converted to land.

North of latitude 32°, the Continental Borderland contains 12 distinct basins. Several others exist to the south, but they are outside California waters and will not be considered here. K. O. Emery has assembled considerable data on these basins. Table 12-1 incorporates some of this information, and Table 12-2 shows the relative importance of the varied topographic features of the area.

328

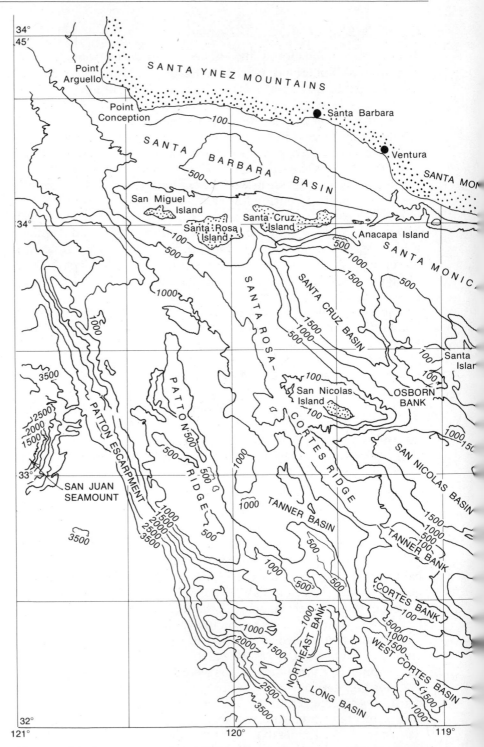

Figure 12-2. Topographic features: southern California shelf. (Source: U.S. Geological Survey)

Bathymetric contours in meters

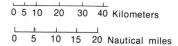

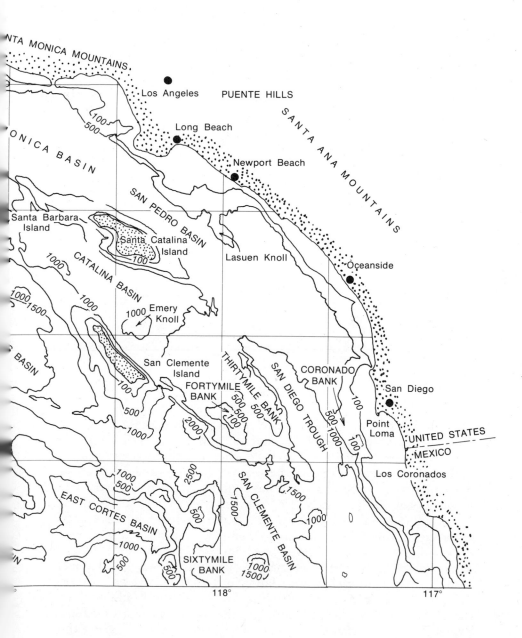

Table 12-1
Characteristics of
Continental
Borderland Basins

Basin	Number of Sills	Bottom Depth		Lowest Sill Depth		Closure		Area of Flat Floor	
		(ft)	(m)	(ft)	(m)	(ft)	(m)	(mi²)	(km²)
Santa Barbara	1	2056	627	1560	476	496	151	420	1090
Santa Monica	1	3078	939	2418	737	696	212	780	2030
San Pedro	2	2994	913	2418	737	576	176	270	700
San Diego	1	3000–4500 900–1400		4500	1400	0	0	680	1770
Santa Cruz	1	6450	1967	3558	1085	2892	882	280	730
Catalina	2	4452	1358	3222	983	1230	375	690	1790
San Clemente	1	6912	2108	5958	1817	954	291	480	1250
San Nicolas	1	6012	1834	3630	1107	2382	727	370	960
East Cortes	1	6492	1980	4644	1416	1848	564	240	620
Tanner	1	5088	1552	3822	1166	1266	386	200	520
West Cortes	5	5892	1797	4470	1363	1422	434	250	650
Long	2	6360	1940	5568	1698	792	242	210	550

Source: After K. O. Emery.

Table 12-2
Areas of
Continental
Borderland
Features

Feature	Areas		Percent of Total Area
	(mi²)	(km²)	
Islands	340	880	1.1
Mainland shelf	1890	4890	6.2
Insular shelves	1390	3580	4.6
Banks	2420	6270	8.0
Basin and trough slopes	19,210	49,700	63.0
Basin and trough floors	5120	13,260	16.8

Source: After K. O. Emery.

The boundaries between relief features are necessarily some-what arbitrary, but it is still evident that the basins and their surrounding slopes constitute a major portion of the province. Admittedly a minor part of the total, the submerged banks and ridges are of particular interest because they have yielded most of the available bedrock samples. Furthermore, they may prove to contain oil reservoirs. Several of the banks are shallow enough to have been emergent during the Pleistocene, when glacially lowered sea levels occurred. A drop in sea level of about 300 feet (100 m), which almost certainly occurred during the last Pleistocene glacial stage, would unite the four northern islands into a single large one, would at least quadruple the area of San Nicolas Island, and would produce three or four new islands along the Santa Rosa—Cortes Ridge. Evidence suggests that there were indeed more islands during the

Pleistocene. Submerged wave-cut benches or terraces have been recognized by their topography and by accumulations of wave-rounded pebbles in water too deep today for these materials to be exposed to wave action.

It is tempting to explain the occurrence of fossil land animals such as the dwarf mammoths of the northern islands by providing a land bridge for them, occasioned by lowered sea level. A gap several miles wide remains, however, even if sea level were lowered 600 feet (180 m)—twice the amount generally acknowledged. How the mammoths actually reached the islands remains a mystery.

Like the northern Offshore area, the southern region also contains submarine canyons. There are at least 32, of which 19 are named. Some are as well known and thoroughly surveyed as any in the world, although none is so large as Monterey Canyon. In addition, much of the investigation into the origin of such features generally has been conducted in this region. Apparently some canyons have been produced by presently operating sea-floor processes, whereas others have resulted from processes that are similar but currently inactive. Six prominent canyons, presumably related to the modern shoreline, occur along the mainland coast. Several others cut the insular shelves. Still others appear to reflect the shoreline and lower sea levels of the Pleistocene.

Rocks

Most submarine rock samples recovered from the Continental Borderland have come from banks, knolls, ridge crests, and steep slopes. As a rule, other sections are blanketed with Quaternary sediments. So far, sampling has revealed a complex distribution of basement rocks without any discernible pattern.

Granitic basement has not yet been detected on the Borderland sea floor, although its presence in the Transverse and Peninsular ranges suggests that it probably occurs at least under the Santa Barbara Basin. Franciscan basement including blue (glaucophane) schist is well exposed on Santa Catalina Island and is present in Palos Verdes Peninsula. Closely similar or identical rocks have been dredged from a sea knoll east of Sixtymile Bank, a low ridge northwest of Santa Rosa Island, and the Patton escarpment. Other rocks of Franciscan type have been obtained from several places including Patton Ridge, the saddle between Santa Barbara and San Clemente islands, southeast of San Nicolas Island, and a knoll in the Tanner Basin. Since Franciscan basement is so widely distributed, some geologists contend that it underlies the entire Borderland.

Upper Cretaceous sedimentary rocks are abundant in coastal southern California. They also are exposed on San Miguel Island and are present at depth on Santa Cruz Island as well as beneath the Santa Barbara Channel. It is probable that rocks of this age occur throughout the area. Much of the upper Cretaceous sandstone is relatively unfossiliferous, though, and may not be positively identifiable as Cretaceous.

Paleocene rocks have been found on San Miguel and Santa Cruz islands, but none has been recovered elsewhere off southern California. The most likely sites for such rocks are the Santa Barbara Channel and the Santa Rosa–Cortes Ridge. These areas contain known Eocene rocks that easily could conceal Paleocene rocks. Eocene rocks are 4000 feet (1200 m) thick on San Miguel, Santa Rosa, and Santa Cruz islands and at least 3500 feet (1050 m) thick on San Nicolas Island.

Because the California Oligocene-Miocene time boundary is disputed, it is difficult to describe Oligocene rock distribution without causing some confusion. If one restricts the Oligocene to Sespe-equivalent rocks, however, it is accurate to say that the Oligocene is so far known only from the Transverse Range part of the Borderland and that there it is entirely nonmarine.

Of all Tertiary rocks, those from the Miocene are most widely distributed both onshore and offshore. They include a wide variety of sedimentary and volcanic rocks, some so distinctive that they often can be identified even when quite unfossiliferous. Middle Miocene rocks are particularly widespread and are distinctive lithologically. They represent three different facies: organic-siliceous shales similar to the Monterey formation; feldspathic sandstones equivalent to those of the Topanga formation in the Transverse Ranges; and blue schist breccias like the San Onofre breccia of the Peninsular Ranges.

The Monterey facies is the most widely distributed. It is missing in only a few places, chiefly where older rocks are exposed. It is probable that most, if not all, of the Borderland was a relatively deep marine environment during the middle Miocene and that little or no land was present.

The San Onofre facies is confined to coastal areas from Los Coronados Islands south of San Diego to the Santa Monica Mountains, Santa Catalina Island, and the islands on the south side of the Santa Barbara Channel. Certain structural reconstructions have been proposed, based on the assumption that San Onofre breccia was deposited in a limited area originally and that its present dispersed occurrence results from fragmentation and considerable lateral fault movement.

Volcanic rocks primarily of middle Miocene age are also widespread and occur on most of the islands. Flows, sills, dikes, and ash beds are represented, some land-deposited but most apparently submarine. Anacapa and Santa Barbara islands are composed almost entirely of such volcanics. Similar rocks are likewise prominent on San Clemente, Santa Catalina, and Santa Cruz islands.

Distribution of Miocene rocks is largely independent of present topography. The implication is strong, therefore, that the system of basins, ridges, and islands distinguishing the region today came into existence only after Miocene strata were deposited. This view is confirmed by the distribution of marine Pliocene strata, which are confined almost exclusively to existing basins and their margins. Pliocene strata reached phenomenal thickness in the nearshore and onshore basins, as demonstrated by well records. It is probable that thick Pliocene strata are present in many other basins as well, because geophysical measurements indicate at least 10,000 feet (3000 m) of sedimentary fill in such basins as San Nicolas. Of course the Pliocene portion of this fill can only be estimated.

Quaternary basin deposits are undoubtedly often continuations of Pliocene patterns. Although Quaternary thicknesses have been established at only a few places, in the Santa Barbara Channel 1600 feet (490 m) of Holocene have been identified. On the islands, deposits of this age are mainly thin nearshore shelf and terrace materials.

Special Characteristics of the Quaternary

Despite some uncertainties about thickness, Quaternary sediments are widespread and varied. Generally basins and slopes are blanketed with fine-grained, often greenish muds. Exceptions occur near the mouths of submarine canyons, where coarser-grained sandy deposits have been carried down from beach and shelf environments. K. O. Emery has divided shelf sediments into five main categories.

1. Detrital materials derived from streams and cliff erosion along shorelines. This is the dominant type.

2. Relict sediments not now in equilibrium with their environment. These are mostly late Pleistocene, coarse-grained deposits submerged by the rise in postglacial sea level.

3. Residual organic debris left from in place weathering of outcrops.

4. Organic sediments, mostly shell gravels.

5. Chemical precipitates, mainly glauconite and phosphorite.

The higher parts of the sea bottom (ridges, banks, and knolls) contain authigenic deposits. These are produced in place by either direct chemical precipitation or alteration of materials already present. Authigenic deposits commonly are associated with organic materials that have accumulated without being overwhelmed by detrital materials from land. On shelves and in nearshore basins, detrital materials greatly dilute authigenic sediments.

Finer-grained materials are usually absent from topographic highs, mostly because of the strength of the tidal currents that sweep across these eminences. Fine deposits can accumulate only in sheltered spots between loose rocks or in cracks and crevices. It is in such environments that much of the authigenic material develops. Two minerals predominate: glauconite (not to be confused with the blue metamorphic mineral glaucophane) and phosphorite.

Glauconite tends to form in oxygen-deficient microenvironments like the interiors of foraminiferal tests and between and beneath the grains of shell gravels. Phosphorite forms more effectively on the better-exposed upper surfaces of bank tops and ridges where oxygenated water is present. Both minerals have potential economic value though neither is being exploited presently. Glauconite is a low-grade source of potassium suitable for some fertilizers because the potassium is released slowly during weathering. Phosphorite is also a potential fertilizer and a source of phosphorous generally. Nodules, lumps, and crusts of marine phosphorite probably cover at least 6000 square miles (15,600 km²) of the Borderland, primarily between depths of 100 and 1000 feet (30–300 m).

Structure

Until quite recently our understanding of the structure of the southern Offshore province was based on island exposures, distribution of earthquake epicenters, and sea-floor topography. The rather striking similarity between Offshore basins and those of the Basin Ranges led to the widely held view that the deep ocean basins were the same kind of down-dropped fault blocks that occur in the Basin Ranges. Distribution of earthquake epicenters seemed to support this theory.

During the past few years, the U.S. Geological Survey has conducted a detailed study of the Continental Borderland. As a result, many basins and ridges are now thought to be the consequence of large-scale synclinal-anticlinal folding rather than faulting. Because of the great structural relief, this conclusion was unexpected. (In this case, structural relief means the difference in elevation between a reference surface under the basin floor and the adjacent ridge or island.) Except for the Ventura–Santa Barbara Basin, maximum submerged structural relief recognized so far occurs between Santa Catalina Island and the San Nicolas Basin—about 20,000 feet (6100 m). On land, the Ventura Basin has structural relief up to 60,000 feet (18,000 m), and the Los Angeles Basin shows at least 40,000 feet (12,200 m).

Folds   Several ridges, such as the Santa Rosa–Cortes Ridge, are anticlinal in general plan. They probably are not simple anticlines, but instead vary in size and orientation. A subordinate anticline well exposed on San Nicolas Island is a broad, nearly symmetrical fold quite unlike the tight, asymmetrical folds occurring on land in such places as the Ventura and Los Angeles basins. There is a general tendency for the folds to become broader and more symmetrical away from shore.

Faults Despite the discovery that it is less important than formerly thought, faulting is still considered significant. The prominent strike-slip faults exposed in the northern islands undoubtedly continue on the adjacent sea floor and may connect with faults on the mainland. Many faults once thought to be long, continuous structures, such as the San Clemente Island fault along the northeast side of the island, are now interpreted as shorter features. Other reasonably well established faults are found along the southwestern side of Santa Catalina Island and on the north and east sides of the San Nicolas Basin.

Gravity Measurements

Because gravity profiles must be interpreted, investigators do not always reach the same conclusions from the same data. For example, a negative gravity anomaly indicates a deficiency of mass that some consider a downbuckle of less dense sedimentary material into more dense basement rock. Others interpret a negative gravity anomaly as a thinning of the underlying crustal material.

Negative anomalies occur a short distance from shore, between La Jolla and Laguna Beach; in San Pedro, Santa Barbara, and Catalina basins; in San Diego trough, southern San Clemente Basin, and southwest of San Clemente Island; and in Arguello Canyon. Gravity highs, generally believed to indicate basement or crystalline rocks close to or at the surface, are found seaward from Point Loma and Del Mar; northwest and southeast of the Palos Verdes Hills and in outer San Pedro Bay; and southwest and west of San Miguel Island and south of Santa Cruz Island.

SUBORDINATE FEATURES

Coastal Sand Transport and Harbor Problems

As man has attempted to manage the shoreline and build harbors, breakwaters, jetties, and so on, he unfortunately has learned about some of nature's processes at great cost to property and financial resources. Perhaps the most depressing aspect of this dreary story is the apparent reluctance to learn from past mistakes or to seek sound advice when it is available. Earlier mistakes are certainly understandable, but recent ones are more difficult to rationalize. In considering the matter, southern California examples are used primarily because this region of the state has experienced the most intensive coastal development. Consequently, the majority of studies have been conducted here.

Santa Barbara   Early in the century, construction of a break-water and harbor at Santa Barbara was repeatedly urged. Santa Barbara had only unprotected coast with a wooden wharf impossible to use in heavy weather. Moreover, there was no sheltered anchorage for boats. Three proposals were reviewed by federal agencies, and each time Santa Barbara was advised not to construct a permanent breakwater. Despite this advice, local interests prevailed and funding was arranged.

In the late 1920s, a breakwater open at both ends was constructed offshore. Sand subsequently accumulated on the protected beach inside the breakwater and inadequate protection from the wind also resulted. The breakwater's western end was extended to the shore to keep sand from entering the harbor and to provide additional wave protection. This greatly improved the harbor—for a while. The longshore current moving east along the shore was forced to deposit its load of sand on the western side of the breakwater (Figures 12-3, 12-4, 12-5). This triangular wedge of sand accumulated for about five years until sand began to pass along the outer

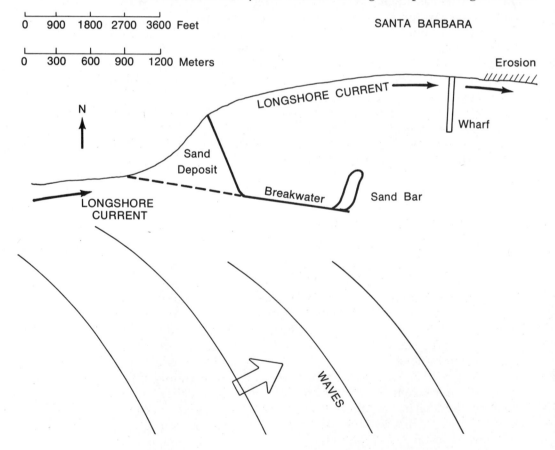

Figure 12-3.   Sketch of wave approach and harbor at Santa Barbara.

Figure 12-4.
Santa Barbara harbor under construction, before sand filling, 1929. (Photo by Fairchild Aerial Surveys, courtesy of Department of Geography, University of California, Los Angeles)

Figure 12-5.
Santa Barbara harbor after sand filling, 1935. (Photo by Fairchild Aerial Surveys, courtesy of Department of Geography, University of California, Los Angeles)

edge of the breakwater and enter the harbor at the east. While the sand was thus being trapped, the longshore current continued its activities east of Santa Barbara. Denied its usual load of sand from the west, it obtained a fresh supply by severely eroding the beaches for 10 miles (16 km) east of Santa Barbara. During a particularly stormy winter in the late 1930s, some beaches were cut away as much as 150 feet (45 m), damaging buildings and narrowing the recreational beaches east of Santa Barbara.

In the middle 1930s, dredging was initiated to counteract the sand moving around the east end of the breakwater into the harbor. The spoil from the dredging was carried east of the wharf and dumped in 18 to 20 feet (5.8–6 m) of water, forming a submarine ridge about 1000 feet (300 m) long. It was expected that the sand from this ridge would be picked up by the longshore current or waves and carried east to the depleted beaches. This did not happen, and erosion of the eastern beaches continued for another two years or so until further dredging of the harbor was needed. By this time, the U.S. Army Corps of Engineers had begun the detailed studies that resulted in landmark reports on the nature of southern California beach processes.

The ten or so dredgings that occurred subsequently have placed the dredged sand on the beach east of Santa Barbara. From here the waves and longshore current can move the sand along the shore, stabilizing if not improving the depleted beaches to the east. The ridge from the first dredging remains today, almost unaltered and in water too deep and far from shore to augment the sand carried by the currents along the beaches.

The studies made by the Corps of Engineers showed that the currents at Santa Barbara moved an average 770 cubic yards (585 m³) of sand daily past any given point (a volume about that contained by 100 full dump trucks). This amount can rise to 4200 cubic yards (3190 m³) daily during stormy winters and drop to 300 to 400 cubic yards (230–300 m³) in the summer. Studies elsewhere in southern California have shown that the volume of sand moving along the Santa Barbara coast is not particularly high; some stretches of coast, such as that between Ventura and Port Hueneme, experience about three times as much sand transport.

Santa Monica In the early 1930s, a breakwater was built at Santa Monica, a site as unsuitable for a harbor as Santa Barbara. The breakwater was left open at each end, on the presumption that the longshore current could move the sand through the sheltered area rather than impounding it on the upcurrent side of an attached breakwater. It was not appreciated, however, that the longshore current and associated beach drifting required normal wave and surf action on the beach. The breakwater was built to eliminate or reduce vigorous wave action, but it thereby also slowed or stopped the currents behind the breakwater. Sand was promptly deposited in the sheltered area and by the early 1940s threatened to fill the harbor and join the breakwater to the land. A dredging program was instituted that is still required to keep the harbor open.

Natural Beach Processes Under natural conditions, what happens to beach sand when it routinely moves along shorelines? Approximate beach equilibrium is maintained by a balance between loss and supply (principally stream and cliff erosion). Sometimes sand is moved inshore by winds to locations beyond the reach of waves. Particularly during storms, waves and currents may carry sand into deeper waters, again beyond the reach of normal transporting mechanisms. In addition, sand deposited into the heads of submarine canyons will be carried by turbidity flows into deeper waters.

Man disturbs any of these processes at some cost. For instance, trapping sand behind inland dams eventually will starve the beaches normally receiving this supply. Similarly, detention of sand because of breakwater construction results in down-current beach erosion, often with associated property loss.

Southern California Submarine Canyons

Eight of the thirteen main southern California submarine canyons cross almost the entire mainland shelf; the other five cross the outer part of the shelf. These five (and presumably others not yet discovered) may have once extended all the way across the shelf almost to the beach, but they have been partially or totally filled with sediment since formation.

Most of the more prominent canyons approach shore where longshore currents are deflected seaward either by the orientation of the coast or by the position of headlands. The pattern is so consistent that it can hardly be accidental. It is well known that longshore currents transport considerable sediment along the coast. In addition, nearshore and shallow-water sediments often move down canyons into deeper water. Evidence for this includes the following elements.

1. Presence of nearshore sediments and remains of shallow-water organisms in the deep-water fan deposits at the mouths of submarine canyons.

2. Direct observation of underwater streams of sand flowing down canyons.

3. Surveys of sedimentary fills in the heads of submarine canyons that prove these materials periodically move down the canyons.

4. Undersea fans that possess levees along the channels crossing the fans.

5. Direct observation of the effects of erosion and abrasion on the walls of submarine canyons.

6. Volumes of fans at canyon mouths are characteristically larger than canyon volumes. This indicates that fan sediments cannot be derived exclusively from erosion of the canyons but require additional material.

For many years it was widely supposed that submarine canyons were cut by land rivers and then drowned by a relative rise of sea level. As these features became better known and their courses were traced to depths of 10,000 feet (3000 m) or more, it became difficult to attribute their existence to sea-level change. Furthermore, other considerations apparently contradicted the possibility of such major sea-level shifts. The close association of canyon heads, present sea level, and coastal configuration seemed to demand an origin related to a presently active marine process.

Although parts of some canyons may reflect the location of land streams during lowered Pleistocene sea levels, most canyons are not clearly associated with land streams. Most are eroded and kept open by periodic flows of turbid water-sediment mixtures that carry near-shore materials into deep water, build fans and levees, and scour canyon walls. Canyons with heads some distance from the present shore apparently formed when lowered sea level changed the coastline. This permitted longshore currents to transport sediments into what are now fossil canyons.

The main obstacle to satisfactorily explaining submarine canyons has been the inability of scientists to observe what happens in these environments. Although turbidity currents had been studied in the laboratory for many years and their existence in submarine canyons was suspected, their erosional competence had not been clearly demonstrated. Consequently, many workers were reluctant to acknowledge that a process with questionable erosional efficacy could account for features the size of the Grand Canyon. Stream erosion, however, was well documented and understood. It was therefore considered the most probable agent, despite the serious problems associated with major sea-level changes. On one hand, an effective process was known that would do the job if only the sea could be disregarded. On the other hand, a likely process appeared to exist, but its presence on the sea floor was doubted and its erosional competence was speculative at best. Nevertheless, as a cause of submarine canyons, it did not require any troublesome change in sea level. Perhaps the choice is obvious today, now that so much is known about events on the sea bottom. The choice certainly was not so obvious even two decades ago, when neither the river-cut nor the turbidity-current adherents could present invincible arguments.

Several of the world's most intensively studied submarine canyons occur off southern California. The best known is La Jolla Canyon, with its branches of Scripps and Sumner canyons. Francis P. Shepard has supervised regular surveying and monitoring of La Jolla Canyon and its sediments. It has been established that La Jolla is one of the longest canyons off southern California, extending about 25 miles (40 km) from the beach to the San Diego trough. Its outer end is about 3300 feet (1000 m) deep.

The deepest southern California canyon is Redondo Canyon in Santa Monica Bay. Although it can be traced only about 10 miles (16 km) to its terminus on the floor of Santa Monica Bay, Redondo Canyon reaches depths of 2000 feet (600 m). Like many others, the head of Redondo Canyon has repeatedly and gradually filled with sediment, only to be abruptly deepened as sediment suddenly moved down the canyon into the basin below. These abrupt flushings have occurred in various types of weather and at irregular intervals. What triggers these outward flows of sediment is not known.

REFERENCES

Anonymous, 1959. Offshore Geology and Oil Resources. Mineral Information Service (now California Geology), v. 12, no. 5, pp. 1–7.

Emery, K. O., 1960. The Sea off Southern California. John Wiley & Sons, Inc.

Hanna, G. D., 1951. Geology of the Farallon Islands. Calif. Div. Mines and Geology Bull. 154, pp. 301–310.

Rusnak, Gene A., 1966. The Continental Margin of Northern and Central California. *In* Geology of Northern California. Calif. Div. Mines and Geology Bull. 190, pp. 325–335.

Shepard, F. P., and K. O. Emery, 1941. Submarine Topography off the California Coast—Canyons and Tectonic Interpretations. Geol. Soc. Amer. Spec. Paper 31.

Vedder, J. G., and others, 1974. Preliminary Report on the Geology of the Continental Borderland of Southern California. U.S. Geological Survey Map MF624 (with text).

Appendix 1

Glossary

Aggradation: Process of building up a surface by deposition.

Alluvial fan: Low, cone-shaped heap, steepest near the mouth of the valley and sloping gently outward with ever decreasing gradient.

Alluvial plain: Plain resulting from deposition of alluvium by water.

Alluvium: Detrital deposits resulting from the operations of modern rivers, thus including the sediments laid down in river beds, flood plains, lakes, fans at the foot of mountain slopes, and estuaries.

Amphibole: Group name for common rock-forming minerals high in iron and magnesium.

Andesite: Volcanic rock composed essentially of plagioclase and one or more mafic constituents.

Anorthosite: Plutonic rock composed almost wholly of plagioclase.

Antecedent stream: Stream whose course was maintained in spite of a barrier lifted in its path by tectonic processes.

Anticline: Strata that dip in opposite directions from a common ridge or axis, like the roof of a house.

Aquifer: Formation, group of formations, or part of a formation that is water bearing.

Armored mudball: Rounded pebble or boulder originally composed of a mud core that became studded with small pebbles as the mass of mud rolled along.

Artesian water: Groundwater that is under sufficient pressure to rise above the level at which it is encountered by a well, but does not necessarily rise to or above the surface of the ground.

Ash: Deposit from volcanic eruption falling from an erupted cloud.

Attitude: General term to describe the relation of some directional feature in a rock to a horizontal plane.

Augite: Common rock-forming mineral belonging to pyroxene group.

Autolith: Fragment of igneous rock enclosed in another igneous rock of later consolidation, each being regarded as a derivative from a common parent magma.

Badlands: Region nearly devoid of vegetation where erosion, instead of carving hills and valleys of the ordinary type, has cut the land into an intricate maze of narrow ravines and sharp crests and pinnacles.

Basal conglomerate: Coarse, usually well-sorted and lithologically homogeneous sedimentary deposit found just above an erosional break.

Basalt: Extrusive rock composed primarily of calcic plagioclase and pyroxene, with or without olivine.

Baselevel: Level below which a land surface cannot be reduced by running water.

Basement: Underlying complex that behaves as a unit mass.

Basic: Refers to igneous rocks that are comparatively low in silica.

Batholith: Massive body of intrusive (plutonic) igneous rock that includes many related but independent intrusives that developed in sequence often over long periods of time.

Bed: Smallest division of a stratified series, marked by a more or less well-defined divisional plane from its neighbors above and below.

Bedrock: Any solid rock underlying soil, sand, clay, and so on.

Biotite: Common rock-forming mineral of the mica group, characterized by its flexible plates and black color.

Bituminous: Containing much organic or at least carbonaceous matter, mostly in the form of the tarry hydrocarbons usually described as bitumen.

Block faulting: Process by which rock is divided into small or large units by faults along which movement has occurred.

Block mountains: Mountains carved by erosion from large uplifted blocks bounded on at least one side by fault scarps.

Brachiopod: Member of phylum of marine, shelled animals with two unequal shells or valves, each of which normally is bilaterally symmetrical.

Brackish: Waters with saline content intermediate between that of streams and sea water.

Breccia: Fragmental rock whose components are angular and therefore, as distinguished from conglomerates, are not water-worn.

Brine: Water strongly impregnated with salt.

Butte: Conspicuous isolated hill or small mountain, especially one with steep sides, or a turretlike formation such as those found in badlands.

Caldera: Large basin-shaped volcanic depression, more or less circular or cirquelike in form, the diameter of which is many times greater than that of the included volcanic vents.

Chert: Compact, siliceous rock formed of chalcedonic or opaline silica, of organic or precipitated origin.

Cienaga springs: Limited area showing growth of water-loving plants appearing sporadically in otherwise arid surroundings, occasionally giving rise to flowing springs.

Cinder block: Volcanic material of large size thrown from a volcano in solid form.

Cinder cone: Conical elevation formed by the accumulation of volcanic ash or clinkerlike material around a vent.

Cirque: Deep, steep-walled recess in a mountain caused by glacial erosion.

Clast: Sedimentary rock fragment that is derived by mechanical transport.

Clastic: Textural term applied to rocks composed of fragmental material derived from preexisting rocks or from the dispersed consolidation products of magmas or lavas.

Clay: Natural substance or soft rock that, when finely ground and mixed with water, forms a pasty, moldable mass that preserves its shape when air-dried.

Coal: Black, compact, and earthy organic rock formed by the accumulation and decomposition of plant material.

Coastal plain: Plain with its margin on the shore of a large body of water, particularly the sea; generally represents a strip of recently emerged sea bottom.

Columnar jointing: Variety of jointing that breaks rock into columns, usually forming hexagonal pattern. Most characteristic of basaltic rocks. Generally considered to be shrinkage cracks due to cooling.

Columnar section: Graphic expression of the sequence and stratigraphic relations of rock units in a region.

Comminuted ash: Finely powdered volcanic ash.

Concretion: Nodular or irregular concentration of certain authigenic constituents of sedimentary rocks and tuffs; developed by localized deposition of material from solution.

Conglomerate: Cemented clastic rock containing rounded fragments corresponding in grain sizes to gravel or pebbles.

Contact: Place or surface where two kinds of rocks come together.

Continental deposits: Sedimentary deposits laid down within a land area in lakes or streams or by the wind, as contrasted with marine deposits laid down in the sea.

Continental shelf: Zone extending from the line of permanent immersion to the depth of about 450 feet (120 m), where there is a marked or rather steep descent (continental slope) toward the great depths.

Continental slope: Declivity from the offshore edge of the continental shelf to oceanic depths. Characterized by marked increase in gradient compared to the continental shelf.

Convection: Process of mass movement of portions of any medium in a gravitational field as a consequence of different temperatures in the medium and hence different densities.

Convoluted bedding: Twisted and rolled stratification in rocks.

Creep: Deformation that occurs along a fault but is not expressed by rupture (offset) along the fault.

Crystal: In mineralogy, the regular polyhedral form, bounded by plane faces (surfaces), that is the outward expression of a periodic or regularly repeating internal arrangement of atoms. In petrology, a crystalline grain in a rock irrespective of the absence of crystal faces.

Crystallization: Process through which crystalline phases separate from a fluid, viscous, or dispersed state.

Current structures: Forms produced by currents of wind, waves, or water in sedimentary rocks as the rocks are deposited.

Deformation: Any change in the original form or volume of rock masses produced by tectonic forces.

Delta: Deposit of sediment at the mouth of a river, resulting in progradation of the shoreline.

Dendritic drainage: A branching tree like pattern of stream courses developed generally on rocks of uniform texture and structure.

Desert pavement: Loose material containing pebbles or larger stones exposed to wind action. The finer dust and sand are blown away and the pebbles gradually accumulate on the surface, forming a mosaic (desert pavement) that protects the finer material underneath.

Desert varnish: Surface stain or crust of manganese or iron oxide that characterizes many exposed rock surfaces in the desert.

Detritus: Material that results from the breaking up, disintegrating, and wearing away of minerals and rocks.

Diapir (piercing fold): Anticline in which a mobile core has injected the more brittle overlying rock.

Diastrophism: Process by which the crust of the earth is deformed.

Differentiation: Process by which different types of igneous rock are derived from a parent magma.

Dike: Tabular body of igneous rock that cuts across the structure of adjacent rocks or cuts massive rocks.

Diorite: Plutonic igneous rock composed of soda-rich plagioclase feldspar and hornblende, biotite, or pyroxene.

Dip: Angle at which a stratum or any planar feature is inclined from the horizontal.

Dip slip: Component of the slip parallel with the fault dip.

Dolomite: Common rock-forming carbonate mineral. Rocks that approximate the mineral dolomite in composition.

Dome: Structural—Roughly symmetrical upfold of strata with beds dipping in all directions, more or less equally, from a point.

Volcanic—Steep-sided protrusion of viscous lava forming a more or less dome-shaped or bulbous mass over and around a volcanic vent.

Drainage basin: Part of the surface that is occupied by a drainage system or contributes surface water to that system.

Dunite: Peridotite consisting almost wholly of olivine and containing accessory pyroxene and chromite.

Earthquake: Groups of elastic waves propagating in the earth, set up by a transient disturbance of the elastic equilibrium of a portion of the earth. Vibration received by waves produced by sudden slippage along a fault.

Elastic rebound theory: Faulting arises from the sudden release of elastic energy that has slowly accumulated in the earth. Just before rupture, the energy released by the faulting is potential energy stored as elastic strain in the rocks. At the time of rupture the rocks on either side of the fault spring back to positions of little or no strain.

Epicenter: Point on the earth's surface directly above the focus of an earthquake.

Escarpment: Cliff or relatively steep slope separating level or gently sloping tracts.

Eugeosyncline: The part of a geosyncline in which volcanic accumulation predominates, for example, continental slope environment.

Eustatic: Real change in sea level resulting from variations in the volume of water in the sea, not a relative change in level resulting from subsidence or elevation of land masses.

External drainage: Drainage to the sea in contrast to internal drainage.

Extrusive rocks: Igneous rocks derived from magmatic materials ejected at the earth's surface, as distinct from intrusive or plutonic igneous rocks that have been injected into older rocks at depth without reaching the surface.

Facet: Nearly plane surface abraded on rocks or a rock fragment; polished surface of a cut gemstone; fault facet on the ridge ends of a fault scarp.

Fanglomerate: Heterogeneous materials originally deposited in an alluvial fan but subsequently cemented into solid rock.

Fault: Fracture or fracture zone along which there has been displacement of the rocks on either side of the fault relative to each other and parallel to the fracture.

Fault block: Body of rock bounded by one or more faults.

Fault scarp: Cliff formed by a fault.

Fault system: Two or more faults or groups of faults that are related in space, movement type, and time.

Fault valley: Valley formed by movement in a fault zone.

Fault zone: Belt hundreds or thousands of feet wide, consisting of numerous interlacing small faults and associated zones at gouge.

Fault-line scarp: Scarp that is the result of differential erosion along a fault rather than the direct result of movement along the fault.

Fault-line valley: Valley excavated by erosion along a fault.

Fauna: The animals collectively of any given age or region.

Feldspar: Group of abundant rock-forming minerals with high proportions of potassium, sodium, or calcium, plus aluminum, silicon, and oxygen.

Flood plain: Portion of level land near river mouth that is periodically flooded with sedimentary deposits.

Flora: The plants collectively of any given age or region.

Focus: True center of an earthquake, within which the strain energy is first converted to elastic wave energy.

Fold: A bend in strata or any planar structure.

Fold axis: Trend of the crest of an anticline or the trough of a syncline (or other fold).

Foliation: Laminated structure resulting from segregation of different minerals into layers parallel to the schistosity.

Foraminifer: Member of a subdivision of the phylum Protozoa; skeletons usually microscopic in size, commonly made of calcium carbonate and more rarely of sand or foreign particles of chitin.

Formation: Lithologically distinct product of essentially continuous sedimentation selected from a local succession of strata as a convenient unit for mapping, description, and reference.

Fossil: Animal or plant remains or traces preserved in the earth's crust by natural methods, excluding organisms buried since the beginning of historic time.

Fumarole: Hole or vent that emits fumes or vapors.

Gabbro: Plutonic rock consisting of calcic plagioclase and clinopyroxene, with or without orthopyroxene and olivine.

Gastropod: Member of the phylum Mollusca, class Gastropoda; usually with an asymmetrically coiled calcareous exoskeleton.

Geomorphology: Branch of science that treats surface features, their form, nature, origin, and development, and the changes they are undergoing.

Geosyncline: Basin of regional extent that subsides over a long time while sedimentary and volcanic rocks accumulate in it.

Geyser: Intermittent eruptive spring in which discharge occurs at more or less regular and frequent intervals, caused by the expansive force of highly heated steam.

Glacier: Mass of ice with definite lateral limits and motion in a definite direction, originating from compaction of snow.

Glass-sand: Extremely pure silica sand useful for making glass and pottery.

Glauconite: Green mineral that is essentially a hydrous potassium iron silicate; commonly occurs in sedimentary rocks of marine origin; a rock with high glauconite content.

Glaucophane: Blue metamorphic mineral of amphibole group containing sodium.

Gneiss: Coarse-grained rock in which bands rich in granular minerals alternate with bands rich in schistose minerals.

Graben: Block that has been downthrown along a fault relative to the rocks on either side.

Graded beds: Stratification in which each layer displays a gradation in grain size from coarse below to fine above.

Granite: Plutonic rock consisting chiefly of potash feldspar and quartz.

Granitization: Production of a granitic rock from sediments, or other rocks without melting.

Granodiorite: Plutonic rock consisting of quartz, plagioclase, and orthoclase, with biotite, hornblende, or pyroxene as mafic constituents; intermediate between quartz monzonite and quartz diorite.

Granulite: Metamorphic rock composed of even-sized, interlocking granular minerals.

Gravel: Fragments of rock worn by air and water, larger and coarser than sand.

Graywacke: Type of sandstone marked by large detrital quartz and feldspars set in a prominent to dominant "clay" matrix.

Hogback: A ridge formed by steeply dipping layered rocks.

Horizontal component: Amount of movement reflecting displacement in the horizontal plane when movement on a fault produces oblique separation.

Hornblende: Most commonly occurring mineral of the amphibole group.

Horst: Block that has been uplifted along faults relative to the rocks on either side.

Igneous: Class of rock formed by solidification of molten or partially molten parent matter.

Inlier: Approximately circular or elliptical area of older rocks surrounded by younger strata.

Internal drainage: Drainage into a closed basin, with no outlet to the sea.

Intrusive rocks: Igneous rocks that, while fluid, penetrate into or between other rocks, but solidify before reaching the surface.

Isocline (isoclinal fold): Anticline or syncline so closely folded that the beds of the two sides have the same dip.

Isotopes: Elements having an identical number of protons in their nuclei, but differing in the number of their neutrons; isotopes have the same atomic number, different atomic weights, and almost the same chemical properties.

Joint: Fracture or parting that abruptly interrupts the physical continuity of a rock mass.

Kernbut: Buttelike hill or buttress on a canyon side or outer, ridgelike edge of a fault terrace or bench; separated from the main hillside by a sag (kerncol).

Kerncol: Sag between a kernbut and the adjoining hillside.

Laccolith: Concordant intrusive body that has domed up overlying rocks, but has a floor that is generally horizontal.

Landslide: Downward sliding or falling of a mass of soil or rock.

Lava: Fluid rock that issues from a volcano or a fissure in the earth's surface; the same material solidified by cooling on the earth's surface.

Left lateral, left slip, left separation: Occurs where the horizontal separation along a fault is such that an observer walking along an index plane (bed, dike, vein) must, upon crossing the fault, turn to the left to find the index plane on the opposite side of the fault.

Limestone: Bedded sedimentary rock consisting chiefly of calcium carbonate.

Lineation: Parallel orientation of structural features in rocks that are lines rather than planes.

Lithology: Study of rocks based on megascopic examination; used loosely to mean the composition and texture of rocks.

Load cast: Mark consisting of a swelling that extends downward into fine-grained softer material.

Lode: Several veins spaced closely enough so that all, together with the intervening rock, can be mined as a unit.

Maar: Relatively shallow, flat-floored explosion crater with walls that consist largely of loose fragments of country rock and only partly of magmatic ejecta.

Mafic: Pertaining to rocks composed dominantly of iron and magnesian silicates.

Magma: Naturally occurring mobile rock material generated within the earth and capable of intrusion and extrusion.

Mantle: Segment of the earth below the crust that lies between the Mohorovicic and Gutenberg discontinuities; regolith material originating from rock weathering.

Mass wasting: Processes by which large masses of earth material are moved by gravity.

Meander: Regular, looplike bend in the course of a stream, developed when stream flows at grade, through lateral shifting of its course toward the convex sides of the original curves.

Melange: Formation consisting of a heterogeneous mixture of rock materials consolidated by tremendous deformational pressure.

Mesa: Flat-topped mountain or other elevation bounded on at least one side by a steep cliff.

Metamorphic rock: Rock formed in the solid state in response to pronounced changes of temperature, pressure, and chemical environment.

Metasedimentary rock: A metamorphic rock derived from sedimentary materials.

Meltavolcanic rock: A metamorphic rock derived from volcanic materials.

Mica: Silicate mineral group with sheetlike structures.

Microcline: Mineral of the feldspar group; common in granitic rocks.

Mineral: Homogeneous inorganic, naturally occurring substance of specific composition, within limited ranges.

Mineralizer: Mineralizing agent; substance that, when present in magmatic solutions, lowers temperature and viscosity, aids crystallization, and permits formation of minerals.

Miogeosyncline: The part of a geosyncline in which volcanic rocks are rare or absent, for example, continental shelf environment.

Mollusks: Members of the phylum of invertebrate animals including gastropods, pelecypods, and cephalopods.

Moraine: Accumulation of drift material within a glaciated region, built chiefly by direct action of glacial ice.

Mudflow: Flow of heterogeneous debris mixed with water and usually following a former stream course.

Mylonite: Fine-grained, laminated rock formed by extreme microbrecciation and milling of rocks during movement along faults.

Natural gas: Mixture of naturally occurring gaseous hydrocarbons; frequently associated with petroleum deposits.

Nose (anticlinal): The end of an elongate fold.

Obsidian: Volcanic glass.

Olivine: Important rock-forming mineral, especially in mafic and ultramafic rocks.

Ooze: Fine-grained pelagic deposit that is more than 80 percent organically derived.

Ophiolite: Basic igneous rocks subsequently altered to rocks rich in serpentine, chlorite, epidote, and albite; currently used to describe a sequence of rocks thought to represent components of the upper mantle.

Orogeny: Process of building mountains, particularly by folding and thrusting.

Orthoclase: Mineral of the feldspar group; common in granitic rocks.

Outcrop: Exposure of unaltered rock above the soil surface of the ground.

P waves: Elastic waves in which displacements are in the direction of wave propagation (in an earthquake).

Paleontology: Study of fossil remains, both animal and vegetable.

Pecten: Member of the pelecypod class of mollusks, with bivalve shells like a clam.

Pegmatite: Coarse-grained igneous rock usually found as a dike associated with a large mass of finer-grained plutonic rock.

Pelecypod: A class of mollusks, characterized by bivalvular exoskeleton with each valve an asymmetrical mirror image of the other.

Peneplain: Land surface worn down by erosion to a nearly flat or broadly undulating plain.

Percolation: Movement under hydrostatic pressure of water through the interstices of rock or soil.

Peridotite: Essentially nonfeldspathic plutonic rock consisting of olivine, with or without other mafic minerals.

Petrification: Process of converting organic matter into stone.

Petrogenesis: Origins of rocks, particularly igneous rocks.

Petrology: Study of the natural history of rocks.

Phenocryst: Relatively large and ordinarily conspicuous crystal of the earliest generation in a porphyritic igneous rock.

Phyllite: Argillaceous rock intermediate in metamorphic grade between slate and schist.

Pillow lava: Lava structure consisting of an agglomeration of rounded masses that resemble pillows; occurs mostly in basic lavas erupted under water.

Placer: Alluvial or glacial deposit, as of sand or gravel, containing particles of a valuable mineral.

Plagioclase: Subgroup of the feldspar family, composed principally of calcium, sodium, and aluminum silicates; common in igneous and metamorphic rocks.

Planation: Widening of valleys through lateral erosion by streams after the streams achieve grade and begin to meander, plus the concurrent formation of flood plains.

Playa: Shallow central basin of a desert plain in which water frequently gathers after a rain and is then quickly evaporated.

Pluton: Mappable body of igneous rock formed beneath the surface of the earth by consolidation from magma.

Plutonic: Class of igneous rocks that have crystallized at depth and have usually assumed a granitic texture.

Porphyry: Textural rock type containing conspicuous phenocrysts in a fine-grained matrix.

Pumice: Cellular, light-weight, glassy lava, generally rhyolitic.

Punky: Spongy, woodlike texture.

Pyroclastic: Detrital volcanic material explosively ejected from a volcanic vent.

Pyroxene: Common rock-forming mineral group with essentially iron and magnesium silicate composition.

Quartz: Silicon dioxide; commonest rock-forming mineral.

Quartzite: Granulose metamorphic rock consisting essentially of quartz.

Quartz monzonite: Plutonic rock containing major plagioclase, orthoclase, and quartz; with increased plagioclase the rock becomes granodiorite and with increased orthoclase it becomes granite.

Radial drainage: Drainage pattern with streams radiating from a central area, like the spokes of a wheel.

Radioactive decay: Change of one element to another by the emission of charged particles from the nuclei of its atoms.

Radiocarbon dating: Determination of the age of a material by measuring the proportion of the isotope C-14 (radiocarbon) in the carbon the material contains.

Radiolaria: Free-living, one-celled marine protozoan characterized by minute shells of opaline silica.

Reentrant: Indentation in a land form, usually angular.

Rejuvenation: Development of youthful topographic features in an area previously worn down to baselevel.

Relative time: Geologic time based on the sequence of oldest to youngest, without regard for specific number of years involved.

Relief: Elevations or inequalities, collectively, of a land surface; difference in elevation between high and low points of a land surface.

Replacement: Process by which a new mineral of partly or wholly different chemical composition may grow in the body of an old mineral or mineral aggregate.

Rhyolite: Extrusive equivalent of granite.

Richter scale: Scale for measuring the magnitude of earthquakes.

Right slip, right lateral, right separation: Occurs where the horizontal separation along a fault is such that an observer walking along an index plane (bed, dike, vein) must, upon crossing the fault, turn to the right to find the index plane on the opposite side of the fault.

Rip-up: Sedimentary structure formed by shale clasts torn by currents from a semiconsolidated mud deposit and transported to a new depositional site.

Rock: Any consolidated or unconsolidated mass of crustal material.

Roof pendant: Older rocks projecting down from the roof into a batholith.

Rubble: Accumulation of loose angular fragments, not waterworn or rounded like gravel.

S waves: Transverse waves that travel through an elastic medium.

Sag pond: Pond occupying a depression along an active fault; depression due to uneven settling of the ground.

Sandstone: Compacted detrital sediment composed predominantly of sand grains.

Schist: Medium- or coarse-grained metamorphic rock with subparallel orientation of its micaceous minerals.

Scoria: Pyroclastic volcanic ejecta, usually of basic composition.

Sea knoll: Submarine hill or elevation of the deep-sea floor, less prominent than a seamount.

Seamount: Submarine mountain rising more than 500 fathoms (900 m) above the ocean floor.

Section: Natural or artificial rock cut; the diagrammatic representation of such a feature.

Sedimentary: Refers to rock formed from sediment, by transportation of fragments from their sources to a depositional site or by precipitation from solution.

Sedimentation: Portion of the rock cycle from separation of particles from their parent rock to and including their consolidation into another rock.

Seismology: Study of earthquakes and the measurement of elastic properties of the earth.

Serpentinite: Rock consisting almost wholly of serpentine minerals, derived from the alteration of previously existing mafic minerals.

Shale: Laminated sediment in which the constituent particles are predominantly clay.

Sheet joints: Parallel jointing, usually in plutonic rocks, that simulate bedded rocks.

Shutterridges: Ridges that by horizontal or oblique fault shift tend to block canyons of streams crossing the fault.

Silica: Silicon dioxide (usually quartz).

Sill: Intrusive body of igneous rock, of approximately uniform thickness but relatively thin compared with its lateral extent, that has been emplaced parallel to the bedding or schistosity of the intruded rock.

Silt: Unconsolidated clastic sediment with specific size range.

Slate: Fine-grained metamorphic rock possessing well-developed fissility (slaty cleavage).

Slip: Relative displacement of formerly adjacent points on opposite sides of a fault, measured at the fault surface.

Soil: All unconsolidated material above bedrock that has been in any way altered or weathered.

Spatter (driblet) cone: Low, steep-sided hill or mound of spatter built by lava fountains along a fissure or around a central vent.

Specific time: Measurement of rock ages in specific years.

Stratigraphy: Study of the formation, composition, sequence, and correlation of the stratified rocks as parts of the earth's crust.

Strato-volcano (composite cone): Volcanic cone, usually of large dimensions, built of alternating layers of lava and pyroclastic material.

Stream capture (beheading): Diversion of the upper part of a stream by the headward growth of another stream.

Strike: Direction or bearing of the outcrop of an inclined bed or structure on a level surface, perpendicular to the direction of the dip.

Strike-slip fault: Fault in which the net slip is practically in the direction of the fault strike.

Subduction zone: In interpretations of plate tectonic theory, a belt along the under margin of a continental plate, where the colliding oceanic plate descends toward or into the mantle.

Superposition: Rule that if a stratiform series is in original relationship of deposition, underlying beds are older than overlying beds; governing principle of sedimentation.

Syncline: Fold in rocks, characterized by strata from both sides dipping inward toward the fold axis.

Synclinorium: Broad regional syncline on which minor folds are superimposed.

Talus: Coarse waste at the foot of a cliff or covering a slope below a cliff.

Tear fault: Strike-slip fault that trends across the strike of the deformed rocks.

Tectonism: Crustal instability; structural behavior and deformation of the earth's crust during and between major cycles of sedimentation.

Terrace: Relatively flat, horizontal or gently inclined surface bounded by a steeper ascending slope on one side and by a steeper descending slope on the opposite side.

Till: Nonsorted, nonstratified sediment carried or deposited by a glacier.

Tilted fault block: Block rotated so that one side is relatively uplifted and the other depressed.

Topographic inversion: Area in which former valleys have been filled and become ridges in a subsequent cycle of stream development.

Topography: Physical features of a district or region, especially relief and contour.

Trace elements: Elements present in minor amount in the earth's crust.

Transgressive unconformity: Progressive pinching out, toward the margins of a depositional basin, of the sedimentary units in a conformable series.

Trap (gas, oil, water): Reservoir rock completely surrounded by impervious rock.

Tremor: Small earthquake.

Trilobite: Primitive, extinct crustacean occurring throughout the Paleozoic, characterized by a segmented body divided into three lobes.

Tufa: Sedimentary rock composed of calcium carbonate or silica, deposited from solution in the water of a spring or lake or from percolating groundwater.

Tuff: Rock formed of compacted volcanic fragments.

Turbidity current: Relatively dense current that moves along the bottom slope of a body of standing water; may also occur as an underflow in a lake or reservoir, in a dust storm, or in a descending cloud of volcanic dust.

Ultramafic: Applies to rocks containing less than 44 percent silica, but rich in magnesium and iron minerals.

Unconformity: Contact between rock units that was once a surface of erosion or nondeposition.

Vein: Crack or fissure filled with mineral matter from underground water solutions.

Vent: Nearly vertical outlet from within the crust.

Vertical component: Portion of a vector that is perpendicular to a horizontal or level plane.

Vesicle: Small cavity in an aphanitic or glassy igneous rock, formed by expansion of a bubble of gas or steam during solidification of the rock.

Volcanic bomb: Mass of magma that is rotated after being tossed into the air and then falls on the ground as a rounded or spindle-shaped unit.

Volcanic glass (obsidian): Glass produced from magma when cooling is too rapid to permit crystallization.

Volcaniclastic: Sedimentary rock formed by materials ejected from a volcanic source.

Wall rock: Country rock into which magma is intruded.

Xenolith: Rock fragment that is foreign to the body of igneous rock in which it occurs; an inclusion.

Appendix 2

Common Minerals and Rocks

Common Minerals

The chemical elements found in minerals are thought to compose all materials of the earth, plus the materials of the other planets in our solar system. Lunar rock studies have so far confirmed this premise, since minerals discovered in lunar samples are similar or identical to minerals found on earth. Nearly 3000 minerals are known, but only about 60 are widely distributed in the earth and only about 25 are at all common.

Minerals are the building blocks of most rocks. In contrast to rocks, however, minerals have specifically organized sets of atomic relationships called *space lattices.* These lattices govern the external forms of mineral crystals and determine, within limited ranges, the physical, optical, and chemical properties of all minerals. Minerals are produced only by nature, and even though many identical chemical compounds can be made in the laboratory, such duplications are not minerals. In addition, materials formed through organic processes, like petroleum and coal, are rocks, not minerals. Organic materials have variable compositions, but minerals have definite compositions within specified limits.

Minerals are most accurately identified by methods that reflect their lattice structures, but involve instruments of limited availability. The lay investigator must rely on external characteristics that primarily depend on the composition of the mineral in question. It is possible to recognize 25 to 50 common minerals by making a few simple tests like those for hardness and color and by noting other distinctive characteristics.

Table A-1 Mineral Systematics

Luster	Hardness ($>$ = greater than, $<$ = less than)	Mineral (and Color)	Other Characteristics
NONMETALLIC Colored minerals	< fingernail	Azurite (blue)	Blue streak; fizzes
		Talc (green, gray)	Soapy or greasy feel and look; very soft; white streak
		Hematite (red, red brown)	Usually harder; wipes brick red on fingers
		Limonite (yellow, yellow brown)	Usually harder; wipes rust brown on fingers
		Malachite (grass green)	Green streak; fizzes
		Graphite (black)	Usually metallic or submetallic luster; marks paper black
		Pyrolusite (black)	Often sooty; usually submetallic luster
		Cinnabar (bright red)	Very heavy; very bright luster
	> fingernail < copper coin	Biotite (black, brown, yellow)	Mica; sheets and flakes flexible and elastic
		Chlorite (black, green)	Sheets and flakes flexible and inelastic
		Serpentine (green, gray green)	Usually harder; greasy look
		Asbestos	Fibrous
	> copper coin < knife	Calcite (brown, gray)	Fizzes; good cleavage (3 ways)
		Dolomite (gray, brownish)	Does not fizz; good cleavage (3 ways)
		Lepidolite (lavender, purple)	Mica; sheets and flakes flexible and elastic
		Sulfur (yellow)	Burns with match (very hot) with odor
		Fluorite (blue, green, yellow, purple)	Does not fizz; good cleavage (4 ways)
		Serpentine (green, gray green)	Greasy look
		Asbestos	Fibrous
		Azurite (blue)	Blue streak; fizzes
		Malachite (grass green, gray green)	Green streak; fizzes
		Chrysocolla (blue green, aqua)	Almost colorless streak; no fizz; enamel-like appearance
		Sphalerite (yellow, dark brown)	Resinous; heavy to feel; looks metallic; much cleavage (6 directions)
		Hematite (red, red brown)	Usually submetallic; heavy
	> knife	Amphibole (green, black)	Cleavage very good and visible; hard to separate from pyroxene
		Pyroxene (green, black)	Cleavage very good, and often not visible; hard to separate from amphibole (amphibole is commoner than pyroxene)
		Chrysocolla (blue green, aqua)	Colorless streak; Enamel-like appearance; often mixed with quartz and therefore tests as though it is very hard
		Epidote (yellow green, pea green)	Hard to identify but a very common mineral in crystalline mats and as coatings, veinlets, and alterations in igneous (and other rocks)

Table A-1 (continued)

Luster	Hardness (> = greater than) (< = less than)	Mineral (and Color)	Other Characteristics
		Olivine (green, black green)	Granular; very common in volcanic rocks
		Garnet (multicolored)	Often in good crystals; massive; heavy for a nonmetallic mineral
		Quartz (multicolored)	Amethyst is purple; rose quartz; citrine quartz (yellow)
		Jasper (brown, red, dark green)	Massive
		Chalcedony (multi-colored but usually white)	Fine-grained quartz
		Tourmaline Pink: rubellite	Gem variety; uncommon but found in southern California
		Black: schorl	Brittle; common; striated with vertical lines on crystal faces
White minerals	< fingernail	Talc	Very soft; feels soapy
		Borax	Powder; coating; sweetish taste
		Kaolinite	"Clay"; odor of earth when moist (breathe on it); compact; very common mineral
		Bauxite	Kaolinite with spherical nodes in it; chief ore of aluminum
		Gypsum	Cleaves readily in two directions unless massive; looks like mica but not "flaky"; very common mineral
		Barite	Usually harder, but noticeably heavy for a white mineral
	> fingernail < knife	Kernite	Uncommon mineral but spread widely from world-famous occurrence at Boron, Calif.; like gypsum, but cleavage is different—breaks into brittle fibers
		Halite	Salt (taste it); often in cubical crystals
		Calcite	Fizzes in acid; 3 directions of perfect cleavage
		Dolomite	Does not fizz; 3 directions of perfect cleavage
		Barite	Glassy luster; does not fizz; heavy to feel for a white mineral
		Fluorite	High glassy luster; usually transparent, but sometimes variegated in color
		Cerussite (Anglesite)	These minerals cannot be separated except by chemical analysis; very high (resinous) luster; very heavy to feel for a white mineral; common ores of lead; associated with galena.
		Scheelite	Commonest tungsten ore; very heavy for white mineral; high luster; hard to distinguish from lead minerals; associated with garnet, epidote, calcite

Table A-1 (continued)

Luster	Hardness ($>$ = greater than / $<$ = less than)	Mineral (and Color)	Other Characteristics
	$>$ knife	Opal	Opalescence

		Feldspar
		   Orthoclase
		   Microcline
			Softer than quartz; often dusted with clay alteration; common rock-forming mineral; widespread in granites; has cleavage; separation of these two not possible by physical properties

		Plagioclase	Plagioclase separated from other feldspars by fine striations (lines) due to twinning of crystals
		Quartz	Commonest and most widely spread mineral in rocks; sand is usually quartz; looks like glass; crystals are striated across faces
		Chalcedony	Chalcedony is finely crystalline quartz in which grains are invisible to eye or lens
METALLIC	$>$ fingernail	Graphite (black)	Submetallic luster; marks paper black
		Pyrolusite (black)	Often sooty; submetallic luster; dendrites are often pyrolusite
		Molybdenite (gray)	Metallic, bright; marks paper gray; heavy; in granites; chief ore of molybdenum
	$>$ fingernail $<$ knife	Stibnite (light gray)	Tarnishes to darker gray; one cleavage; blades and needles; lighter weight and color than galena; chief ore of antimony
		Galena (dark gray)	Very heavy; cleavage (3 directions, forming cubes); chief ore of lead
		Chalcocite (black or steel gray)	No cleavage; strong fracture; brittle; not so heavy as galena; chief ore of copper
		Bornite (peacock blue)	Tarnishes to many colors; associated with chalcocite and chalcopyrite; ore of copper
		Sphalerite (black to brown)	Much cleavage; high luster; chief ore of zinc; not so heavy as lead or copper minerals
		Chalcopyrite (brass yellow)	Softer and yellower than pyrite
		Chromite (black, brownish black)	Brown streak; nonmagnetic; softer than ilmenite and magnetite
	$>$ knife	Magnetite (black)	Highly magnetic; important ore of iron
		Ilmenite (bluish black)	Less magnetic than magnetite
		Pyrite (pale brass yellow)	Fool's gold; lighter in color than chalcopyrite; harder than chalcopyrite; common mineral
		Hematite (steel gray)	High luster; common iron ore; red brown streak

Mineral Recognition It is helpful to keep the following points in mind when trying to recognize minerals.

1. The geology of the region in which specimens are found can give important clues regarding what minerals will *not* be found in the area.

2. Most common minerals are white, but white minerals can be difficult to recognize because dozens of uncommon minerals are also white. Frequently occuring white minerals are: quartz and the common varieties of the quartz group (agate, jasper, chalcedony, chert, flint, rock crystal, and onyx), feldspar, calcite, gypsum, kaolinite, barite, and fluorite.

3. The most frequently found minerals are physically hard and durable and usually chemically stable. Quartz and feldspar are good examples.

4. About 100 chemical elements compose the minerals of the earth. Eleven elements make up 95 percent of the crust, with the result that the common minerals tend to be chemical compounds involving these elements. The dominant elements are aluminum, calcium, hydrogen, iron, magnesium, nitrogen, oxygen, potassium, silicon, sodium, and carbon.

5. The common minerals are mostly silicates, compounds involving silicon and oxygen.

6. All physical, chemical, optical, and other properties of minerals are constant within prescribed ranges. Although there are some limitations, this constancy permits the use of physical properties to recognize common minerals. Luster, color, hardness, cleavage, and specific gravity are the properties most often considered; occurrence and habitat are also important.

Items normally used to help determine the physical properties of minerals are a knife blade, magnifier, magnet, weak acid, copper coin, and an unglazed porcelain plate. Procedures are as follows.

1. Examine specimen with eye and magnifier. Note texture and uniformity. If it is nonuniform, it may be more than one mineral, or a rock. Note color and color variation. Is color throughout or a coating? Is it a white or colorless mineral, stained?

2. Determine luster (reflection) of mineral on a fresh surface: metallic (like a metal) or nonmetallic.

3. Determine hardness (always examine with magnifier to see whether in fact a scratch has been made): harder than knife blade—over 5.5; harder than copper coin, but softer than knife—between 3 and 5.5; harder than fingernail, but softer than copper coin—2.5 to 3; softer than fingernail—2.5 or less.

4. Put small drop of weak acid on specimen. Is there a reaction (fizz)?

5. Locate mineral in recognition chart and name specimen.

Common Rocks

Rocks are composed primarily (but not entirely) of minerals, and it is the minerals of a rock on which the name of the rock is usually based. Since rock names are numerous and have been assigned over many years by many petrologists, precise naming is often complicated and can be resolved only

Table A-2   Recognition of Rocks: Their Properties

	Appearance	Minerals Visible	Name	Variety
IGNEOUS* (from cooling of magma)	Glassy (very rapid cooling)	None	**Obsidian** (gray to black)	Basalt glass (shiny cinders on volcanos)
	Fine grained (rapidly cooled)	Feldspar Quartz	**Rhyolite** (lava; pink to white)	None
		Feldspar Dark minerals: (1 or more) Biotite Amphibole Olivine Pyroxene No quartz	**Basalt** (lava; black to brick red)	None
	Coarse grained (slowly cooled)	Feldspar Quartz	**Granite** (pepper and salt appearance; sugary texture)	None
		Feldspar Dark minerals No quartz	**Gabbro** or **diorite** (dark and colored minerals)	None

*Notes on igneous rocks: Pumice is froth from obsidian—light weight, porous, and light-colored. Porphyry is an igneous rock with discrete large grains (crystals) in a massive or glassy matrix. Dark minerals are often called "ferromagnesian" because of their iron and magnesium content.

SEDIMENTARY (from erosion, precipitation, accumulations of other rocks, minerals, and organic matter)

CLASTIC: Made of rounded particles of other rocks and minerals

Material	Coarse Grain	Medium Grain	Fine Grain	Very Fine Grain
Unconsolidated material	Gravel	Sand	Silt	Mud (clay)
Consolidated material	**Conglomerate**	**Sandstone**	**Shale** (Siltstone)	(Mudstone)

ORGANIC: Made of organic materials

Material	Rock Name
Shells	**Coquina** (fizzes with acid)
Shells and limy mud	**Shelly limestone** (fizzes with acid; chalk, ooze)
Diatoms	**Diatomite** (does not fizz)
Plants	**Coal**

CHEMICAL: Formed from water solutions

Material	Rock Name
Salt	**Rock salt** (salty taste)
Limy mud (as on inside of tea kettle)	**Limestone** (fizzes with acid; dolomitic or calcareous)

Table A-2 (continued)

		Original Rock	Metamorphic Rock	Properties
METAMORPHIC (rocks changed from original nature by heat, pressure, chemical action)		Shale	Slate	Many colors; splits easily
		Sandstone	Quartzite	Very hard, tough; rock grains hard to see; many colors; makes pebbles
		Slate	Schist	Splits easily; rich in mica flakes
		Granite	Gneiss	Looks like granite, but minerals arranged in crude bands
		Limestone	Marble	Coarse grained; many colors; fizzes with acid; knife scratches it; when dark colored, and hit with hammer, gives off odor
		Basalt	Serpentinite	Green to black; shiny, slippery feel; soft

by microscopical study. The problem may be somewhat simplified for the general student, however, if some basic guidelines are followed. First, recognize the process by which the rock of interest originated and, second, note the minerals in the specimen (most rocks are composed of more than one mineral).

Igneous Environments   Some igneous rocks are formed when volcanos erupt molten material (magma) that congeals on contact with the atmosphere. Such congealed products may take many forms, but all are volcanic igneous rocks. Magma congealed within the earth's crust forms plutonic or hypabyssal (intermediate in depth) igneous rocks. The specific names assigned depend on cooling rate and on the original minerals that crystallized from the magma.

Sedimentary Environments   Once a rock is exposed on the surface of the earth, the atmosphere begins to change it by alteration (weathering) and transportation (erosion). Eroded products ultimately reach the sea as rounded fragments, gravel, sand, or mud, which are then consolidated into conglomerates, sandstones, shales, and mudstones. All these belong to the large class known as sedimentary rocks, the environment of formation being that of sedimentation. The source of such sediments can be any material, including previously formed sedimentary rocks, igneous rocks, or recycled metamorphic rocks.

Metamorphic Environments   Under the impact of pressure, heat, and chemical change, any material buried in the crust of the earth can become a new (metamorphic) rock type. By definition, however, the heat involved cannot be great enough to change the metamorphic rock to magma. Metamorphic rocks have complex mineralogy, but fortunately they are named according to external characteristics. Five common types are recognized: gneiss (often metamorphosed plutonic rock), schist (often metamorphosed shale or volcanic rock), slate (metamorphic rock with good rock cleavage), quartzite (metamorphosed sandstone), and marble (metamorphosed—"crystalline"—limestone).

Table A-3   Igneous Rock Classification (primarily for hand specimen recognition)

Characteristic (essential) minerals	CHIEF FELDSPAR ALKALI Potash Feldspars: Microcline or Orthoclase		ALKALI and SODA-LIME Orthoclase-Plagioclase		CHIEF FELDSPAR SODA-LIME Plagioclase			FELDSPARS ABSENT
	Quartz: 0–5%	Quartz: greater than 5%	Quartz monzonite	No quartz	Quartz	No quartz	No quartz ±Olivine	No quartz / Olivine / Other Fe-Mg minerals
Percentage of silica	66–55%	75–65%	75–65%	65–50%	70–62%	65–50%	60–45%	50–30%
Family or clan	Syenite	Granite	Quartz monzonite	Monzonite	Quartz diorite	Diorite	Gabbro	Peridotite
Characteristic texture 1. Granitic (megacrystalline) **[PLUTONIC]**	Syenite	Granite / Alaskite	Quartz monzonite / Granodiorite	Monzonite	Quartz diorite	Diorite	Gabbro / Anorthosite / Norite	Peridotite / Dunite
2. Porphyritic — Ground mass: a. megacrystalline, b. microcrystalline; Phenocrysts **[HYPABYSSAL]**	Syenite porphyry (a. mega-)	Rhyolite porphyry (b. micro-)	— Porphyries —					
3. Felsitic **[VOLCANIC]**	Trachyte	Rhyolite	Quartz Latite	Latite	Dacite	Andesite (less than 50% dark minerals)	Basalt (greater than 50% dark minerals)	Various terms in wide use

Glassy Rocks

Obsidian: Bright vitreous luster; usually black
Pitchstone: Dull, pitch-like obsidian
Perlite: Gray, black, pearly, with small and large spherical modes layered like an onion
Scoria: Vesicular basaltic lava
Pumice: Froth of obsidian; strong cellular structure

Pyroclastic Materials

Unconsolidated: Blocks, bombs, lapilli; ash
Consolidated: Volcanic breccia; agglomerate; tuff

Appendix 3

Geologic Sequence and Time

Geologists use two approaches when considering the vast dimension of geologic time: relative or qualitative and specific or quantitative. Relative measure is not a true measure, but rather a sequence, since a given event occurs before or after another event. Specific measure is a true although not exact measure. Geologists think primarily in terms of a sequence of events, each event being relative to every other event, sequentially from oldest to youngest. Specific ages are in some ways incidental to the basic task of the geologist. Hence geologic events are arranged in chronologic sequence, based on stratigraphic and paleontologic data. The magnitude of each known event is estimated in both time and space; relative magnitudes are acknowledged by assigning designations with varying degrees of emphasis. Thus the term *era* designates major time units, separated from one another by major events; *period, epoch,* and *age* refer to time spans between events of progressively lesser magnitudes.

A standardized set of labels for recording geologic history is the geologic time scale, which has been accepted by international agreement. Continental and regional designations may vary, however, particularly those of lesser magnitude. Estimates of specific time are commonly inserted.

Current techniques for establishing specific measurement of geologic time involve pairs of chemical elements that have "mother-daughter" relationships. Elements such as uranium-lead, rubidium-strontium, and potassium-argon occur widely, though in exceedingly small proportions, in some of the commonest rock-forming minerals. Age determinations from the elements in these minerals may yield measurements up to many millions of years. The date of mineral crystallization from magma or even the date of later metamorphism can often be determined. In general, igneous rock-forming minerals give the best results. Carbon-14 measurements are only valid to 50,000 years before the present, so this method is of limited geologic use.

Table A-4 summarizes the isotopes currently used for age dating. For example, potassium breaks down into calcium and argon at a known rate. Thus, if the amounts of each of these elements in a certain mineral is known, the time involved in the change can be calculated.

Geochronologic techniques appear to suggest solutions to some fundamental problems. Present applications provide: (1) an independent test of the relative time scale, particularly in the post-Precambrian; (2) greater sophistication in determining rates of sea-floor spreading and continental drift; (3) refinements of dating reversals in the earth's magnetic field; (4) increasing definition and division in the tremendous span of Precambrian time; and (5) assistance in meteoritic studies, which promise insight into pregeologic history of the earth.

In another use of the radioactive decay process, the rubidium[87]-strontium[86] system has proved especially helpful in establishing rock origins. Rubidium is known to be concentrated in the continental crust, where rubidium-strontium ratios are higher by a factor of at least ten than they are in rocks derived from the upper mantle. Furthermore, rubidium-strontium ratios for rocks derived from the lower mantle contrast with those derived from the continent or upper mantle.

Table A-4   Common Isotopes Used in Radiometric Age Dating

Materials Used	Isotopes		Half-life of Original (years)	Maximum Dating Range (years)
	Original (parent)	Decay Product (daughter)		
Uraninite, pitchblende, zircon in rocks	Uranium-238 Uranium-235	Lead-206 Lead-207 (not terrestrial)	4.5 billion > 10 million	10 million to 4.6 billion
Micas (muscovite and biotite); feldspar: micro-cline; some metamorphic rocks	Rubidium-87	Strontium-87	47 billion	10 million to 4.6 billion
Micas (muscovite and biotite); amphibole (horn-blende); some volcanic rock	Potassium-40	Argon-40 Calcium-40	1.3 billion	100,000 to 4.6 billion
Wood, peat, and various plant materials; charcoal, some bones, cloth, shells; waters, both ground and ocean; deposits like stalactites and stalagmites	Carbon-14	Nitrogen-14	5730 ±	100 to 50,000

Appendix 4

Some Theories Pertinent to California Geology

Within 15 years, the wide acceptance of continental drift made the formerly conventional view of stable continental and oceanic masses untenable to most geologists. Corollaries of continental drift include sea-floor spreading, polar wandering, and plate tectonics. California's position at the edge of a continent confers on the state a geologically important position as a testing ground for many new and exciting ideas in the earth sciences.

Geosynclines and Geosynclinal Theory

Continental bodies are subject to periodic marine invasions that develop elongate basins, primarily along the edges of the continents. The seas thus established have seldom been deeper than 1500 feet (450 m), but they frequently accrued great thicknesses of marine sediments as the basin floors sank under the loads of deposition. *Geosyncline* is the term for the large depression (fold) that sedimentary deposits create as they settle on the underlying crust. The geosynclinal theory of mountain building contends that marine waters are eventually expelled from the geosyncline by compressive or vertically activated forces that typically produce a mountain chain of parallel folds. The Appalachian chain is a classic North American example.

The geosynclinal theory has substantially influenced North American geology and California geology in particular. Prior to development of plate tectonic theory, the standard explanation for the Sierra Nevada and Coast Ranges involved their evolution from continentally based geosynclines. About 1940, however, it was suggested that geosynclines are primarily marginal to continents rather than continentally based. A further point is that sedimentation within the geosyncline can be divided into an oceanward eugeosynclinal zone, containing predominantly volcanic materials, and a continentward miogeosynclinal zone, containing predominantly clastic sediment and limestone. The current view is that geosynclines form along continental plate margins over both oceanic and continental crusts.

Isostasy

Isostasy has been accepted as a fundamental premise of earth science for more than 100 years. Initially it was asserted that high-standing rock masses like the Himalayas and the Andes deflected the plumb bob (in measurements of gravity) less in amount than the mathematical calculation of gravitational variance required for the supposed mass of the mountain block in question. We now know that this anomaly occurs because mountains have deep roots and a mountain mass with roots has a lower density (mass) than the adjacent, crustal units. Constant shifting is required to maintain balance between high-standing (mountain) blocks and low-standing (basin) blocks as erosion transfers material from the mountains to the lowlands.

The principle of isostasy applies to small and large areas alike, although because of the general rigidity of rocks the results may not be measurable if the area is small. Larger areas may show deformation appropriate to the volume (mass) transfer on the surface. Isostasy has been frequently suggested as the causal mechanism of mountains, both those originating from geosynclines and those formed by block faulting. A variety of geologic events could trigger isostatic adjustment: widespread lava outpourings like those of the Columbia and Modoc plateaus, the formation of deltas or continental ice masses, and response to erosional and depositional processes in general. Changes in sea level often reflect isostasy, for epicontinental seas may be created or removed through isostatic adjustment. Large vertical movements such as those that produced the Tibetan and Colorado plateaus have been attributed to isostasy.

Overthrust Faults

Horizontal movement represented by overthrusting on flat-lying fault planes was once considered an impossibility by most physicists, who questioned the concept that blocks of rigid crust rode upward and outward across nearly horizontal planes. Physicists generally maintained that forces could not move bodies of such rigidity across miles of underlying, also rigid bodies without substantially shattering overlying rock masses. Nevertheless, geophysicists M. King Hubbert and W. W. Rubey demonstrated that the supporting pressure of fluids in pore spaces of rocks could permit rock masses to ride miles across underlying units with minimal friction and without causing the severe shattering predicted for such brittle bodies. In fact, it appears that vertical movements may well be incidental to the worldwide horizontal movement of rock bodies called for by plate tectonics. Although horizontal displacement on strike-slip faults like the San Andreas was traditionally assumed to have occurred on vertical or nearly vertical planes, the cause of such horizontal shift was difficult to explain until plate tectonic models introduced the transform fault as an element in plate boundaries.

Rock Magnetism and Paleomagnetism

It is generally accepted that the earth's changing magnetic field has impressed its past orientations on the iron minerals found in rocks. Accordingly, plotting of magnetic polar positions from orientation of minerals in ancient rocks supports the contention that continents have changed their positions with respect to the magnetic poles. Furthermore, we now know that magnetic field reversals are common over both short and long periods of geologic time; there is evidence that reversals have occurred as many as nine or more times in the past 3 million years. Establishment of the time-polarity calendar has become possible because of advances in the potassium-argon dating method as applied to volcanics and deep-sea sediments.

Continental Drift

The idea of drifting continents was first advanced seriously in 1910 by F. B. Taylor, who did not pursue the concept. It remained for Alfred Wegener to articulate and develop the theory with his first paper in 1912. Wegener's ideas were eventually synthesized in his 1915 treatise entitled *The Origins of Continents and Oceans*. His view was that an original single continent became slowly but continuously dissociated by horizontal shift of its adjacent parts, creating suture lines and rifts or trenches between adjacent masses. The discrete parts evolved by slow drift into the major continents we know today. Wegener presumed that the continents were supported in a substrate of mantle or subcrustal material. Although American geologists were slow to accept Wegener's hypothesis, today drift is accepted by almost all earth scientists as the basic explanation for shape and position of the continents.

Sea-Floor Spreading

In the early 1960s, the view was advanced that midoceanic ridges like the Atlantic ridge are the places from which magmas rise from the mantle and spread laterally to produce new crust and, eventually, new continental plates. As plates migrate laterally, they collide. The deep trenches existing in many ocean basins (often impinging on continental borders) are now thought to be the result of the more dense oceanic plates riding under the lighter continental bodies.

Evidence from polarity reversals in deep-sea sediments seems to support the view that new crust develops by mantle extrusion from midoceanic spreading centers. It is of particular interest for California geology that spreading centers along the East Pacific rise may extend into the Gulf of California, and possibly beneath the Imperial Valley and even farther north (with the concomitant development of the San Andreas fault).

Plate Tectonics

In 1967, it was first proposed that the earth's crust is divided into immense plates, five or six of which are rigid, relatively thin continental plates. These plates are thought to be moving continually, primarily horizontally with lesser vertical components of movement. Downward movements tend to occur along plate margins, where spreading centers cause rifting or where oceanic plates impinge on the forward edges of continental plates in down-sweeping subduction zones. Upward movements tend to occur where the forward edges of continental plates float up over oceanic units, or where continental plates collide or impinge on each other. According to plate tectonic theory, California's mountains have been produced chiefly by the collision of the east-moving Pacific plate with the west-drifting North American plate. Moving plates presumably have at least three kinds of margins: (1) sutures like the mid-Atlantic ridge, where new material is being added to the trailing edge of a plate; (2) boundaries where moving plates impinge on one another, usually with one plate moving below the other (subduction); and (3) boundaries in which two plates slide past one another (shearing), rather than one passing beneath the other.

Transform Faults

Transform faults, or simply transforms, are special cases of strike-slip faulting that separate major structural elements of tectonic plates. A North American example would be the interpretation of the San Andreas fault as a partial boundary between the Pacific and North American plates. The transform was originally proposed to describe the relative motion between discrete units of oceanic crust upwelling at oceanic rises, where magma repeatedly fills a central suture and shoves the blocks of oceanic crust away from the rise. The upwelling zones in the central rise are themselves offset, so the blocks on either side of the faults in some places move in the same direction and in other places move in opposite directions, with normal strike-slip offset. These are ridge-ridge transforms, in contrast to the boundaries where plates collide or move past one another by shearing. Both are correctly termed strike-slip faults, however.

Index